MECHANISM, MENTALISM, AND METAMATHEMATICS

SYNTHESE LIBRARY

STUDIES IN EPISTEMOLOGY,

LOGIC, METHODOLOGY, AND PHILOSOPHY OF SCIENCE

VOLUME 137

JUDSON CHAMBERS WEBB
Boston University, Dept. of Philosophy, Boston, Mass., U.S.A.

MECHANISM, MENTALISM, AND METAMATHEMATICS

An Essay on Finitism

D. REIDEL PUBLISHING COMPANY
DORDRECHT : HOLLAND / BOSTON : U.S.A.
LONDON : ENGLAND

Library of Congress Cataloging in Publication Data

Webb, Judson Chambers, 1936–
Mechanism, mentalism, and metamathematics.

(Synthese library; v. 137)
Bibliography: p.
Includes indexes.
1. Metamathematics. I. Title.
QA9.8.W4 510'.1 79-27819
ISBN 90-277-1046-5

Published by D. Reidel Publishing Company,
P.O. Box 17, 3300AA Dordrecht, Holland

Sold and distributed in the U.S.A. and Canada
by Kluwer Boston Inc.,
190 Old Derby Street, Hingham, MA. 02043, U.S.A.

In all other countries, sold and distributed
by Kluwer Academic Publishers Group,
P.O. Box 322, 3300 AH Dordrecht, Holland

D. Reidel Publishing Company is a member of the Kluwer Group.

Printed in The Netherlands

TABLE OF CONTENTS

PREFACE

This book grew out of a graduate student paper [261] in which I set down some criticisms of J. R. Lucas' attempt to refute mechanism by means of Gödel's theorem. I had made several such abortive attempts myself and had become familiar with their pitfalls, and especially with the double-edged nature of incompleteness arguments. My original idea was to model the refutation of mechanism on the almost universally accepted Gödelian refutation of Hilbert's formalism, but I kept getting stuck on questions of mathematical philosophy which I found myself having to beg. A thorough study of the foundational works of Hilbert and Bernays finally convinced me that I had all too naively and uncritically bought this refutation of formalism. I did indeed discover points of surprisingly close contact between formalism and mechanism, but also that it was possible to undermine certain strong arguments against these positions precisely by invoking Gödel's and related work. I also began to realize that the Church-Turing thesis itself is the principal bastion protecting mechanism, and that Gödel's work was perhaps the best thing that ever happened to both mechanism and formalism. I pushed these lines of argument in my dissertation with the patient help of my readers, Raymond Nelson and Howard Stein. I would especially like to thank the latter for many valuable criticisms of my dissertation as well as some helpful suggestions for reorganizing it in the direction of the present book. Innumerable discussions with Jean van Heijenoort on the history and philosophy of mathematical logic have also been invaluable to me in writing this book. It would not exist at all, however, much less in its present form, but for Ilona Lappo: she has been its typist, its editor, its sounding board, and its muse.

INTRODUCTION

This is an essay on the significance of various metamathematical theorems, notably those of Gödel and Church on incompleteness and decidability, for both psychology and the philosophy of mathematics. It is widely believed, for example, that Gödel's incompleteness theorems refute the formalist philosophy of Hilbert. A less prevalent but not uncommon view – one which Gödel himself inclined towards – sees in such theorems a refutation of mechanism in psychology, i.e. as showing that man is not, or could not be adequately modelled by, a machine. I shall criticize both of these views at length, and argue to the contrary that both the formalism of Hilbert and Bernays and the mechanism of Hobbes, La Mettrie, and Turing are on balance supported and strengthened in various ways by these so-called 'limitative' theorems. Incompleteness, I shall argue, cannot coherently be interpreted as our inability to formalize our notion of *number* as such, but rather as our inability to completely describe *the behavior of certain machines.*

I shall also explore the close relation, pointed out by N. Wiener and G. Kreisel and implied by their common metamathematical support, between formalism and mechanism. Indeed, J. Lucas conceded, in a rejoinder to my earlier criticisms [261] of his attempt to refute mechanism using Gödel's incompleteness, that

> It is inherent in my program that I cannot prove to the formalist that he cannot be represented by a formal system, but only enable a reasonable man to see that in fact he is not. But that, however achieved, is enough to refute mechanism ([177], p. 312).

I hope to show, however, that formalism is a reasonable position, and that, insofar as Hilbert was a formalist, he was a very reasonable man, especially in the light of Gödel's work. Prior to this work Poincaré had objected to Hilbert's formalism that his formal systems for number theory *would* be complete and hence sterile for research, while Brouwer had objected that, since Hilbert's mechanical formal systems represented only the hollow shell of mathematical language and not mathematical thought itself, their study could lead to nothing of mathematical value; in particular, he felt that a proof of their consistency would be so easy as to be trivial! This is not unrelated to classical objections to mechanism on the grounds that machine behavior, being predictable in principle, was no match for the unpredictability of human behavior.

A full understanding of Hilbert's formalism is impossible without a close study of his work on the foundations of geometry, which contains the germ of all his later ideas for the foundations of mathematics. In fact, Hilbert developed a program for the study of geometry which was in some ways more extensive than his later one for arithmetic and set-theory. At the heart of his mathematical philosophy lies a certain '*Parallelisierung*' of geometry and arithmetic with which he opposed the prevalent belief of his day that true rigor resided only in arithmetic and analysis. And we shall see that the issue of mechanism arises in a more full-blooded and interesting way in Hilbert's geometry than it does for the arithmetical theories of his later program.

Another criticism leveled against Hilbert was that his later program required clarification of such hopelessly unmathematical ideas as 'decidability in a finite number of steps'. Most of us would nowadays accept such analyses as Turing's as showing that such notions can be treated in a purely mathematical way. In fact, Church's thesis, especially in Turing's form, plays a central role in my whole present approach. In my previous criticism of Lucas, I suggested that since the generality of Gödel's incompleteness depends on Church's thesis, one could insulate mechanism against the incompleteness argument simply by giving up that thesis. After all, some rejected it in any case. But this suggestion was quite misguided, not only for having underestimated the evidence for Church's thesis, but also for having not fully appreciated the force of Church's thesis *as a restricted form of mechanism itself*. Indeed, this thesis implies that however much creative insight may have gone into the discovery of an algorithm, and however abstract the concepts or objects be in terms of which it is defined or specified, as long as it is effective it will be mechanizable after all. In the words of Post, this thesis is our "axiom of reducibility for finite operations". In fact, it was only when Post came to appreciate this aspect of Church's thesis that he gave up his own incompleteness argument against mechanism! The thesis does not say, of course, that all mental operations *are* effective, but by implying the mechanizability of those that are, it shows that mechanism cannot be *effectively* refuted. I will spell this out in various ways in subsequent chapters.

By focussing my analysis on Turing's thesis I try to avoid having to consider such implausible constructions as 'the Gödel sentence of a man', supposed to be a machine, which hang over this literature like a fog, lending it an unreal and forced air. Individual Gödel-sentences of a formalism are, after all, of no special significance in and of themselves: what is important is that the whole *set* of them is not recursively enumerable, the significance of which is formulated already by Church's thesis itself. In short, only the 'mass phenomena' of undecidability, if anything in logic, is of ultimate importance for mechanism. Thus any role played by Turing's computability theory in psychology will have to be somewhat analogous to that played by thermodynamics in physics. That is, just as this branch of physics deals very generally with both capabilities and limitations of machines, such as energy conversion and dissipation respectively, which are established independently of the physical details of their realizations, so Turing's theory established both the existence of universal machines and the non-existence of any other machine predicting their halting behavior, all from very general finiteness assumptions. I shall elaborate this point in Chapter I by comparing L. Szilard's attempt to exorcise Maxwell's demon with Turing's exorcism of what I will call 'Hilbert's demon'.

As I have already mentioned, not everyone accepts Church's thesis or Turing's argument for it. In fact, Kreisel has complained of the undeserved 'glamour' which surrounds Turing's analysis, arguing that he only analyzed mechanically effective and not *humanly* effective functions. Specifically, both Kreisel and Gödel have claimed that Turing's finite-state hypothesis is unjustified for the latter notion. The claim that the mind of a human computer has somehow an infinite number of states has also been used by R. Abelson to refute the mind-body identity thesis. Part of my criticism of these infinitistic assumptions and arguments, as well as my defense of the finiteness principles of Hilbert and Turing at which they are directed, turns on a close analysis of Dedekind's classic work on number, recursion, and infinity. The essential point is that, although it built profitably upon his admirable analyses of number and recursion, modern logic – including Brouwer – could not accept Dedekind's ingenious 'proof' for the existence of an infinite set based upon a consideration of the structure of thought. We have long since replaced his proof by an *axiom* of infinity. Moreover, a careful examination of the difficulties of Dedekind's abortive proof leads one to see similar and related difficulties in the arguments of Kreisel, Gödel, and Abelson for the infinity of mental states. Given the failure of all attempts so far to prove the infinity of either our numbers, thoughts, or mental states, finitism may yet win out without ever firing a shot. In this connection we may do well to ponder a remark of Nelson Goodman. Goodman distinguishes his brand of nominalism from finitism which he admits is nevertheless "a friendly companion of nominalism". In answering a challenge of Church that he show his nominalistic syntax adequate to salvage the main results of 'non-finitist syntax', including Gödel's incompleteness theorem, Goodman remarks that

> Admiration for Gödel's incompleteness theorem does not make the result any more welcome. If incompleteness depends essentially upon those aspects of arithmetic or its supposed foundations that are not finitistic or nominalistic, this argues *for* rather than *against* finitism or nominalism. Gödel's result stands as a *reductio ad absurdum* giving notice of an anomaly that calls for correction. Incompleteness is no more to be cherished for the sake of Gödel's theorem than is crime for the sake of detection; banishment of crime and incompleteness to the realm of fiction would hardly be a matter for regret ([104], p. 154).

As for myself, I have come to cherish incompleteness for the support it lends to mechanism generally, and to Turing's thesis in particular. But in trying to work out the details of this striking development I have more than once been struck by the make-believe air in which parts of the subject flourish. Perhaps indeed, it cannot be made to satisfy finitist scruples completely. If not, we may yet, in all honesty, have to banish it to a realm of fiction where, though shorn of truth, its beauty may still shine.

Finally, a word about my historical approach. Originally, I had planned to treat these matters entirely systematically, especially in view of the modern technicalities involved in this material. But the more I tried to sort out and understand the arguments, to sift claim and counterclaim, the more I found that most of the central figures, however original they might have seemed, had really gotten key ideas from their teachers and predecessors – sometimes very obscure ones at that. And, most importantly, time after time I found that, *because* of my ignorance of these antecedents, *I had not, nor could have, really understood those ideas.* All the logical analysis in the world will not reveal the intentions behind ideas, and without these intentions one all too easily misunderstands and misjudges the ideas and theories of a writer no longer living. Moreover, I found that many of the arguments I thought had necessarily arisen in connection with modern developments had in fact already been made long ago, and in contexts in which their strengths and weaknesses stood more nakedly exposed. On the other hand, one also finds that current ideas and results can illuminate older and crustier ideas. The lesson seems to be this: we cannot fully understand our own conceptual scheme without plumbing its historical roots, but in order to appreciate those roots we may well have to filter them back through our own ideas. This is somewhat analgous to the relation of mutual dependence between quantum theory and classical physics urged by Bohr. In any case, it has become clear to me that many of the positions and ideas dealt with below, especially with respect to mechanism and formalism, have been badly misunderstood and misrepresented by those who have ignored their history. The 'genetic fallacy' is only fallacious in straw-man cases.

CHAPTER I

MECHANISM: SOME HISTORICAL NOTES

> When your Daemon is in charge, do not try to think consciously. Drift, wait, and obey.
>
> R. Kipling

1. MACHINES AND DEMONS

The concept of machine has dual origins in mechanics and human fabrication. A machine was often taken as any device whose behavior can be explained solely by mechanical laws, and often as any man-made device for performing some task, and often as both. Machine behavior was assumed to be completely determined by such laws, and as a consequence of this, also *predictable* in principle, for the laws allowed one to calculate its behavior. A paradigm machine having all of these properties was the cuckoo clock which Descartes adopted as a model for living organisms. But insofar as the behavior of organisms is often *de facto un*predictable and they remain beyond human fabrication – and biology irreducible to mechanics – any mechanistic thesis that organisms are machines will require a concept of machine which uproots and generalizes its origins, and suitably modifies or rejects the properties of determinism and predictability that were based on them. Thus it was only by assuming that God made the universe that Descartes could claim it was only "a machine in which everything happens by figure and motion". And when Newton discovered the laws by which the figures in Descartes' 'universal machine' actually move, Laplace could only embrace it by admitting the possibility of an 'intelligence' or *demon* who could use them to calculate its whole future course from sufficient data about its current state: "for it, nothing would be uncertain". Pascal, however, had already demurred here: precisely because he felt nature to be *un*predictable, he argued that it could not be a machine.[1]

Laplace's demon distills the core of the classical mechanist view of nature, which is generally taken to be its predictability from deterministic laws. Bohm has argued, however, that the essence of mechanism, which is indeed implicit in Laplace's demon, is rather the view that

> ...the great diversity of things that appear in all of our experience ... can be reduced completely ... to nothing more than consequences ... of a final set of qurely quantitative laws determining the behavior of a few kinds of entities ([27]' p. 37).

Later I shall discuss the 'finitism' that underlies this interpretation of mechanism. For now we note that Bohm also rejects the view that the rise of field theories already overthrew mechanism: they admittedly strained it inasmuch as "fields, being continuous, required a non-countable infinity of variables for their mathematical expression" ([27], p. 46). But Laplace's demon survives, for one can imagine it

> ...endowed with the power of dealing with a non-countable infinity of variables, so that he could then calculate the future of the universe with complete precision, although the labor involved would obviously be infinitely greater than for a universe consisting of nothing but bodies (*ibid.*).

Descartes' view of organisms as machines, meanwhile, led eventually, through its impact on physiology, to the 19th century idea that organisms were heat machines governed by the two thermodynamic laws of the conservation and dissipation of energy, and in fact these laws turned out to be irreducible to mechanics alone: new assumptions about *probabilities* are needed. In examining this crucial development, especially for how it affected our concept of machine, it will be convenient to begin with W. Coleman's useful formulation of two senses of 'machine' current at the end of the 17th century:

(1) "A contrivance ... for the application of power to a particular task".
(2) A device operating "mechanically, that is, without occasional intervention or sustained by voluntary conscious action" ([55], p. 122).

Sense (2) goes back to Descartes' view that machines don't think, for they involve only the motion of figures which are but modifications of *extension*, which he assumes to be fundamentally different from thought. But Descartes also believed that human thought *does* effectively intervene in the human *bodily machine*, which would conflict with (2) unless some kind of intervention could be imagined which does not compromise the autonomy of machines implicit in (2). But since he assumed the mind to be *un*extended and not subject to mechanical law, its causal intervention in any extended bodily machine, in whatever sense of machine, is bound to be rather mysterious, prompting Ryle to speak of a 'ghost in the machine'. Henceforth, we will simply speak of such a mind as 'Descartes' demon'.[2]

Sense (1), meanwhile, led to the problems of vitalism, which held that organisms supplied their own power by special non-physical means, an unassailable position until combustion in living organisms was understood in physical terms. Some clues came from the steam engine showing the convertibility of heat into mechanical work and then other devices demonstrating the interconvertibility of other known physical processes, all leading to the idea of a common measurable basis for such conversions, energy. R. Mayer conjectured the chemical process of oxidation as the source of an organism's energy, and extensive research confirmed not only this but also the conservation of energy itself. From a physiological standpoint, organisms were chemical heat machines which convert and conserve energy.

This view of organisms also extended the machine concept itself, for machines were no longer devices which only apply power or transfer energy as in sense (1), but now actually transformed energy in the sense of converting one form of it into another. Moreover, the thermodynamic laws governing these conversions are not derivable from mechanics, on which the classical machine concept had rested, even in its cartesian sense (2). But these laws also impose certain *limitations* on what machines can do in principle.[3] Thus, the conservation law implies that machines cannot produce perpetual motion by creating energy. And insofar as human bodies are viewed as chemical heat machines, this also seems to exorcise the cartesian demon, for

> If the mind acted on the body, either it was itself a physical system that exchanged energy with the body, or it was not a physical system, and its interaction with the body violated the conservation principle ([110], p. 205).

But this line of argument, which drove Leibniz to monadology, has not escaped criticism.[4] In any event, it might still be the case that organisms could create perpetual motion by using their *psychological* functions to violate the limitations imposed on the efficiency of heat machines by the second thermodynamic law of the dissipation of energy (law of increasing entropy).[5] This possibility is raised by another famous demon introduced by Maxwell to emphasize the deep difference between the second law and the laws of mechanics, in particular, the statistical nature of the former.

Maxwell's demon would create perpetual motion by using its intelligence and perception to effectively but effortlessly distinguish the fast from the slow molecules of a gas in thermal equilibrium and thereby reduce its entropy by operating a trap door to sort its molecules without the physical expenditure of energy required by the second law. Indeed, J. Larmour could still write in 1910 that this possibility

> . . . shows that the principle of the dissipation of energy has control over the actions of those agents only whose faculties are too gross to enable them to grapple individually with the minute portions of matter which are the seats of energy ([163], p. 401).

Intelligent beings might also be limited by "the rudeness of the machinery . . . at their disposal for controlling the behavior of those ultimate portions of matter" (*ibid.*, p. 401). But apparently by nothing else.

This puzzling situation, sometimes called 'Maxwell's paradox', was clarified in 1929 when L. Szilard argued that the energy gained by Maxwell's demon would be balanced by the energy it would have to spend to get the *information* needed for deciding when to open the door. However, in order to make his analysis part of physics, Szilard replaced the demon with a suitable mechanical substitute, arguing that since "biological phenomena are beyond physics in the strict sense", a 'direct test' of demons was impossible. Nevertheless, we can still understand the essential problem posed by the demon, for

> We may be sure that intelligent beings – insofar as we are dealing with their intervention in a thermodynamic system – can be replaced by non-living devices whose biological phenomena one could follow ([242], p. 302).

Szilard justified this by analyzing in physical terms exactly what is required, in certain simple situations, to effectively decrease entropy and found that, in general "a measurement procedure underlies the entropy decrease effected by the intervention of intelligent beings" (*ibid.*, p. 307). Specifically, they need a 'coupling' of a value y of the measuring system with the simultaneous value x of a fluctuating parameter of the system being measured: such couplings provide "a sort of *memory* faculty, in the sense that one can recognize by the state parameter y what value another parameter x had on an earlier occasion" (*ibid.*, pp. 302–3). And since only such measurement is essential, "we need not construct any complicated models which imitate the intervention of living beings in detail" (*ibid.*, p. 307). Indeed, Szilard specified simple mechanical devices which, in certain typical situations, effect the required measurement and could thus replace the demon. Contrary to popular belief, however, he did not, nor could he, *prove* that such a device exists for every *conceivable* situation – no more than Turing could prove that a mechanism always existed which simulated every calculation of his 'human computer' – and consequently did not infallibly demonstrate the impossibility of Maxwell's demon.[6] His analysis cannot exorcise a demon, say, with extrasensory perception – if such exists – but only one whose perception is based on known physical processes and whose measurements are effective in the sense of being based on systematic physical procedures. He himself did not analyze perception explicitly in physical terms, but only the information one gets from certain measurements. L. Brillouin carried Szilard's analysis a step further by analyzing physically certain processes of perception, which he claimed to be presupposed by any exercise of intelligence. Like Szilard, he concludes that such processes increase entropy by more than the information subsequently gained by them allows one to decrease it. But, as in Szilard's analysis, one may wonder whether the 'intelligence' of the demon has really been exploited. In effect, all Szilard required was a certain kind of *mechanically realizable memory*. Thus W. Ehrenberg has objected to Brillouin's analysis that it

> . . . eliminates the intelligent portion of the being by demonstrating the existence of a physical entropy-producing link in the relay action this intelligent being represents. As a result the agent does not rely on his intelligence, since he needs physical means to obtain the information – but given the physical means we do not need the agent any longer because we can replace him by a machine![7] ([75], p. 109).

Later we shall see that similar objections, inspired by intuitionism, have been raised against Turing's analysis of 'effective human computability' in terms of his machines.

It is still possible that the demon might accidentally decrease entropy by lucky guesswork, or even that the gas molecules just might separate spontaneously, for Boltzman had already shown that the entropy law must be interpreted as a probability law: it is only highly probable that entropy always increases. Entropy itself is now interpreted as a function of the number of microstates of the gas consistent with a given macrostate and can thus be interpreted as 'missing information' about the real state of the gas. In confronting the random fluctuations of a gas in equilibrium the demon is faced with the problem of finding an effective 'gambling system' and the second law now says essentially that there *are* none. For the first time laws of probability took a place alongside of mechanical laws as laws of nature, yet we must admit with Cassirer that

> It is evident . . . that in viewing the entropy law as a probability law there had been intro-

duced into the very concept of law itself a dualism wholly foreign to its original meaning. This dualism is illustrated by Maxwell's 'demon', who, without violating a single dynamic law, without acting counter to the laws of mechanics or the principle of conservation of energy, could reverse the direction of world events and abrogate the law of entropy ([47], p. 288).

The question is whether this dualism has compromised mechanism itself. Does mechanism have room for chance and real randomness that is not simply, as it was for Laplace, a reflection of our ignorance?

We now resume our discussion of Bohm's philosophy of nature, for he has argued forcefully that real randomness not only does not compromise mechanism, but rather it *enhances* mechanism. It is true that mechanism must now part ways with determinism:

But in doing this it has conserved and in fact enhanced the essential characteristic of mechanism, namely that everything in the universe can be reduced completely ... to nothing more than the effects of a set of mechanical parameters undergoing purely quantitative changes ([27], p. 63).

The loss of detail entailed by the incorporation of randomness doesn't make things less mechanistic, but only signifies a shift from one mechanist paradigm to an even more mechanistic one:

Indeed, the introduction of absolute arbitrariness and lawlessness into a theory is analogous to taking as a model of the world, not an idealized frictionless machine of the type envisaged by Laplace but, rather, an idealized roulette wheel that would give an irregular distribution of results *depending on nothing else at all* (*ibid.*).

We thus arrive at what Bohm calls 'indeterministic mechanism', the foundation of which was laid by von Mises' theory of probability, based on the notion of a '*Kollektiv*', his term for a sequence of events or objects to which the laws of probability apply. *Kollektivs* are required to be 'genuinely random', in the sense that "there are no causal relationships at all, and the distribution is completely 'lawless' " (*ibid.*, p. 64). The question is; *are* there such distributions? Von Mises characterized the randomness property by the impossibility of an effective 'gambling system', that is, a method for specifying a subsequence of the *Kollektiv* having a different limit of the frequency of favorable cases than the *Kollektiv* itself. This method must clearly be *effective*, for otherwise we can simply define nonconstructively such a subsequence.[8] So the question now becomes: what is an 'effective' method? It was precisely the attempt to clarify this notion that led to our current abstract and most general concept of machine: the Turing machine. And the best way to motivate the analysis by which Turing introduced this concept is to see it as the exorcism of yet another demon, one which we will call 'Hilbert's demon'.

Hilbert's attempt to clear up certain puzzles and contradictions in the foundations of mathematics concerning infinity had led him to introduce the notion of a *formal system* as a 'finite picture' of infinity, and to formulate an *Entscheidungsproblem* for it, the problem of deciding whether a given sentence is provable or refutable in it. Though such systems may 'refer' to infinity, like a perspective painting with its vanishing points, they themselves contained only a *finite* number of symbols and rules of the kind that one could effectively enumerate all their proofs, and hence also the theorems and their negations. Thus one could apply the 'museum' method to their *Entscheidungsproblem*: just search through these two enumerations until the given sentence is found in one of them.[9] Hilbert discovered that one of his formal systems, call it *P*, was a sort of logical womb into which virtually all of mathematics fed in the sense that a solution of its *Entscheidungsproblem* could be used to settle virtually any mathematical problem whatever. Now the museum method can hardly be applied by a mortal mathematician, but one can easily imagine a demon to carry out the search without having to endow it with any notion of infinity. For as O. Becker insisted in connection with Hilbert's problem:

> Der Mensch ist gewissermassen so wesenhaft sterblich, dass ihn auch eine unbegrentze Lebensdauer nicht 'unsterblich' machen würde; auch der 'Dämon' als Grenzbegriff des Menschen kann zwar belibig grosse Endlichkeiten, auch keine aktuale Unendlichkeiten erfassen ([8], p. 342).

Though such a '*dämonische Mathematiker*' could never itself be sure that it could prove all theorems

> Wohl aber kann er schliesslich (im Gegensatz zum Menschen!) jeden belibigen ihm vorgelegten Satz beweisen oder widerlegen, denn ein bestimmter Satz hat einen bestimmten minimalen Beweisumfang (*ibid.*, p. 343).

Prima facie there was even less objection to Hilbert's demon than to Maxwell's, and yet, just as the latter would have brought energy-free work, Hilbert's demon would have brought *thought-free knowledge*. Both demons offended common sense, which was nevertheless powerless to exorcise them.

Turing's exorcism is based on a close connection he found between Hilbert's formal systems and his own abstract computing machines, which was not surprising since, as Wiener observed in his introduction to cybernetics:

> . . . it has become quite evident, both to the nominalists like Hilbert and to the intuitionists like Weyl, that the development of a mathematico-logical theory is subject to the same sort of restrictions as those that limit the performance of a computing machine ([272], p. 13).

Specifically, Turing showed that a given sentence S is provable in Hilbert's system P just in case one of his machines eventually halts when fed a description of S and P. So Hilbert's demon need only solve the 'halting problem' for these deceptively simple machines, and Turing had only to show that he cannot. However, Turing could no more directly deal with Hilbert's demon than could Szilard with Maxwell's: as physics says nothing about perception per se, mathematics says nothing about *thought*. But just as Szilard argued that inanimate devices could make the measurements essential for reducing entropy, Turing argued that his machines could simulate any *effective* procedure for solving the halting problem. So one of his machines could solve it if it was solvable at all. He then refined Cantor's diagonal argument to show that none of his machines could solve it: another demon exorcised. But this exorcism depends essentially on Turing's fundamental thesis:

(T) A procedure (function) is 'effective' just in case it can be simulated (computed) by a Turing machine.

which by its nature cannot be proved. The crucial assumption behind (T) is that Hilbert's demon, insofar as its cognitive procedures are to be effective, must be capable of only *finitely* many 'states of mind' for otherwise, Turing argued, "some of them will be 'arbitrarily close' and will be confused". He also showed that all known effective functions are in fact computable by his machines, but the decisive case was his *universal machine U* which actually computes the function $u(n, x) = \phi_n(x)$, where ϕ_n is the unary function computed by the nth machine M_n in the enumeration

(3) $M_1, M_2, M_3, \ldots, M_n, \ldots$

of all his machines. Whereas *Turing* could certainly compute $u(n, x)$ effectively, this was seemingly only *because he knew the code that he himself had quite arbitrarily set up in (3)*: it was by no means obvious that one of his machines could 'break' his code, nor, having broken it, could then go on to convert the program for M_n into computation.[10] Indeed, it seems rather like a miracle of pre-established harmony that U could be so in tune with all the conventions required for simulating all other machines.

So all Hilbert's demon really had to do was to decide whether or not

U itself would halt. Surely he can do this. After all, *U* is a single perfectly deterministic machine capable of only a finite number of internal states, so its behavior ought to be just as effectively predictable from its input and program as was Descartes' universal machine by Laplace's demon. If *U* does halt, Hilbert's demon will find out by patient search. But if *U* is not going to halt, the demon will have to suspend his search and *prove* this fact. There doesn't seem to be any reason why he can't do this effectively too, for Hilbert and other mathematical logicians had developed very strong formal systems *F* which prove theorems of higher mathematics, and in which one could *also* prove theorems of the form

(G) *U* will never halt when fed the numbers *n* and *x*.

So given *n* and *x*, the demon need only search through the theorems of *F* until he either finds one of form (G), or *U* halts. This not only seems to effectively solve *U*'s halting problem, but also to refute (T)! But another miracle blocks the demon and saves (T): Gödel discovered that in every formal system *F* of the kind required, there were true sentences of the form (G) *not provable* in *F*. No formal system can provide the demon with all the information he needs to say whether *U* will halt. This is all the more amazing since *F* can be chosen to prove all the strongest known theorems of higher mathematics and set theory, and yet it must fail to completely follow the behavior of this single machine *U*. So far from refuting mechanism, as many have claimed, Gödel discovered a kind of invisible protective shield encasing the foundation for modern mechanism, namely Turing's thesis (T). This is indeed a very strong thesis, for it says that *any* effective procedure whatever, using whatever 'higher cognitive processes' you like, is after all mechanizable.

R. Laing has recently shown how to simulate any Turing machine computation by a 'Maxwell demon system', call it *D*, which he introduces as a generalization of classical thermodynamic systems inclusive of their demons. Roughly speaking, the 'demon proper' of *D* corresponds to the finite-state control unit of the machine, the 'gas chambers' of *D* to machine memory tapes, and the 'molecules' of *D* to machine symbols. Maxwell's paradox is reformulated for *D* and Szilard's analysis is invoked to show that the apparently 'free computation' provided by *D* must 'cost us at least as much as whatever energy is necessary for the Demon to carry out his particle detecting and sorting activities" ([161], p. 176). We can also invoke Turing's analysis to show that the demon system *D* which simulates *U* must be effectively unpredictable: in this case, there is no effective method for telling whether, when started with some initial distribution of molecules in its chambers, *D* will eventually reach some desired 'halting state', say with all the fast molecules in one chamber and the slow ones in another. This might seem understandable for *D*, recalling the random fluctuations associated with classical thermodynamic systems, were not *U* itself a purely deterministic machine. Is there something 'random' about *U* also?

This question is clarified and answered by a new theory of algorithmic information theory and randomness developed by G. Chaitin and others, which also throws new light on the incompleteness discovered by Gödel. The key idea is that of the 'program size complexity' of a set *A*, roughly, the length of the shortest program that a suitably chosen universal Turing machine can use to generate *A*. Roughly, *A* is random if its shortest program is approximately as long as *A* itself: for such *A* there is no more efficient description than to just present *A* itself. This idea of randomness

> . . . is based on the observation that the information embodied in a random series of numbers cannot be 'compressed', or reduced to a more compact form ([50], p. 47).

Unlike von Mises' theory of randomness, this theory applies to *finite* as well as infiite sets.[11] Chaitin [49] defines all of these ideas very precisely and shows that the resulting concept of program size complexity has *all the formal properties of the entropy concept of information theory*. He also defines 'the probability that *U* will halt' – when its programs are chosen by coin-flipping – and shows that this probability is random, i.e. the binary sequence representing it has maximal complexity or entropy. Finally, Chaitin shows that in any formal system there are random sets whose randomness cannot be proved in that system, which might clarify the 'true source' of incompleteness.[12] In any case, such results show that Bohm was hasty when he claimed that von Mises' idea of applying probability only to 'completely lawless' distributions "can never be given a clear meaning in any specific problem or application" ([27], p. 65).

Laing suggested that the embeddability of arbitrary Turing machines in the physical universe might result in unsolvability and other machine properties becoming "possible properties of the universe" (*ibid.*, p. 174). Specifically, if the universe contains a realization of *U*, the unsolvability of its halting problem will pose serious difficulties for any predicting demon, for, as Bernays once remarked, "even a divine intelligence can *effectively* calculate nothing more than we are able to effectively calculate". Are there prediction methods that are not effective? Suppose Laplace's demon is given a state and input pair $\langle q, s\rangle$ of U as well as its program (its 'laws of motion'), and tries to calculate whether or not U will ever reach a halting state. Turing's analysis shows that any effective general method he could possibly adopt for arbitrary $\langle q, s\rangle$ must either give wrong answers or none at all. But might not the demon possibly find, for each $\langle q_0, s_0\rangle$, a separate method telling only whether U will halt from $\langle q_0, s_0\rangle$? Church suggested that this question was meaningless "on the ground that the universal quantifier which it contains is intended to express a mere infinite succession of accidents rather than anything systematic" ([61], p. 107). One can perhaps interpret Church's remark in the light of the randomness of the halting probability of U, which underscores the complete lack of any connection between those pairs $\langle q, s\rangle$ leading U to a halt. Laplace's demon could conceivably find these infinitely many separate methods, but surely not by calculation. Nor could it, as Laplace had claimed, embrace everything in a *single formula*. Rather must it proceed by divination, a process that can only be assumed, not actually described. Recall that Bohm had even granted the demon the ability to survey an uncountable set, which vividly shows how far one is obliged to go with 'pure spirit' or 'intelligence': exorcism only seems to be feasible for demons which are supposed to perceive, as was Maxwell's, or ones which are supposed to proceed effectively, as was Hilbert's. This brings us back again to Bohm's critique of mechanism in the philosophy of nature and his exorcism of Laplace's demon from the universe as he sees it.

Indeterministic mechanism culminates in quantum mechanics. Many would invoke its uncertainty principle to exorcise Laplace's demon, but disagreement over its interpretation is deep enough to decisively affect the result.[13] In any case, Bohm does not accept this principle as final, and looks rather for 'hidden variables' to causally explain the randomness which the Copenhagen interpretation sees as "completely arbitrary and lawless, having no causes at all" ([27], p. 102). Such variables would provide the 'deeper substratum of continuous motion' for the subatomic domain denied by quantum mechanics. But such denials, Bohm admits, enable quantum mechanics to *save the essential features of mechanism*, namely the description of nature by a "set of purely quantitative laws of probability fitting into . . . a general physical and mathematical scheme that is absolute and final" (*ibid.*). In order to contrast this 'quantum mechanism' as sharply as possible with Bohm's own philosophy, let us recall Poincaré's formulation of the original hypothesis of Planck on which it was built:

(F) "A physical system is capable of only a finite number of states; it jumps from one of these to another without going through a continuous series of intermediate steps"[14] ([207], p. 85).

To classical mechanism, which treated nature 'in terms of a finite set of qualities and laws', and quantum mechanism based on (F), Bohm opposes 'the qualitative infinity of nature' which asserts that

(I) The possible variety of basic properties and qualities existing in nature is infinite.

This is understood to imply the existence of enough levels in nature to allow for explanations of randomness encountered on a given level by appeal to factors existing at a deeper level: since on every level some events depend on factors of another it follows that "If the universe has an infinite structure, then . . . data taken at a certain level will necessarily give imprecise predictions because of random fluctuations" ([27], p. 22). Thus (I) tries to explain the appearance of a relative randomness which (F) eventually forces one to accept as absolute. Classical mechanism also admitted 'blind chance', but saw the probabilities needed to analyze it as purely subjective, which is just what Laplace illustrated with his demon: to say that nothing was uncertain for it was his way of saying that probabilities do not occur in nature, but only *in our finite minds*. Now since the exorcism of a predicting demon is *prima facie* easier from worlds manifesting some kind of real randomness, such as the indeterministic one of quantum mechanism or even a deterministic one realizing a universal Turing machine, *we see that the existence of such a demon is by no means essential to mechanism in general*. But these exorcisms rest ultimately on the finiteness assumptions (F) and (T) respectively, so we may ask if suitable such assumptions might not be applied more directly. Thus D. Hawkins would exorcise Laplace's demon by making him part of the world whose state S he must obtain and then assuming the world to have only a finite amount of energy. The demon's measurement of S will require an amount of energy proportional to the number of discriminations he must make, and an infinite number of discriminations will be needed for perfect accuracy. Thus he is limited to a finite number of discriminations and can only measure S with finite accuracy. How then is the demon to be exorcised from Bohm's infinite world of (I)?

In (I) the demon must do "an infinity of different kinds of experiments each of which depends significantly on the laws of a given context" just to know the totality of nature's laws, and for this he needs to know not only the existing laws, "but also all the *new* laws that are expressible only in terms of the infinity of new qualities, new entities, and new levels that are going to come into being into the infinite future" ([27], p. 159). Bohm claims that even a 'super-being' will never be able to predict the future of such an infinite universe *if* he

> ... resembles us to the extent of obtaining his knowledge through a series of partial investigations of the universe, and not ... by divine revelation or by *a priori* intuitions which he finds by plumbing the depths of his mind (*ibid.*).

Now without explicit finiteness assumptions of some kind, this is not so clear. However, we have already seen that the existence of the demon is not essential to mechanism: what is important is the sense in which Bohm's world is no longer 'mechanical'. According to him, in this world, over an infinite period of time

> ... the determination of ... the essential features of an effect is evidently not purely mechanical, because it involves not only an infinite number of contingent factors but also an infinity of kinds of qualities, properties, laws of connection, all of which themselves undergo fundamental changes with the passage of time (*ibid.*, p. 160).

It would thus seem as if the essential conflict between mechanism in the philosophy of nature and its opponents must center on the *finitism* inherent in mechanism: to go beyond mechanism requires very strong infinitistic assumptions about nature itself. It does no good to merely exorcise Laplace's demon for that would require constraints on his mind, and it is not easy to think of ones *which would not as easily allow his exorcism from mechanistic worlds as well.* Mechanists meanwhile, as we shall see, have questioned whether we really have a coherent conception of infinity at all.

We now conclude this section by summarizing the development of the machine concept and its relation to the exorcism of our various demons, and then considering a criticism of our current machine concept.

Laplace's demon was a spiritual calculator who would predict with certainty the future of Descartes' universal machine. Since this machine only passes from one state to another by determinate laws, no difficulty of principle seemed to stand in this demon's way, assuming it could completely measure a given state.

Descartes' demon was a spiritual thinker who would causally interact with the cartesian bodily machine to improve its performance. When this machine was discovered to conserve its energy, this demon could no longer intervene in it without becoming material; in particular, Descartes' 'proof' of the spirituality of this thinker had to be re-examined to see whether or not it could be consistently incorporated into the bodily machine. Meanwhile, the efficiency of the energy conversions of this machine were seen to be limited by the dissipation law: its energy tends to become less available for work.

Maxwell's demon was a very small physical observer who would direct, on the basis of measurement procedures, the molecular motions constituting energy to reverse its dissipation and make it available for work again. But this demon must deal with the randomness of these motions: he must somehow take advantage of fluctuations in which he can observe no pattern or redundancy. Szilard showed that the measurements typically required could be made by mechanical devices capable of a certain kind of memory, but in a manner consistent with the dissipation law.

Hilbert's demon was an idealized mathematician who would effectively solve the halting problem for Turing's universal machine – and thereby virtually every question in mathematics without further thinking. Turing showed that this could be effected by a Turing machine if it could be done at all, but that no such machine could do this. The immediate difficulty is Gödel's discovery that no formal system can have complete information about U, which may in turn be seen as a consequence of the *randomness* of the set of situations $\langle q, s \rangle$ from which U halts. U's behavior is effectively unpredictable. Hence, if U is embedded in the universe it will pose serious problems for Laplace's demon. Thus, suppose he is given the 'complete state' of the universe. We are now inclined to regard his remaining problem as 'purely mathematical', essentially solving differential equations. But if we let M stand for the system of methods he uses for solving them, there will exist situations $\langle q, s_M \rangle$ for which U will halt just in case M is *inconsistent*. So by another discovery of Gödel, if the demon's mathematics M *is* consistent, then U will never halt for $\langle q, s_M \rangle$, but our demon will not be able to prove this by M alone: he will have to find *new* mathematical methods. We can, of course, assume 'mathematical omniscience' for the demon, but this possibility depends on Hilbert's belief that all mathematical problems are 'solvable in principle'. If there are 'absolutely unsolvable' ones, mathematical sentences which are neither true nor false 'in reality', then even an omniscient demon must be uncertain about *them*. I shall examine Hilbert's thesis at length in Chapter III.

The preceding discussion depended on the assumptions that U is embeddable in the universe and that the 'complete state' of the universe is available to Laplace's demon, but both can be questioned. Embedding

U in the universe introduces memory into it: unlike Descartes' universal machine, which merely passes from state to state, there is for *U* a distinction between its *internal* state and what it does, and its states constitute an internal memory in that they depend on previous states and inputs. But these 'inputs' may be outputs of *U* itself which were stored in *U*'s environment, its so-called 'tape', which thus constitutes an external memory. Turing's thesis (T) limits *U* to a finite internal memory, but his unsolvability theorem depends on a potential infinity of available external memory: *U*'s tape must be allowed to grow without limit, and thus is not embeddable in all universes, and perhaps not ours.

The second assumption, that the 'complete state' of the universe be available to Laplace's demon, can be questioned on the basis of quantum mechanics, which reciprocally restricts the measurement of conjugate physical variables in accordance with Heisenberg's uncertainty relations. Thus, if Laplace's demon, like Maxwell's, is supposed to obtain state information by its own measurements, then he cannot obtain the 'complete state' *as it was classically conceived.* If 'in reality' every particle in the universe has a position and momentum at every point in time, then only the knowledge of this state is limited. But it would still exist to be given to the demon, and the universe still viewed as Laplace did. But if there is no 'reality' corresponding to the classical state, then the restricted quantum state *is* the complete state, only it is defined probabilistically. But on this view it is no longer clear that any *limitation* is imposed on Laplace's demon: it would now be absurd, for example, to say that he was limited to 'only' probability distributions of particles appearing on a scintillation screen but was unable to predict the whereabouts of individual particles in transit, if these particles *have* no well-defined location at each point in time; in particular, none when Laplace's demon tries to measure it. Anyone still wanting a classical state can follow Bohm's search for hidden variables. Now von Neumann and others have proved that such variables could only be introduced into the quantum formalism at the cost of physically strange consequences, e.g. forms of action at a distance. But such results cannot preclude the existence of formulas in some other formalism which give effectively the arrival times of individual particles at a screen. In this respect, the limitations imposed on methods of particle prediction by quantum mechanism are less stringent than those of Turing on methods for halting predictions; for in the latter case the claim is not merely that no such effective method is definable in this or that formalism, but that it cannot be defined in any conceivable formalism – indeed, that such a method does not exist at all. As Gödel has observed, Turing's computability concept is 'absolute' in the sense of being independent of any particular formalism.

H. Pattee [198] has argued that the Hilbert-Gödel-Turing theory of formal systems and computability has ignored thermodynamic dissipation and quantum indeterminacy to remain unjustifiably wedded to "the classical ideal of machines and symbols". Specifically, whereas thermodynamics should have taught logicians that the "writing of . . . symbols is an inherently irreversible classification process, and its physical representation . . . therefore probabilistic", they have continued with Post to assume "the permanence of symbols", presumably because statistical mechanics still appeared consistent with determinism. But with quantum indeterminacy "the machine analogy to physics broke down", even though the "machine concept remains an ingrained part of our thought." The upshot of this is that "we are unable to clarify the basis of either formal or mechanical logics", because of having tied the concept of computability too closely to the classical machine. Indeed, Pattee claims that

> . . . in spite of enormous efforts to clarify the foundations of logic and mathematics we still find the intuitive ideas of a 'formal system' of symbols based on the entirely classical concept of a strictly deterministic 'mechanical procedure' . . . even though this is a physical impossibility ([198], p. 315).

He admits, however, that quantum mechanics itself seems to require the classical concepts for its very application, and nicely reformulates this "apparent paradox of quantum and classical concepts" in terms of machines and symbols. In any case, two critical comments are necessary here. First, the quantum laws are as deterministic as classical ones: probabilities enter only into the *state* concept. But this is not inconsistent with Turing's thesis (T): the only assumption about machine states which it appeals to is that they are finite in number, which means that the limitations discovered by Turing are independent of whatever way these states are physically realized or defined. Second, Turing's theory seems to be quite in accord with thermodynamics. In particular, the unsolvability of the halting problem corresponds to the law of dissipation: both are equivalent to the impossibility of effective gambling systems in the face of randomness. But the nature of this impossibility is, if anything, strengthened and clarified in the newer logical theories of computability and randomness. Indeed, in opposition to Pattee, I would say that, in spite of enormous efforts to clarify the foundations of thermodynamics and probability, we still found that the intuitive idea of 'randomness' was based on hopelessly vague notions like 'lack of order' and 'lawlessness'. The new algorithmic theory of randomness which avoids these is based on Turing's theory. Pattee quotes Planck on the 'unfathomable gulf' between an improbability and an 'absolute impossibility'. But the new theories based on (T) show precisely what is 'absolute' in connection with such improbabilities, namely the impossibility of *effective* methods of gambling and decision. I shall examine the 'absoluteness' of Turing's concept in Chapter IV.

This concludes our discussion of mechanism in the philosophy of nature. Our general result is that mechanism here is a kind of finitism, of which there are several forms and which owes both its success and the development of its concepts to the exorcism of certain demons. None of these exorcisms can be logically air-tight, but can only be made morally certain: demons die hard. But the crucial demon is Descartes' mind which interacts with his body, and its exorcism is the most problematic of all. The most direct approach is to extend Descartes' mechanism of the body to man as a whole, as was first attempted by La Mettrie and more recently by Turing himself.

2. MACHINES AND MEN

Descartes' view of animals as machines involved serious difficulties. One of the most conspicuous features of animals as living organisms is their ability to reproduce themselves, but how could mere machines accomplish this? One can conceive of one machine producing another, but only, it would seem, a *less complex* machine. Indeed, Descartes' basic axiom of causality states that a cause must always have more reality and perfection than its effect, so that a 'parent machine' would always have to produce more and more imperfect offspring, contrary to what we observe in nature with real animals. Nevertheless, Descartes hoped that his mechanistic physiology could be extended to cover self-reproduction, and he struggled considerably with the problem in his later years:

> In consequence I could perhaps enable you to understand how, from the humour there [in the organs of generation] assembled, another machine can be formed similar to the parent machine ([237], p. 353).

But he failed to make machine self-reproduction comprehensible, so his mechanistic thesis

(α) Animals and human bodies are machines.

could only appear very implausible at first.

Man was more than a machine: he was a thinking mind 'united' with a bodily machine fundamentally different from it. Machines are material and extended, while souls are immaterial and unextended. Machines also differ from minds in being divisible, while a mind is an 'indivisible whole' without parts. One might object that minds were divisible into their faculties, or even their thoughts, but Descartes seems to have had something more like the self in mind which, let us grant, has a certain 'unity'. The main problem for cartesian philosophy is to explain the nature of mind or self; in particular, the crucial point is to *prove its immateriality*, and the classic argument given by Descartes turns on the logic of doubting:

> I am able to doubt whether I have a body, nay, whether any body exists at all; yet I have no right to doubt whether I am, or exist, so long as I doubt or think ... Hence I who doubt and think am not a body ([69], p. 80).

Hobbes objected that a thinking mind *is* corporeal, for granting the validity of 'I think, therefore I am', the knowledge of 'I think' must itself come from something corporeal, for

> It is not by another thought that I infer that I think ... we cannot think that we are thinking, nor similarly know that we know. For this would entail the repetition of the question an infinite number of times. Whence do you know, that you know, that you know, that you know ([69], p. 62).

Descartes' replies that it is indeed obvious that "one thought cannot be the subject of another thought", but that this is irrelevant: thoughts are the subject of the mind, which is immaterial. Descartes justifies the premises of the above argument for this by appeal to "clear and distinct ideas", which criterion in turn depends on God whose existence is established by a proof "based explicitly upon the necessary connection of infinity and existence: God exists, and exists as a *causa sui* in virtue of his infinity" ([152a], p. 197). Hobbes objected that we *have* no "conception of anything we call *infinite*", whereas Descartes used the very thesis that a finite mind cannot produce an idea of the infinite to prove that an infinite being must have put it in our minds. Again we see that mechanism leads to problems about infinity.

A more operational argument advanced by Descartes against viewing man as a machine was that a machine could never be built to use language as 'appropriately' as man: some of a machine's 'verbal replies' might be appropriate, but not all of them.[15] A related claim is that a machine's behavior cannot have 'sufficient diversity' to simulate all of man's. To these arguments it may be replied first that not all the verbal responses of man are 'appropriate' either, and, secondly, that if men are machines, such machines need not be built. To prove that men are not machines it must be shown that machines cannot be *conscious*, which brings us back to animals.

Descartes' thesis (α), in fact, leads not only to the problem of mechanical self-reproduction, but to others as well, including the problem of consciousness. Besides begging the question of the difference between man and animals, (α) commits Descartes to denying *all* reason to animals, and moreover, at the cost of having to admit an inequality of reason among men.[16] His basic claim is that, quite apart from language, animals just don't think, but this requires him to account for animal sensation without any appeal to consciousness: he must say that their sensations only "occur to them" without their explicit awareness of having them. This amounts to denying *self-consciousness* to animals, for as Gibson points out, for Descartes

> ... the only kind of consciousness is self-consciousness, as discovered first of all in the act of doubt; so that where there is no self-consciousness there is only mechanism ... when I think of myself walking I know that I am a spiritual being, but when I simply walk, I am obeying a mechanical propulsion ([98], p. 204).

We may summarize Descartes' position as follows. Animals are machines, whose self-reproduction is, however, miraculous; whereas man is a thinking being miraculously united to a machine. So there is a basic difference between them, unless machines *can* think and be self-conscious. Descartes denied this, but having never clarified just what thought is, he left open the possibility that thought itself is some kind of mechanical process, a line pursued by Hobbes.

All thought or reasoning, according to Hobbes, is just *ratiocination*. This nominalism is inspired by the Latins who

> ... called accounts of money *rationes*, and accounting *ratiocinatio*; and that which we call *items*, they called *numina* – that is, *names;* and thence it seems ... that they extended the word *ratio* to the faculty of reckoning in all other things ([131], p. 42).

In support of this nominalism Hobbes notes that the Greeks had but one word, *logos*, for both speech and reason, because they believed that reason presupposed speech. In his objections to Descartes' *Meditations*, he opposes just this nominalistic view of thought as the manipulation of symbols, names, and words to Descartes' more conceptualistic view of thought as operating somewhow directly with ideas and conceptions. Hobbes had meanwhile become convinced that

(H) "Reason is nothing but *reckoning* – that is, adding and subtracting – of the consequences of general names agreed upon for the *marking* and *signifying* of our thoughts" (*ibid.*, p. 46),

for which he apparently found support in Pascal's construction of the first calculating machine.[17] Indeed, according to Pascal's sister, her brother's machine

... was an entirely new thing in nature, by which he reduced to mechanism a science which resides in the mind, and by which he found the means of carrying out all the operations with a complete certainty *without recourse to reasoning* (Quoted in [137a], pp. 22–3).

So whereas for Pascal and Descartes nothing which can be 'reduced to mechanism' requires reasoning, for Hobbes reasoning is just a computational process whose mechanical nature is confirmed by such constructions as Pascal's machine. However, Hobbes was in no position to show that all computation can be reduced to mechanism. Nor could he show that, because a machine can do what had previously required a mind, a mind then is a machine. To Pascal it seemed obvious that the mind was not a machine, that there was an infinite gulf between thought and body, yet for proof he relied essentially on Descartes' divisibility argument. Indeed, this was still the principal argument that greeted La Mettrie's extension of (α) to man himself.

Previous philosophy, according to La Mettrie, had failed to explain the nature of the soul. Cartesian dualists gratuitously assumed two distinct substances in man "as if they had seen them, and positively counted them", whereas Leibnizian monadology was simply an "unintelligible hypothesis". Neither viewpoint explains how we can ever "define a being whose nature is absolutely unknown to us". But this does not argue for materialism, for according to La Mettrie, "we can have no idea of the essence of matter". Our ignorance of the ultimate nature of both soul and body, however, does not prevent us from establishing the dependence of the soul on the body, for medical science experimentally shows that

(C) ". . . the diverse states of the soul are always correlative with those of the body" ([162], p. 97).

La Mettrie does not commit himself to either the identity or difference of soul and bodily states, even though he thinks of the latter as causing the former.[18] Whereas the emphasis in cartesian philosophy is on the mind acting on the body, La Mettrie emphasizes the effect of the body upon the mind. But in view of (C), La Mettrie proposes that the best way of studying the soul is to extend Descartes (α) to man himself: he is a machine, one which even "winds its own springs, a living image of perpetual motion". However, La Mettrie argues this extension of mechanism requires a compromise because

(M) "Man is so complicated a machine that it is impossible to get a clear idea of it beforehand, and hence imposible to define" (*ibid.*, p. 89).

This is why all *a priori* investigations of man have failed to explain his nature, which can only be discovered *a posteriori*, though never with certainty, "by trying to disentangle the soul from the organs of the body" (*ibid.*). La Mettrie finds no entity as such to disentangle, but rather that

> ... the faculties of the soul depend to such a degree on the proper *organization* of the brain ... that apparently they *are* but this organization itself (*ibid.*, p. 128; italics mine).

In view of the equation of the soul with the organization of the brain – which echoes the Platonic view of the soul as the form of the body and anticipates the cybernetic view of mind as a program for a machine – La Mettrie grants that "the soul is clearly an enlightened machine". But if a machine can have only finitely many states – as Turing's analysis and quantum mechanism say – then La Mettrie's (C) and (M) will imply that the soul also is capable of only finitely many states. This point shall loom large in later discussions.

Anti-materialists had argued that matter by itself has no tendency to move at all, much less intelligently, and appealed to the soul as the source of bodily movement. La Mettrie appealed rather to the newly discovered irritability of living tissue. In fact, he tried to combine the basic ideas of vitalism and materialism: he doesn't deny the 'vital' aspects of organisms, but argues that they are best studied as aspects of his concept of a self-moving, self-sufficient 'organic machine'. But he warns that we shall never scientifically explain how the soul arises from matter in motion.

Nor should we seek the origins of man, for then we shall "lose ourselves in the infinite" of which we have not the least idea, nor of its personification in God. The argument from design is our only reasonable approach to God, but "perhaps man was created by chance". This is hard to accept, but then God is only a 'logical entity' himself. La Mettrie proposes a 'creative nature' as a way out of this dilemma: it may be that nature has "causes hidden in her breast that might have produced everything".

The main difficulty here, as with the cartesian mechanism (α), was self-reproduction, but Trembly had at least found the polyp "to contain in itself the causes of its regeneration". This shows, according to La Mettrie, that "to destroy chance is not to prove the existence of a supreme being". Of course, Darwin's later mechanistic explanation of man's origins had to include a stiff dose of chance after all in the form of random variation, but even it still presupposed the self-reproductive function which had troubled Descartes and La Mettrie. Chemical and mechanical accounts of self-reproduction have only become possible fairly recently. Of particular interest is von Neumann's existence proof for self-reproducing machines: after generalizing Turing's concept of universal simulation to universal *construction* machines, he further refines the diagonal method to prove the existence of machines which construct replicas of themselves – with or without 'mutations' – which include the instructions for self-reproduction. His 'child machines' are still reproductive. The possibility of overcoming the old problem of the 'degeneracy' of machine construction, which had blocked Descartes' (α), was implicit in Turing's universal machine U which could simulate machines arbitrarily more complex than itself.

La Mettrie seems to follow Hobbes' (H) in interpreting thought as "the effective manipulation of symbols".[19] All knowledge is but a "huge mass of words and figures" which cause our brains to imagine their objects by exercise of the faculty of imagination. When the brain is well organized the imagination is able to "deduce . . . a long chain of consequences" from its inputs. Such is the "generation of intelligence", but Luzan objected to La Mettrie that it was inconceivable how a machine, dependent wholly on its organization, could draw any consequences at all. In short he seems to have been asking how La Mettrie's organic machine could also be a *logical* machine.

3. MACHINES, ARITHMETIC, AND LOGIC

Some of the rashness of Hobbes' (H) may well have resulted from his having taken too seriously Pascal's unjustified claim that his machine executes "in controlled movement *all* the operations of arithmetic" ([236], p. 168). In fact Pascal's machine itself only adds and subtracts – even this is stretching things a bit – and can only be used to multiply and divide by a human operator who himself iterates these processes a suitable number of times. Hobbes claims that though men name other operations, such as multiplying and dividing, than those mentioned in (H), they are nevertheless the same as them, "for multiplication is but the adding together of things equal, and division but the subtracting of one thing as often as one can" ([131], p. 45). Actually, multiplication and division were only reduced to mechanism some thirty years later by Leibniz, who very clearly pointed out that a quite different construction than Pascal's is needed to perform these operations *directly*.[20] The same would be true if one tried to reduce exponentiation to mechanism.

Another seeming weakness in (H) is the apparently narrow arithmetical focus on 'adding and subtracting', but Hobbes rightly points out that

> ... these operations are not incident to numbers only, but to all manner of things ... For as arithmeticians teach to add and subtract in *numbers*, so the geometricians teach the same in *lines, figures, angles, proportions, times*, degrees of *swiftness, force, power*, and the like (*ibid.*, p. 45).

From this it appears that by 'reason' Hobbes meant scientific reasoning. Finally, by reference in (H) to 'consequences of general names', he tried to incorporate *logic*, which itself was traditionally taken to comprise the 'laws of thought':

> ... the logicians teach the same in *consequences of words*, adding together two *names* to make an *affirmation*, and two affirmation to make a *syllogism*, and many syllogisms to make a *demonstration*, and from the *sum* or *conclusion* of a syllogism they subtract one proposition to find another (*ibid.*, p. 45).

One easily sees that this ingenious reduction of logical relations to arithmetical ones – and hence to mechanism – strained Hobbes' arithmetical metaphor far beyond what could be literally justified in his time: indeed, one could say that it was not justified explicitly until Gödel's arithmetization of logic. Two specific questions raised by (H) were these then: (i) Are all possible arithmetical operations reducible to mechanism? (ii) Is logic reducible to mechanism?

The first of these questions was answered affirmatively in principle by the 'Analytical Engine' of Charles Babbage in 1840, which is the theoretical forerunner to Turing's universal machine and the practical forerunner of general purpose computers. Prior to this no essential progress in calculating machines had been made beyond those of Pascal and Leibniz for the specific operations of addition and multiplication, and even these machines still left important control functions in the hands of human operators. Babbage's basic new idea of *inserting a program* in a machine allows these functions to be brought entirely under the control of the machine itself: whereas his predecessors had mechanized specific arithmetical operations, Babbage mechanized the more mentalistic idea of a plan or design for a calculation. This idea also enables a finite machine to encompass an infinity of arithmetical operations, even though it obviously cannot 'include infinity'. Though his machine could not occupy unlimited space,

> ... it is possible to construct finite machinery, and to use it through unlimited time. It is this substitution of the *infinity of time* for the *infinity of space* which I have made use of to limit the size of the engine and yet to retain its unlimited power ([188], p. 60).

Interestingly enough, the inspiration for Babbage's key technical innovation – namely, punched cards inserted into his engine to control the course of its calculations – came from the Jacquard loom which used such cards to fully automate and mechanize the weaving of complex patterns which had hitherto required many human hands. Such cards controlled the arbitrary repetitions of subroutines in the weaving process; more generally, they represented the designs one wished to weave. Indeed, Babbage claimed, "it is known as a fact that the Jacquard loom is capable of weaving any design which the imagination of man is capable of conceiving" ([188], p. 55). Reflection on this 'universality' of Jacquard looms and the applicability of their punched cards to arithmetical calculation engendered in Babbage a "shadowy vision" in which "the whole of arithmetic now appeared within the grasp of mechanism", (*ibid*., p. 52) and which crystallized in his analytical engine, a machine "of the most general kind". In fact, his expositor Menabrea declared the engine able to

(B) "... reproduce all the operations which the intellect performs in order to obtain a determinate result, if these operations are themselves capable of being precisely defined" (*ibid*., p. 241).

Now this description of the powers of Babbage's machine is nothing less than a form of the main thesis of cybernetics generally, and more specifically, a form of Turing's thesis (T) itself.[21] Not surprisingly, Babbage's vision was hardly understood or appreciated in his day, much less the "glimpses" it gave him of the "Engine eating its own tail", and even being unpredictable.[22] The realization of these visions in Turing's work undercuts old arguments against mechanism on the score of the unpredictability of man's behavior. The roots of this realization in Gödel's work on incompleteness are discussed in detail in Chapter IV.

Another interesting attempt to mechanize arithmetic was made by

Husserl [137], who attempted a complete development of the algorithmic conception of arithmetic, which required "*die logische Untersuchung des arithmetischen Algorithmus*". The notion of algorithm, Husserl felt, had to be bound up with that of a 'mechanical process': in fact, Husserl regarded arithmetic itself as the '*wunderbarsten geistigen Machine*'. He saw the scientific value of arithmetic continually increasing through the replacement of laborious "conceptual operations" by the "arts of mechanical calculation" rendering the former superfluous. He sets out to analyze the concept of calculation taken in its widest sense to include "*jede Herleitungsweise von gesuchten Zahlen aus gegebenen*", where '*Zahlen*' are certain "*sinnlichen Zeichen*". Husserl's theory of calculation has some of the flavor of Church's calculi of λ-conversion: 'systematic numbers' (e.g. arabic numerals) result from a series of rule governed 'reductions' of 'unsystematic numbers' (terms compounded out of numerals with function symbols), also called '*symbolische Bildungen*'. Assuming the potentially infinite set of numerals and the four arithmetical operations, Husserl poses the problem – *die erste Grundaufgabe der Arithmetik* – of developing algorithms for reducing all possible symbolic constructions to their 'normal forms'.[23] Husserl's conception of the '*Mechanismus der symbolischen Methodik*' itself is very similar to Post:

Man kann ihn nämlich fassen als jede geregelte Art der Herleitung von Zeichen aus Zeichen innerhalb irgendeines algorithmischen Zeichensystems nach den diesem System eigentumlichen 'Gesetzen' – oder besser: Konventionen – der Verknüpfung, Sonderung, und Umsetzung (*ibid.*, p. 258).

Observing the rich possibilities for extending any set of arithmetical operations leads Husserl to conclude that only an 'exact study' could ever discover a 'closed totality' of all possible such operations. Especially remarkable is his suggestion that the question whether an arbitrary '*Rechnungsaufgabe*' is always defined (*eine Bedeutung haben*) for any number will require a deep analysis.[24] Husserl sums up the goal of his study as follows:

Ein für allemal gebildete Regeln ersparen die immer wieder erneute Mühe schwierig Überlegungen und gestatten auch hier, ein rein mechanischen Operieren an Stelle des wirklichen Denkens zu setzen. *Auch die Regeln für die Verbindung, Anordnung, und Umsetzung der Operationen schliessen sich zu einem lückenlosen Rechenmechanismus zusammen* sowie die spezielleren Regeln für die Auswertung der einzelnen Operationen (*ibid.*, p. 280).

This replacement of 'real thought' by mechanical operations is clearly of a piece with the formulation (B) of the powers of Babbage's engine, and

Husserl's idea here of a '*lückenlosen Rechenmechanismus*' pretty much corresponds to the universal engine itself. Indeed, when we add in Husserl's requirement that all the rules which '*schliessen sich*' in this mechanism be compatible with our '*endlichen Geistesfähigkeit*', we see that he had pretty much tackled the problem only later solved by Church, Turing, and others of explicating effectiveness. They only succeed because of having the basic discoveries of Gödel to guide them.

Attempts to mechanize logic go back to Leibniz's dream of a 'universal characteristic', a logical language into which every clear question could be translated and then settled by calculation. Explicit mechanization in logic finally emerged in the construction of logical machines by Jevons (1870) – his 'logical piano' – and Marquand (1881), which set off many philosophical discussions of the nature of logical reasoning and its relation to mechanism, the most penetrating being those of Bradley and Peirce.

The machines of Jevons and Marquand were not designed to check whether a given logical argument is valid, but to indicate implicitly conclusions that *could* be drawn from given premises which alone were fed into them. Thus, if fed $A \rightarrow B$ and $B \rightarrow C$ they did not give the conclusion $A \rightarrow C$ outright, but rather indicated the four valuations in which both of these premises were true. One could then easily infer $A \rightarrow C$ by observing that it was also true in those cases. However, Bradley pressed the point that "the process is not finished when the machinery stops; and the rest is left to be done by the mind. What is called 'reading' the conclusion is to some extent *making* it" ([33], p. 384). He denied that machines *can* perform inferences, but as he could see that Jevons' machine clearly did perform logical operations, he conceded that it "performs mechanically an operation which, if performed ideally, would be an inference" (*ibid.*, p. 383). To be 'performed ideally' is to be performed in the mind – but what goes on there during logical inference is rather obscure, and Peirce urged that "needed light on the nature of the reasoning process" was bound to result from studying just "how much of the business of thinking a machine could possibly be made to perform, and what part of it must be left for the living mind" ([200], p. 165).

However, Peirce insisted that no such light could illuminate logic itself, for it doesn't really depend on "the nature of the reasoning process", but rather on *truth*, which is independent of any thinking process. He supports this attack on psychologism in logic by arguing that its laws apply equally well to human or mechanical reasoning. If it be objected that machines do not think, Peirce's reply is that thinking "has nothing to do with logical criticism, which is equally applicable to the machine's performance and to man's" ([201], p. 33). Indeed, we must admit that the machines of Jevons and Marquand, as well as Babbage's engine, all perform inferences, for

> If from true premisses they always yield true conclusions, what more can be desired? Yet those machines have no souls that we know of. They do not appear to think ... in any psychical sense; and even if we should discover that they do so, it would be a fact altogether without bearing on the logical correctness of their operations ... (*ibid.*, p. 32).

Thus did Peirce, as did Husserl after him, *use logical machines to argue against psychologism in logic*. Those who insisted, with Bradley, that machines do not really reason or infer were guilty of psychologism. But if calculating machines also reason – and Babbage's engine, according to Peirce, "would perform considerable feats in mathematics" – where do we draw the line? Must we admit also that "a steam engine would be working out, at every revolution, its problem in thermodynamics"? What exactly is a 'logical' machine?

To answer such questions we must understand what Peirce calls the "secret of all reasoning machines", namely that

> ... whatever relation among the objects reasoned about is destined to be the hinge of a ratiocination, that same relation must be capable of being introduced between certain parts of the machine ([200], p. 168).

Thus to reason, say in the Barbara syllogism, a machine need only

> ... have a connection ... such that when one event *A* occurs in the machine, another event *B* must also occur. This connection being introduced between *A* and *B*, and also between *B* and *C*, it is necessarily virtually introduced between *A* and *C* (*ibid.*).

But now, Peirce maintains, we must admit that *every* machine is really a reasoning machine, for in every machine certain relations between its parts will "involve other relations that were not expressly intended" (*ibid.*). Indeed, a so-called logical machine reasons no more than any other machine, but differs from others

> ... merely in working upon an excessively simple principle which is applied in a manifold and complex way, instead of upon an occult principle applied in a monotonous way ([201], p. 32).

The special truth-conditions satisfied by logical machines are not the exclusive mark of all reasoning, but this is largely a terminological issue for Peirce: he insists only that

... a man may be regarded as a machine which turns out ... a conclusion, the man-machines having been fed a premiss. Since this performance is no more than a machine might go through, it has no essential relation to the circumstance that a machine happens to work by geared wheels, while a man happens to work by an ill-understood arrangement of brain cells (*ibid.*, p. 33).

But his attack on psychologism does not lead Peirce to accept either La Mettrie's (M) or Hobbes' (H), for although he expressly agrees with Hobbes that computation is reasoning, he denies that all reasoning is computation. This brings us to what Peirce sees as the essential difference between human and machine reasoning.

Every reasoning machine, according to Peirce, has two "inherent impotencies". First, "it is destitute of all originality, of all initiative. It cannot find its own problems . . . It cannot direct itself between different possible procedures" ([200], p. 168). As evidence for this Peirce invokes our incredulity that a machine could "automatically thread its way through such a labyrinth" as von Staudt's long proof for Desargues' theorem. And even if it could it still "would only do the special thing it had been calculated to do" (*ibid.*, p. 169). Thus it seems that an essential part of what Peirce meant by the 'lack of originality' is really the lack of *universality*: no more than anyone else in his day does he seem to have grasped the fact that Babbage had conceived, if not built, a machine which could execute *all* 'possible procedures', and not just some "special thing". Since then, Babbage's vision of universal machines which can "direct themselves" between different programs has been realized. On the other hand, Turing's argument that a universal machine must be considered to have 'originality' if any machine has it begs the question whether any machine *does* have it.[25]

The second inherent impotence Peirce finds in any machine is that "the capacity of the machine has absolute limitations; it has been contrived to do a certain thing, and it can do nothing else" (*ibid.*, p. 169). Peirce's explanation of this second impotence – which at first sounds like the first one all over again – is simply that existing logical machines can only deal with a limited number of symbols. In fact,

The unaided mind is also limited in this as in other respects; but the mind working with a pencil and plenty of paper has no such limitation. It presses on and on, and whatever limits can be assigned to its capacity today, may be overstepped tomorrow (*ibid.*, p. 169).

But this is pretty clearly a matter of external memory capacity here, which, as we have seen, must also be assumed to be potentially infinite for a universal machine.

Again, despite the "considerable feats" he would have expected from Babbage's machine, Peirce failed to appreciate the full force of (B). This failure shows up in another way in Peirce's otherwise admirable discussions of the importance of distinguishing between 'non-relational' (monadic) logic and the 'logic of relatives' (full predicate calculus). He points out that the former has misled logicians into believing that necessary reasoning was all a matter of following rigid rules, i.e. rules which infer just *one* conclusion from given syllogistic premises, and hence that machines might carry out *all* such reasoning. But this expectation breaks down in relative logic where "from any proposition whatever . . . an endless series of necessary consequences can be deduced" ([202], p. 407). Peirce even saw the logic of relatives as a refutation of Kant's view of logic as analytic in the sense that it "only elicits what was implicitly thought in the premisses": for in relative logic "Matter entirely foreign to the premisses may appear in the conclusion" (*ibid.*, p. 408). Elsewhere he admits that relative logic has failed to "completely eradicate" the idea that all necessary reasoning could conceivably be left to a machine, although "it does show that much *unexpected truth* may often be brought to light by the repeated re-introduction of a premiss already employed" ([202], p. 506; italics mine). Nowadays one would try to make this point precisely by appealing to the undecidability of the *Entscheidungsproblem*. But we know of this only because of Turing's demonstration that a machine such as Babbage's is able to find every such truth while remaining *itself* undecidable. Any method for predicting what such a machine will do in general will be confronted with 'unexpected truths' about its halting behavior, so unexpected indeed that it will not predict them at all. In fact, Peirce did not see any real difficulty in constructing a machine to "work the logic of relatives", even with many terms:

> But owing to the great variety of ways in which the same premises can be combined to produce different conclusions in that branch of logic, the machine, in its first state of development, would be no more mechnical than a hand-loom for weaving in many colors with many shuttles. The study of how to pass from such a machine . . . to one corresponding to a Jacquard loom, would be likely to do very much for the improvement of logic ([200], p. 170).

This was a beautiful formulation of what was needed, but as we have seen, Babbage had already done quite literally this by applying the punched cards of the Jacquard loom to the problem of mechanizing calculation, while Turing later established that it could indeed "work the logic of relatives".

If Peirce's claims and diagnoses of machine impotence no longer bear scrutiny, his use of logic machines to argue against psychologism is a lasting and brilliant achievement, and is in fact somewhat clearer than some of Husserl's later and better known arguments against this view. However, Husserl saw clearly that the refutation of psychologism depends on the objectivity of truth as opposed to the subjectivity of thought, and tried to establish this directly by showing that logical laws could not be interpreted in terms of '*Denkakten*', but rather required the '*Sätze an sich*' of Bolzano. Bolzano had used them to prove the existence of infinite sets, and indeed, Frege would later see in Dedekind's refinement of Bolzano's argument an existence proof for the 'objective thoughts' (*Gedanken*) by which he also claimed to refute psychologism. Again, our discussion of mechanism, this time as applied to man as opposed to nature, has led to problems about infinity.

We may summarize the three sections of this chapter as follows. Mechanism in the philosophy of nature was originally associated with determinism but eventually parted company with it upon the introduction of probability laws. Its essential feature was then seen to be, in general, a finitistic approach to the description of nature. An especially pointed example is Poincaré's formulation (F) of Planck's quantum hypothesis, as opposed to the infinitism of Bohm's (I). Mechanism thrives on the introduction of any kind of real randomness into nature and the consequent exorcism of certain demons from it. A general principle that lies behind these exorcisms is Turing's fundamental thesis (T) that every effective process can be finitely mechanized. Indeed, (T) implies the existence of a universal machine *U* with only finitely many internal states, which nevertheless can simulate any other machine whatever. *U* turns out to hold one key to the original difficulty facing Descartes' mechanism (α) in biology, namely of conceiving self-reproduction purely mechanically. *U* also provides relief for a difficulty facing La Mettrie's extension of mechanism to man himself, namely his unpredictability. This in turn supports the claim in (M) that the machine that man is cannot be studied *a priori*. Meanwhile, an argument against Hobbes' mechanistic interpretation (H) of thought itself was blunted by Babbage's (B), which is the generalized precursor of Turing's (T). However, part of this argument did succeed in refuting psychologism, and attempts to find more general arguments against this view have led, as we shall see, to arguments which seek to refute La Mettrie's correlation hypothesis (C), upon which his (M) rests, by showing the mind to have infinitely many states. Indeed, whereas Descartes' refutation of (M) would have rested ultimately on his existence proof for an infinite God, others have more recently tried to base their refutation of mechanism on Dedekind's existence proof for an infinite world of thoughts.

Finally, any straightforward mathematical attempt to find counter-examples to the 'pearl of mechanism', Turing's thesis, would seem to stumble over Gödel's incompleteness of formal systems, for they would otherwise lead to straightforward refutations of (T). This suggests that incompleteness really provides just the opposite of what it is normally supposed to: namely, that it supports mechanism. I shall explore this suggestion at greater length in later chapters.

NOTES

[1] See Baird [5], especially pp. 312–14.

[2] Not to be confused with the so-called 'malignant demon' invoked momentarily by Descartes to motivate his doubt about the existence of the external world.

[3] The idea of general limitations on machines, which can be established independently of their particular physical constitution, goes back to the pioneering thermodynamic studies of S. Carnot.

[4] See in particular C. D. Broad [34], pp. 105ff.

[5] The 'dissipation' of energy refers to a decrease in the amount of *free* energy, available for doing work. This in turn may be expressed as an increase in the 'disorder' or entropy of the system.

[6] As Ehrenberg [75] observes, the value of Szilard's work was to have made plausible through examples a certain *postulate* relating entropy to information.

[7] This interpretation is strikingly elaborated in Laing's recent work discussed below on computation by 'Maxwell demon systems'. Daub [59] has suggested, however, that Maxwell himself never considered the necessity of energy consumption in the perception of his demon because he conceived it in "the very image of the Newtonian God who has ultimate dominion over the world and senses the world in divine immediacy" (p. 224).

[8] See Martin-Löf [180] for a survey of attempts to clarify randomness along these lines, and Burks [41] for an exhaustive discussion of different senses of randomness.

[9] This assumes, of course, that such systems are complete, which they are not! See below.

[10] Von Neumann was especially interested in developing analogies between thermodynamics and the logic of computation. Burks [40], by way of motivating some of his ideas, has remarked that

> A program is really a state of an automaton, so the interchangeability of programs and automata is really an interchangeability of state and system. One is reminded here of the analogy between kinetic and potential energy. A finite automaton M is active — a form of kinetic energy — while its description $D(M)$ recorded in a succession of cells is passive — a form of potential and kinetic energy. In both cases some activity (kinetic energy) is required in the system initially . . . (*ibid.*, p. 113).

[11] It was Popper who stressed this weakness of von Mises' theory and who first tried to construct a theory of randomness for finite sequences.
[12] See Chaitin [48]. Kleene [61] suggested his hierarchy theorem for arithmetic predicates as such a source "because the complexity of the predicates soon exceeds the limited form representing the concept of provability in a stated formal system" (p. 279).
[13] The main conflict is between those who hold that (i) particles don't really *have* simultaneous positions and momenta to measure, and those who only claim that (ii) they really have such, only we cannot simultaneously *measure* them.
[14] What Poincaré meant here was that, as a consequence of Planck's hypothesis of a quantum of action, there could only be a finite number of ways that a finite total energy of a system in thermal equilibrium could be distributed throughout that system. This is related to the circumstance that a thermal equlibrium of a system represents a state in which no more 'shuffling' of its energy is possible. Eddington formulated the point by analogy with shuffling a deck of cards:

If after shuffling the pack you tear each card in two, a further shuffling of the half-cards becomes possible. Tear the cards again and again; each time there is further scope for the random element to increase. With infinite divisibility there can be no end to the shuffling. The experimental fact that a definite state of equilibrium is rapidly reached indicates that energy is not infinitely divisible, or at least that it is not infinitely divided in the natural process of shuffling. Historically this is the result from which the quantum theory first arose ([74], p. 78).

See Lodge [171] for the scepticism concerning continuity in nature which was invading science from all sides at the turn of the century.

[15] See Gunderson [107a] for a thorough discussion of this argument.
[16] See Gibson [98], pp. 207ff.
[17] See Bowden [30], pp. 333–4.
[18] But he denies, à la Hume, that men can ever understand the relation between cause and effect.
[19] Vartanian [256], p. 28.
[20] See Smith [236], Vol. I, p. 174.
[21] I am assuming here that 'effective' behaviors are the same as those which can be 'precisely defined'. See Wilkes [275] for a detailed discussion of the sense in which Babbage did and the sense in which he did not anticipate our modern notion of a 'program'.
[22] See Morrison [188], p. xx.
[23] *Ibid.*, pp. 260–3.
[24] *Ibid.*, pp. 277–8.
[25] See Turing [251], p. 21.

CHAPTER II

MIND, NUMBER, AND THE INFINITE

In science nothing capable of proof ought to be accepted without proof.

R. Dedekind

Our ideas are not always proofs of the existence of things.

J. Locke

1. THE OBLIGATIONS OF INFINITY

Proclus spoke for antiquity in branding the infinite as incomprehensible and unknowable, remarking that "the imagination recognizes the infinite by not understanding it." Philosophers long maintained the incomprehensibility of the infinite – according to Descartes, "incomprehensibility is comprised within the formal concept of the infinite" – though they later tried to explain it by reference to the *limitations* of the human mind. Hume still spoke for many when he said that "the capacity of the human mind is limited, and can never attain a full and adequate conception of infinity", where by 'adequate conception' he meant a clear and distinct image in the mind. Hobbes had already elaborated this thesis as follows:

Whatsoever we imagine is *finite*. Therefore there is no idea or conception of anything we call *infinite*. No man can have in his mind an image of infinite magnitude, nor conceive infinite swiftness, infinite time, infinite force, or infinite power. When we say anything is infinite, we signify only that we are not able to conceive the ends and bounds of the thing named, having no conception of the thing but of our own inability ([131], p. 36).

Descartes argued against atheism precisely that "we must consider our minds . . . as finite and limited, and God as . . . incomprehensible and infinite" ([69], p. 138). Indeed, Descartes tried to erect modern philosophy on the consequences of this very distinction. Since our minds are finite, they are subject to the gravest *doubt*: they can doubt everything except doubt itself, more generally, everything except *self-consciousness. Cogito, ergo sum*. But how then can such pure subjectivity reclaim anything indubitably objective? According to Descartes, only through the necessary connection between finite self-consciousness and the consciousness of an infinite God. "By the name God I understand a substance that is infinite . . ." ([69], p. 165), he begins, and then argues that God must exist, for though I am guaranteed an idea of substance by being a substance, he thinks, "nevertheless I should not have the idea of an *infinite* substance – since I am finite – if it had not proceeded from some substance which was veritably infinite" (*ibid.*, p. 166). Hobbes and Gassendi objected that our idea of infinity was purely negative, arising only from the exertions of our own minds by the negation of our limitations, but Descartes had tried to anticipate this objection by claiming that "in some way I have in me the notion of the infinite earlier than the finite – to wit, the notion of God before that of myself" (*ibid.*). This priority of the infinite is explained in a letter to Clerselier in which he claims that, as opposed to our merely negative idea of what he would call the 'indefinite',

> . . . the notion I have of the infinite is in me before that of the finite because from it alone I conceive *being* or *that which is* without thinking whether it is finite or infinite . . . but in order that I should conceive a finite being, it is necessary to remove something from this general notion of being, which consequently must precede (quoted in [69], p. 267).

Our ideas of number, space, and divisibility are only indefinite, while only God is truly infinite. But when Descartes explains this 'positive' infinity as that "in which nowhere limits are to be found" ([69], p. 17), it sounds just as negatively conceived as the indefinite. Descartes tries to save his distinction between the indefinite and the infinite – which is essentially that between potential and actual infinity – by further distinguishing between the 'formal notion' of the infinite and the infinite object itself:

> . . . for as for infinity, even though we understand it to have as much positive reality as may be, yet we understand it only in a certain negative fashion, . . . namely, that we perceive no limitation in the thing; but the thing itself which is infinite is indeed positively understood, though not adequately, i.e. we do not comprehend the whole of what is intelligible in it ([69], p. 17).

But Gassendi and others objected that if we really have an idea, we ought to *comprehend* it. Having granted that we don't comprehend an infinite God, Descartes saves his idea of it nevertheless by denying any obligation to comprehend our ideas: a finite spirit cannot comprehend anything infinite, but that does not prevent such a spirit from 'apprehending' it, "just as one can touch a mountain without being able to embrace it" (*ibid.*, p. 130). And after all, Descartes could argue, his idea of infinity was not of his own making but was put innately in him by his creator. He could at most be expected to comprehend ideas that he had himself constructed out of his own experience. This principle that we ought to completely comprehend our own ideas will turn up again later in our discussion of the mathematical philosophies of Gauss and Dedekind, and will loom large in subsequent discussion of the significance of incompleteness.

Both the priority of the infinite over the finite – in which Descartes anticipates modern set theory – and its innateness were challenged by Locke, who defended experience as the source of all our ideas and knowledge. Specifically, he held all our ideas to derive from either sensation or *reflection*, which occurs when the mind

> . . . turns its view inward upon itself, and observes its own actions about those ideas it has, takes from thence other ideas, which are as capable to be objects of its contemplation as any of those it received from foreign things ([169], p. 159).

Such reflection, which Fraser called "self-consciousness intensified", must somehow account for our idea of infinity which Locke saw as a crucial test case for his empiricism. The finite offers no difficulty, according to Locke, for the "obvious portions of extension that affect our senses, carry with them into the mind the idea of finite" (*ibid.*, p. 277).

The idea of infinity presents a difficulty, however, "since the objects we converse with come so much short of any approach . . . to that largeness" (*ibid.*). But if one reflects on any one idea of spatial length, say a foot, one "finds that he can repeat that idea". Or one may just begin 'doubling' it, whereupon one discovers that

> . . . after he has continued his doubling in his thoughts, and enlarged his idea as much as he pleases, he *has no more reason to stop*, nor is one jot nearer to the end of such addition, than he was at first setting out: the power of enlarging his idea of space by further additions . . . he hence takes idea of infinite space (*ibid.*, pp. 277–8, italics mine).

But, "since our ideas are not always proofs of the existence of things", Locke emphasizes that such reflection gives only the *ideas* of infinite space and time, not existing infinite things themselves. In general, these ideas of infinity are due to "the power we observe in ourselves of repeating, without end, our own ideas" (*ibid.*, p. 279). However, when we "pursue" our idea of infinity into space and time, it is only *number*, Locke claims – our "simplest and most universal idea", the one "most intimate to our thoughts" – that keeps our mind from losing itself in a "confused heap". For, being "the clearest and most distinct idea of infinity we are capable of" (*ibid.*, p. 283), number enables us to individuate the repetitions of lengths and times that would otherwise become "confused". Locke even suggests that, when applied to space and time, "our idea of infinity . . . seems to be nothing but the infinity of number" (*ibid.*, p. 274). In any case, "The infinity of numbers, to the end of whose addition everyone perceives there is no approach, easily appears to anyone that reflects on it" (*ibid.*, p. 281). Locke nevertheless maintained the absurdity of "the actual idea of an infinite number". Thus our idea of number, though more clear and distinct than those of space and time, can no more provide us with actually existing infinities than they can.[1] The same holds for the infinite divisibility of space: since "our thoughts can never arrive at the utmost divisibility, there is an apparent infinity . . . also in that, which has the infinity also of number" (*ibid.*, p. 284–5). Locke insisted finally, as had Hobbes and Gassendi, that our ideas of infinity are only negative, that we are unable to conceive any "positive idea of infinity", but the main question is whether Locke has shown how to derive *any* notion of infinity solely from experience.

The crux of Locke's account of our idea of infinity lies in the "power" he finds in the mind for endless repetitions of certain of its ideas. Now if by 'endless' repetitions are meant infinitely many of them, then Locke's account would, as some have claimed, be either circular or tantamount to Descartes' doctrine of the innateness of our idea of infinity: it would simply find our idea of infinity in our *power* of infinite repetition. But if we look at Locke's account of the mind's powers, his position is seen to be more resourceful. He points out that the mind's idea of 'active powers' comes less from external sensation than "from reflection on its own powers" (*ibid.*, p. 313). And of the crucial power of a 'free agent', namely volition, Locke asks

> . . . is that faculty anything more than . . . the power of the mind to determine its thought, to the producing, continuing, or stopping any action, as far as it depends on us? (*ibid.*, p. 320).

But if the mind has *this* power, why must Locke speak of endless repetitions "of which there *can* be no end" (*ibid.*, p. 280)? Can't one simply choose to 'stop' them? Locke's editor Fraser claimed that he never recognized his power for endless repetition for what it really was, namely, a 'mental obligation', or even an 'intellectual necessity'. But any such obligation or necessity would still have to be reconciled with our mental or intellectual *freedom*, also clearly pointed out by Locke. When Locke speaks of repetitions 'of which there *can* be no end', he need only mean, given his view of volition, that there would be no end if we chose to continue them. In any case, it is puzzling when Leibniz claims, about Locke's mental process of doubling mentioned above, that "it is never *possible* to stop the process" ([165], p. 163). Can a mental process go on endlessly by itself even if we will it to stop? Locke only said of this process that one has no *reason* to stop it, which makes more sense than saying one *cannot* stop it.[2] But he could also have said that one has no reason to continue it, beyond, say, any specific counting need. What could be more pointless than to do so? Elsewhere Locke says of the mind's power of adding numbers that it "leaves always to the mind *room* for endless additions" (*ibid.*, p. 285, italics mine), which makes more visible the modal aspect of infinity: the mind 'leaves room' for as many additions as possible. The basic modal fallacy committed by Leibniz, Kant, and Fraser lies in assuming that since it is possible for some repetition to continue endlessly, it somehow must actually *do* so. When Fraser complained that a "merely empirical 'repetition' . . . does not explain the *intellectual need* for *continuing without end* the process of repetition" (*ibid.*, p. 278), he not only committed this fallacy, but also begged the question of intellectual need and obligation. After all, our greatest intellectual obligation is to *consistency* and Locke was only too aware of the "perplexities and contraditions" which the idea of infinity had led the mind into.[3]

D. J. O'Connor has observed that Locke gave no "correct definition" of an infinite number, crediting this to the 19th century. This is true enough, but a correct definition of infinity is little better than a vague idea of it unless one can use it to *prove the existence* of something infinite, and as we shall see, the existence proof for infinite sets which Dedekind tried to base on his correct definition of them failed rather badly, leading instead to perplexity and contradiction. As Locke said, "our ideas are not always proofs of the existence of things." In fact, our examination of Dedekind's proof will bring us back later to the very endless repetitions of our thoughts that Locke's discussion centered on. Indeed, Poincaré rejected Dedekind's entire theory along with its reduction of mathematical induction to logic, but his explanation of the synthetic *a priori* character of induction could well have been written by Locke:

> It is only because it is the affirmation of the power of the mind which knows it can conceive of the indefinite repetition of the same act, when the act is once possible. The mind has a direct intuition of this power, and experiment can only be an opportunity of using it, and thereby becoming conscious of it ([205], p. 13).

But if Dedekind's definition does not enable us to prove the existence of anything infinite, we shall see that it does show why mere 'endless repetitions' do not by themselves lead to anything infinite either. As in the case of the parallel postulate, we now realize that the existence of infinite sets cannot be proved, but must be assumed as an *axiom*: their existence can no more be settled by pure thought than could the uniqueness of parallels by pure intuition. The same is also true of the axiom of continuity which Dedekind also first correctly formulated, though the impossibility he claimed of proving space to be continuous is somewhat less generally appreciated. Indeed, the independence of the continuity axioms, though a mathematically simpler result, was on the whole even harder to accept than that of the parallel axiom. It is instructive to review its history.

Zeno's paradoxes showed that a consistent account of continuity in geometry would be difficult. Aristotle thought he had resolved them by allowing the points on a line to be only potentially, not actually, infinite, but he admitted that geometry was less exact than arithmetic because it had to deal with continuous as opposed to discrete quantity. Leibniz still claimed that no one had solved "the labyrinth of the continuum", that is, the problem of whether the line was infinitely divisible or composed of small discrete parts. Still, he adopted the 'Law of Continuity' as a basic principle of his philosophy, one formulation being

(ℓ) "Whenever two instances approach each other continuously, so that one passes over into the other, it is necessary for their consequences or their results to do so also" ([164], p. 539).

Sometimes he simply says that "nature does not make leaps". Leibniz claims that (ℓ) "has its origins in the *infinite* and is absolutely necessary in geometry" (*ibid.*, p. 539), but that it will enable the geometer to "complete all his demonstrations" without ever having to enter the labyrinth of the continuum himself. When challenged to prove (ℓ), Leibniz replied that anyone who rejected continuity was committed to saying that "motion is nothing but a succession of leaps . . . which . . . are recreations by God in separate places" (*ibid.*, p. 847). But his proof of (ℓ) uses "the principle of order, with the aid of supreme reason, which does everything in the most perfect way" (*ibid.*, p. 847). Needless to say, his proof was no more acceptable than Dedekind's for infinite sets.

It was Hume who clearly saw that "no geometrical demonstrations for the infinite divisibility of extension can have so much force" that we must accept it, for "geometry fails of evidence in this single point, while all its other reasonings command our fullest assent" ([133], p. 58). Still, the early Kant was confident that the geometer can prove the infinite divisibility of space, for if he starts a process of division of a line, "he recognizes with the greatest certainty . . . that this division *must* go on endlessly" ([142], p. 9). But later he admits that "the continuous is . . . unrepresentable according to the laws of intuitive cognition" (*ibid.*, p. 48). This did not make continuity impossible or inconsistent, even though, according to Kant, it had been "expelled in disgrace from a number of schools" (*ibid.*, p. 49). But it forced one to describe continuity as an "object of pure reason" which transcended any concrete intuition. This view, as we shall see later, deeply influenced Hilbert.

Lambert was already defensive about his use of continuity in the theory of parallels; and some misgivings about the non-euclidean geometry of Bolyai and Lobachevski were due to its heavy use of continuity just to prove the existence of hyperbolic parallels. This may also have futher inhibited Gauss from publishing his similar theory of parallels, for he felt strongly that the method of exhaustion should be entirely banned from elementary geometry. In any case, when the archimedean axiom on which this method depends was finally formulated explicitly in 1880 by Stoltz, it was soon doubted by some of the greatest geometers including Hilbert, whose *Festschrift* to Gauss' way of thinking would indeed eliminate it from elementary geometry. Cantor claimed to have shown the impossibility of 'nonarchimedean' geometries, but this was only because, like Leibniz, he regarded a still stronger continuity axiom – essentially the one first formulated by Dedekind – as a *logical necessity*.[4] But the work of von Staudt, Fano, and Veblen led even to the construction of *finite* geometries having only finitely many points and lines. Originally having only incidence relations, these finite geometries have been extensively developed through the introduction of suitable axioms of order and congruence by Järnefelt and Kustaanheimo, who have proposed their use in physics to solve certain paradoxes due apparently to the use of continuous variables.[5] They have constructed an infinite sequence G_1, G_2, G_3, . . . of these geometries, having ever larger finite numbers of points which has euclidean geometry as its limit. Insofar as euclidean geometry is approximately correct, it is clear that one can find finite geometries G_i which are as experimentally verifiable as euclidean geometry without making its unjustifiable assumptions that every figure has exact scale models as small as one pleases.[6] These finite geometries clearly show the futility of trying to demonstrate the necessity of infinite divisibility. Such a result must be postulated, whatever obligation one may feel to continue dividing a line in one's thoughts. However, even the eventual use of finite geometries in physics would not necessarily relieve us of a need for some infinitistic mathematics, for our ignorance of which finite geometry actually applied would create a need for theorems about all such geometries. For example, Desargues' theorem implies Pascal's in every finite geometry, but the only known proof of this uses Wedderburn's theorem that every finite field is commutative, which in turn requires arithmetical induction on the integers. Of course, we are assuming here that there *are* infinitely many integers: it remains to be seen to what extent this assumption can be rationally justified.

2. MIND AND PHILOSOPHY OF NUMBER

The philosophies of infinity just discussed lead to basic problems about number. Unless we assume with Descartes that God gives us some idea of infinity innately, it is hard to see it as anything but a purely mental construct of our own. And yet, classical philosophers all agreed that we cannot completely comprehend this idea. Shouldn't we be able to completely comprehend an idea purely of our own making? And since, as Locke pointed out, our clearest idea of something infinite is of the numbers themselves, shouldn't they also be seen as of our own making? And thereby also as completely comprehensible?

The idea that number is, as Berkeley put it, "entirely the creature of the mind", is common in philosophy, and has long been used by philosophers to distinguish arithmetic from geometry, which has usually been regarded as having a subject matter external to ourselves. Proclus had already come to the conclusion that numbers are "purer and more immaterial" than the magnitudes of geometry and also "free of any extraneous figure". Leibniz claimed that "Number is perceived by all external senses, but because it is perceived by the internal sense, and even more so, arithmetic is more rightly subordinated to metaphysics" ([164], p. 557). According to Kant, "the science of number . . . is a pure intellectual synthesis which we represent to ourselves in thought", whereas geometry elaborates the structure of outer sense. The idea that arithmetic deals, in a way that geometry does not, with an internal or mental subject matter was expressed pregnantly, and in philosophically neutral language, by Gauss:

> I am profoundly convinced that the theory of space occupies an entirely different position with regardl to our knowedge *a priori* from that of arithmetic; that perfect conviction of the necessity and therefore the absolute truth which is characteristic of the latter

is totally wanting in our knowledge of the former. We must confess in all humility that number is *solely* a product of our mind. Space, on the other hand, has also a *reality outside* our mind, the laws of which we cannot fully prescribe ([217], p. 17).

We isolate from this the thesis, later made famous by Dedekind

(D0) Numbers are our own mental creations or constructions.

which will figure prominently in what follows. Also important is the further view implied by Gauss that since number is merely a product of our own mind, we can *write down all its laws a priori and completely*. We formulate the general principle needed to justify this view as

(CM) We can in principle completely describe our own mental constructions.

as well as an essentially stronger principle, surely not intended by Gauss but apparently held by Kronecker:

(DM) We can in principle completely decide the properties of our own mental constructions.

These principles are both rather vague, but there are conceptions of the mental for which they are plausible, for example, certain phenomenological views. (D0), (CM), and (DM) have all figured importantly in arguments we shall have to consider and have played a large part in shaping views about the foundations of mathematics and science. Philosophies such as positivism, phenomenology, phenomenalism, or idealism that try to base science on something certain and immediately accessible, such as protocol statements, perceived eidos, elementary experiences, or thought, respectively, must face the questions (CM) and (DM) when they turn to mathematics. If, as Brouwer, for example says, "*Mathematik ist identisch mit dem* exakten *teil unseres Denken*", the question arises why we should not be able to completely decide all mathematical questions. Is anything more under our complete control and observation than the exact part of that to which we have the direct access of introspection? In fact, versions of (CM) and (DM), formulated as principles of 'phenomenological access', inspired attempts by Husserl and Kaufmann to prove the completeness and decidability of number theory.[7] Since it is now clear that number theory is undecidable, we may ask what this implies for (D0) and (CM). Thus Gödel used (DM) to argue for platonism and against (D0). His main premise was that either mathematical objects are external to us or they are our own mental constructions. But we have proved that they are undecidable. Therefore it follows from (DM) that they are external to us. Gödel also, of course, presupposed Turing's thesis in his argument.[8] However, Kreisel has criticized this argument on the grounds that (DM) is implausible. As we shall see, Dedekind defended (D0) at length, but explicitly rejected (DM).

One may also ask whether the discovery of the formal incompleteness of arithmetic does not make (D0) and (CM) incompatible. And hence also, in view of the success of Hilbert and others in axiomatizing various geometries, whether modern logic and axiomatics do not weigh against Gauss' view that the laws of number could be completely prescribed whereas those of space could not. Thus, after recounting the seeming failures of the axiomatic method in arithmetic and set-theory, Kac and Ulam hasten to add that

> Work on the axiomatic foundations of geometry, on the other hand, had a different and much more constructive effect. It turned out that much of our intuition about space can be deductively codified . . . One is almost tempted to conclude that in some undefined way there is a deep distinction between the problem of axiomatizing a branch of mathematics that originated in *external* stimuli and the problem of axiomatizing *internal* processes of thinking ([139], p. 706).

But this is still compatible with Gauss' view quoted above. With respect to geometry Gauss was only asserting that experience rather than deduction would be required to settle the *parallel* postulate. On the other hand, there is a clear mathematical sense, due to Gauss' student Dedekind, in which one *can* say that number has been 'completely described' despite the formal incompleteness discovered by Gödel. But also like Dedekind, Gauss would surely have rejected (DM), being too deeply impressed with the difficulty of finding proofs for even simple theorems in 'higher arithmetic'. His third proof for his theorem of quadratic reciprocity is prefaced by these words:

> The questions of higher arithmetic often present a remarkable characteristic which seldom appears in general analysis and increases the beauty of the former subject. While analytic investigations lead to the discovery of new truths only after the fundamental principles of the subject have been mastered; on the contrary in arithmetic the most elegant theorems frequently arise experimentally as the result of a . . . stroke of good fortune, while their proofs lie so deeply embedded in the darkness that they elude all attempts and defeat the sharpest inquiries ([235], p. 112).

Many years later Hilbert complained along the same lines of the "monstrous difficulty" of finding proofs for theorems of arithmetic easily discovered by induction. Later we shall examine Hilbert's attempt to reconcile this difficulty with his belief that every mathematical problem is solvable. Philosophers often say that even in so 'simple' a domain as elementary arithmetic we must accept undecidability, but this was precisely where mathematicians would expect such a result. The opposite result would have made it indeed difficult to see how mathematical logic could have any relevance to actual mathematical experience. Still, undecidability results are very theoretical and have not yet been shown to really 'explain' the difficulties mentioned by Gauss and Hilbert in any explicit way.[9]

Closely related to the problem of decidability is that of pure existence proofs. When, for example, the search for certain kinds of effective methods for solving algebraic equations of degree higher than four had bogged down, mathematicians temporarily had little choice but to try to prove that the solution of such an equation was not *logically impossible*. The classic example is due to Gauss who gave the first rigorous proof of the fundamental theorem of algebra. Many fallacious proofs of this theorem were published before Gauss, and he took pains to diagnose their common error: they all tacitly assumed the existence of the roots of a given equation, while concentrating only on its form.[10] Thus, at certain stages in the development of mathematical theories, pure existence theorems are the natural outcome of logical rigor. Meanwhile, the search for effective ways of actually finding the roots whose existence Gauss had proved culminated in Sturm's theorem of 1829 which provided, for any equation of degree n, n effectively computable polynomials, the calculation of which allowed one to determine the number of roots in any given interval. Sturm's result, which decisively influenced Kronecker's views, shows how one may eliminate the need for continuity assumptions in Gauss' proof of the fundamental theorem of algebra. Far from a relic in the history of mathematical algorithms, this result is intimately related to some of the most important results of modern logic, namely Tarski's proofs of the decidability of the elementary theories of algebra, geometry, real numbers, and algebraic and real closed fields.[11] Sturm's result stimulated the search for other algorithms in higher mathematics, and was one of the first algorithms for which Babbage wanted to program his analytical engine.

Returning to elementary arithmetic, we recall that it had always been traditionally conceived in terms of algorithms and calculation. Even Gauss' *Disquisitiones Arithmeticae* of 1801, which shifted the focus of arithmetic to proof of theorems in his congruence theory, had as an essential goal the explanation of why certain algorithms work as well as proofs that they always will work. In the middle ages, 'algorithm' was defined to be the arithmetic which calculates with Indo-arabic numerals. Conversely, the 1771 edition of the *Encyclopedia Britannica* defined 'Arithmetick' in terms of the notions of number and algorithm. Number was defined, following Euclid, as "either a unit, or a multitude of units", and once a person has "the idea of number in his mind", he is ready for the "science of arithmetic":

(A) "Notation, Addition, Subtraction, Multiplication, and Division are . . . the whole of arithmetic, as every arithmetical operation requires the use of some of them, and nothing but a proper mixture of them is necessary in any operation whatever; and these are called Algorithm" (*ibid.*).

(A) formulates what we will call the 'algorithmic conception of arithmetic'. It regards the basic operations of arithmetic as algorithms rather than as functions in the modern sense, i.e. as *rules* rather than sets. We can contrast (A), which concentrates on the construction of arithmetic operations and has a practical orientation, with the more theoretical approach of Frege and Dedekind, which reduces the notion of number to concepts of pure logic and then concentrates on the proofs of arithmetical propositions. We could express their respective ideals of completeness as 'algorithmic completeness' – which would require an account of all the 'proper mixtures' of basic operations in (A) – and 'deductive completeness', which would require an account of all logical axioms. The algorithmic conception tends to stress formalisms and concrete symbols while the deductive conception stresses concepts and abstract objects.

An early proponent of logicism was H. Grassman [106], who held that while geometry depended to some extent on spatial intuition, number depended wholly on the law of thought. He was the first mathematician both to approach arithmetic axiomatically and to employ recursive definitions for the basic arithmetic operations. The two are related, for if one decides to use, say, the recursive 'definition' $a + (b + 1) = (a + b) + 1$ for addition, then the fact that it does not in general enable one to eliminate the plus sign, together with the difficulty of replacing it with anything more basic or obvious, obliges one to take it as an 'axiom'. On the other hand, if one wants to use the axiomatic method in arithmetic, then, if one has no set theory available and eschews the mere verbal definitions for addition and multiplication that used to be common, it is hard to see how else one can introduce these operations except recursively. Recursive definitions for the basic arithmetic operations began to appear frequently in the literature after Grassman, and were seen as raising two kinds of problems, which we can classify roughly as logical and mathematical.

The first to raise logical objections was Frege, who objected that Grassman's recursive definition of addition violated a basic principle of logic which forbids *circular* definitions. Somewhat reminiscent of Gauss, Frege demanded that such definitions be justified by existence proofs of unique functions satisfying them.[12] There are actually two different kinds of 'existence' problems connected with such definitions. Thus if we demand with Frege an existence proof that the equations

$$f(n, 0) = n,$$
$$f(n, m + 1) = f(n, m) + 1,$$

can be 'solved', we are not looking simply for an ordinary arithmetic or algebraic formula, such as $n + m$, but rather for something formulated in a wholly new language, such as Frege's 'logic', which we regard as describing directly the basic logical entities that 'really exist'. In short, to demand this in the case of the basic arithmetic operations is to demand a *reduction* of arithmetic to something else. However, recursive definitions were used well before Grassmann, who was only the first to so define the *basic* arithmetic operations. Thus, if we demand a proof that a solution exists for the Fibonacci equations:

$$(1) \qquad f(0) = 1,$$
$$f(1) = 2,$$
$$f(n + 1) = f(n) + f(n - 1),$$

we are probably just looking for some arithmetic or algebraic formula expressing f explicitly as a function of n. In this case, the method of difference equations yields the solution

$$(1)' \qquad f(n) = \frac{5 + 3\sqrt{5}}{10}\left(\frac{1 + \sqrt{5}}{2}\right)^n + \frac{5 - 3\sqrt{5}}{10}\left(\frac{1 - \sqrt{5}}{2}\right)^n,$$

which can be proved by mathematical induction to satisfy (1).[13] (1)′ is an explicit description of a function only implicitly defined by (1). The construction of (1)′ and the proof that it satisfies (1) justifies the apparent 'circularity' of the latter. However, we have no general way of finding algebraic solutions such as (1)′ for arbitrary 'recursions' like (1), or even any assurance that they always exist. This more mathematical problem, including the characterization of 'arbitrary recursions', proved to be difficult and could not be generally solved until the work of Gödel appeared. For example, it is not obvious which sets of equations, having the 'recurrence' property that the function being defined recurs in the definition, will have solutions at all. Even for the simple equations

$$f(1) = 1,$$
$$f(2n + 1) = n^2 - n + 1,$$
$$f(3n + 1) = 2n + f(2n + 1),$$

one must calculate several values to realize that no function can satisfy them, i.e. that they are contradictory, since their inconsistency does not show up until one tries to calculate $f(13)$.

A very attractive feature of recursive definitions was seen in the fact that, despite not being explicit, one could still use them to effectively calculate a function: as Cantor had already put it in 1888, "*Zur praktischen Berechnung bieten sich Rekursionsformeln dar*" ([44], p. 434). This lead eventually to the most difficult problem of all: whether *every* function capable of 'practical calculation' could be defined by a suitable kind of 'recursion formula'. This problem was only settled by the cumulative work of Dedekind, Skolem, Hilbert, Herbrand, Gödel, Kleene, and Church, much of which we shall examine.

3. DEDEKIND'S THEORY OF ARITHMETIC

Dedekind tried to build a logical foundation for arithmetic and our intuitive notion of natural number, where by 'logic' Dedekind still meant the 'laws of thought'. This implies for Dedekind that our number concept is *independent of intuitions of space and time,* that numbers are rather "free creations of the human mind"; for Dedekind (D0) is an essential consequence of logicism. Since he proves that, in a certain sense, his theory is complete, we may ask whether he accepted (CM): being convinced of (D0), did he not then feel that numbers must thereby admit in principle of a complete description? We know that he argued conversely: he regarded his completeness proof as justifying (D0). More specifically, the abstract characterization of 'natural number series' justifies (D0) – "With reference to this freeing the elements from every other content (abstraction) we are justified in calling numbers a free creation of the human mind" ([63], p. 68) – and the completeness proof justifies the abstract characterization. He carefully avoids anything like (DM) however. In fact, he only published his essay to publicly oppose Kronecker's demand that only decidable definitions be used in mathematics and to defend his own use of sets ('systems of thought') for which membership was only "determined, not actually decidable":

> In what manner this determination is brought about, and whether we know a way of deciding it, is a matter of indifference for all that follows; the general laws to be developed in no way depend upon it; they hold under all circumstances. I mention this expressly because Kronecker . . . has recently endeavored to impose certain limitations upon the free formulation of concepts in mathematics which I do not believe to be justified ([63], p. 45).

Dedekind believed that our "inner consciousness" is incapable of providing a foundation for arithmetic and thought that by providing proofs for propositions once held as self-evident to this inner consciousness he had shown that our belief was never really based on that faculty at all: rather this shows that belief to be "always gained only by more or less complete repetition of the individual inferences". Dedekind's scientific credo is:

(D7) In science nothing capable of proof should be accepted without proof (*ibid.*, p. 31).

A year after Dedekind's death Weyl, already well into his intuitionistic phase, rejected (D7) in favor of an immediate insight based on '*das Erlebnis der Wahrheit*'. Meanwhile Brouwer was pressing the intuitionistic position that in a sound mathematics there could be no room for anything like the distinction drawn by Dedekind between 'determined' and 'decidable'.[14]

For discussion, we list six major achievements of Dedekind's essay:

(D1) Definition of an infinite set.
(D2) 'Proof' that infinite sets exist.
(D3) Definition (using chains) of simply infinite systems and natural number series.
(D4) Proof that all simply infinite systems and natural number series are isomorphic.
(D5) Justification (set theoretic derivation) of proof by induction.
(D6) Justification (set theoretic derivation) of *definition* by induction (recursion).

These are the highlights of his analysis of the 'intuitive natural number sequence':

(N) $0, 1, 2, 3, \ldots, n, n', \ldots,$

which, as he tells us in his letter to Kefferstein, did not come easily:

How did my essay come to be written? Certainly not in a day; rather, it is a synthesis constructed after protracted labor, based upon a prior analysis of the sequence of natural numbers just as it presents itself, in experience, so to speak, for our consideration. What are the mutually independent fundamental properties of the sequence N, that is, those properties that are not derivable from one another but from which all others follow? And how should we divest these properties of their specifically arithmetic character so that they are subsumed under more general notions and under activities of the understanding *without* which no thinking is possible but *with* which a foundation is provided for the reliability and completeness of proofs and for the construction of consistent notions and definitions? ([255], p. 99).

Let us follow Dedekind in his "prior analysis" of the intuitive sequence (N) "as it presents itself, in experience, . . . for our consideration". It is a classic case of what Carnap called an explication of an '*explicandum*', an intuitive inexact concept belonging to what Carnap called "a previous stage in the development of scientific language". Although we are never "presented in experience" with more than a finite segment of (N), our intention, expressed inexactly and intuitively in our second group of dots, is that (N) should go on infinitely. We must express this intention for (N) more exactly. Dedekind's answer in (D1) is that any set N is called infinite if there exists a 1–1 function which maps N into a proper subset of itself. We will formulate this in current logical terminology:[15] a *model* is a triple $\langle M, a, g \rangle$ where M is a set, a an element of M, and g a function on N. Evidently the system (N) can be regarded as a model, viz. $\langle N, 0, {}' \rangle$. Dedekind's explanation of the infinity of N now is that the model $\langle N, 0, {}' \rangle$ satisfies the conditions:

(P1) $x' \neq 0$, for all x,

(P2) $x' = y' \rightarrow x = y$, for all x, y,

where the variables range over N. These conditions express the intention we have about (N) that it should begin with 0 and then go on forever always producing something *new*. In short, they express our intentions for $'$ in (N). But not yet completely, for our dots are supposed to point to only numbers, i.e. only to things we can get by applying the $'$ operation an arbitrary finite number of times beginning with 0: we must exclude what Dedekind called "alien intruders". Dedekind emphasized this as the heart of the problem: how to banish the intruders without taking the circular route of explaining the locution 'n is a number' by simply requiring that n be generated from 0 by a finite *number* of applications of $'$. Dedekind's solution in (D3) is to further require the system (N) to be 'simply infinite', which we can formulate by saying that our infinite model $\langle N, 0, ' \rangle$ should be a *Peano model*, i.e. that it should further satisfy

(P3) $\{(A)\ [A \subseteq N \ \&\ 0 \in A \ \&\ (x)(x \in A \rightarrow x' \in A)] \rightarrow A = N\}$.[16]

Dedekind then defines a 'natural number series' as the result of "entirely neglect[ing] the special character of the elements" of any Peano model, retaining only properties of them implied by (P1) – (P3), whose logical consequences become "the first object of the science of arithmetic". This also amounts to a *thesis*, namely,

(DT) The intuitive number system (N), viz. $\langle N, 0, ' \rangle$, is a Peano model.

Before discussing Dedekind's justification (D4) for this thesis, let us pause to consider the general problem of proving any explication adequate. Our ultimate concern will be Church's thesis, which is often said to be unprovable because it relates an 'intuitive idea' to a mathematical one. In short, Church's explication can never be proved adequate. Is this true of explication generally? But Dedekind's (DT) is really no less a thesis than Church's: it relates an informal intuitive mathematical idea with a more formal abstract one. And though Dedekind's analysis leading up to (DT) is convincing enough for most mathematicians, there is room for doubt that Peano models adequately describe (N) as long as we do not know which Peano model is meant: the correctness of (DT) may depend on which one we pick. Dedekind tried to remove any such doubt by his isomorphism result (D4): since any two Peano models are isomorphic, they all have intrinsically the same structure, and hence, apart from matters of notation, (DT) completely characterizes (N) – showing that explications of intuitive concepts *can* be proved adequate. In short, the conditions (P1)–(P3) defining a Peano model comprise, when regarded as axioms, a categorical system. This is also Dedekind's justification for saying that "the first object of the *science of numbers* or *arithmetic*" will be

(A)′ ". . . the relations or laws which are derived entirely from the conditions . . . [defining Peano models] and therefore are always the same in all . . . [Peano models], whatever names may happen to be given to the individual elements" ([63], p. 68).

As we have already noted, Dedekind regarded his isomorphism result as justifying the thesis (D0) also. Now for Dedekind the essential meaning of (D0) consists of the following two denials:

(D0)′ The system (N) need not be represented in *intuition.*
(D0)″ The numbers of (N) cannot be known as *things-in-themselves.*

This is the twofold sense in which our number creations are 'free': unlike Kant, Dedekind did not require them to be represented in intuition, and unlike Frege and Russell, he did not think it was possible to identify any unique object as the 'real number one', for example. Now (D4) does seem to justify (D0) when the latter is understood in this twofold sense: it justifies (D0)′ insofar as it seems to show that for the scientific purposes of (A)′, *no new laws are gained by representing (N) intuitively*; and it justifies (D0)″ because it shows that our numbers *cannot be identified scientifically as the elements of any unique model.* Essentially this view of the significance of isomorphism has been argued with particular eloquence by Weyl:

A science can only determine its domain of investigation up to an isomorphic mapping. In particular, it remains quite indifferent as to the 'essence' of its objects. That which distinguishes the real points in space from number triads or other interpretations of geometry one can only *know* (*kennen*) by immediate intuitive perception. But intuition is not a blissful repose never to be broken; it is driven on toward the dialectic and adventure of cognition. It would be folly to expect cognition to reveal to intuition some secret essence of things hidden behind what is manifestly given by intuition. The idea of isomorphism demarcates the self-evident insurmountable boundary of cognition. This reflection has enlightening value, too, for the metaphysical speculations about a world of things in themselves behind the phenomena. For it is clear that under such a hypothesis the absolute world must be isomorphic the the phenomenal one . . . Thus even if we do not *know* the things in themselves, still we have just as much *cognition* about them as we do about the phenomena ([269], pp. 26–7).

Now this expresses an essential tenet of *formalism*: that, being unable to intuit or know the objects of science in themselves, we must settle for the formal laws they satisfy. This is important to remember in connection with the incompleteness difficulties which may arise if we apply isomorphism results to explicitly formalized theories, for otherwise we are easily misled into thinking that they refute formalism. But as we shall see, formalism was never the view that such isomorphism proofs could be, as Hilbert put it, always "finitistically transformed" into syntactical completeness: it rather claimed that isomorphism, finitizable or not, would be the *best* we can do. It represents, to repeat Weyl's words, "the self-evident insurmountable boundary of cognition". This, in fact, was just the issue which divided Hilbert and Frege.[17]

As it happens, Dedekind's isomorphism result follows almost immediately from his justification (D6) of definition by induction, i.e. what we have been calling 'recursive definition'. We also need (D6), of course, to define various functions on N in terms of which we can express the laws about numbers we wish to prove, that is, we need (D6) to get started with what Dedekind called "the science of arithmetic". In terms that we have been using, the general theorem behind (D6) – which we will call *Dedekind's recursion theorem* – says the following.

Let $\langle N, 0, ' \rangle$ be a Peano model, and let $\langle M, a, g \rangle$ be any model. Then there exists a unique function f satisfying the conditions

$$(2) \qquad f(0) = a, \quad f(n') = g(f(n)).$$

In case $\langle M, a, g \rangle$ is actually a Peano model then the two such f's which the theorem gives, one in each direction, can be composed and then easily shown to yield an isomorphism between the two models. But we are really more interested in the proof of Dedekind's recursion theorem itself, as well as its application in building up arithmetic, and above all, its relation to mathematical induction and intuition. Let us consider the special case where $M = N$; hence a is a number, and g is a numerical function. Dedekind shows that (2) can be used to meet Frege's demand for an existence proof for Grassman's recursion equations for addition: taking g as $'$, there exists by the theorem, for each m, a unique function f_m satisfying

$$f_m(0) = m, \quad f_m(n') = f_m(n)',$$

and we can then explicitly define addition by $m + n = f_m(n)$, constructing a binary function from an infinite sequence of unary functions.[18] Taking next $a = 0$, and $g(n) = n + m$, which has now been defined, the theorem yields for every m a unique function h_m satisfying

$$h_m(0) = 0, \quad h_m(n') = h_m(n) + m,$$

and we can explicitly define multiplication as $m \cdot n = h_m(n)$. Taking now $a = 1$ and $g(n) = n \cdot m$ yields for each m a k_m satisfying

$$k_m(0) = 1, \quad k_m(n') = k_m(n) \cdot m,$$

and we define exponentiation by $m^n = k_m(n)$. Dedekind uses these recursions and induction to give perhaps the first rigorous proof of basic arithmetical laws holding for these functions. What is especially interesting is that Dedekind is now using recursion to provide a precise mathematical basis for the systematic introduction of new arithmetical operations, a process which he had described in his 1854 Habilitation lecture [62], and again in the preface to his memoir on continuity, quite mentalistically as a series of 'creative acts' in which each operation arises by a '*Zusammenfassung*' of the arbitrary repetition of the previous one, starting with the successor as the 'simplest arithmetic act'.[19] Thus, if we let f_1, f_2, f_3 stand respectively for $m + n$, $m \cdot n$, m^n as each was just recursively introduced with the help of the previous one and $f_0 = {}'$, we see that this process can be extended to yield an infinite sequence of new binary arithmetical operations by the inductively defined scheme

$$\text{(3)} \qquad \begin{aligned} f_{z+1}(m, 0) &= b_{z+1}(m), \\ f_{z+1}(m, n + 1) &= f_z(f_{z+1}(m, n), m). \end{aligned}$$

Of course, none of these f_z beyond f_2 have any use in the market place, as already by f_4, called 'elevation' by Husserl, we encounter a growth rate so steep as to make its calculation a practical impossibility for all but the smallest arguments. By f_5 even our standard notational conventions begin to buckle, while for f_6 we presumably will have to remain forever content with a recursion formula of (3) as our only feasible description of it.[20] Thus, very early in the sequence (3) one is led to wonder whether, and in what sense, we can be talking about 'arithmetical' operations at all. We also begin to appreciate the great difficulties that faced a theoretical elaboration of the claim (A) of algorithmic arithmetic that every arithmetical operation can be generated by a 'proper mixture' of the four elementary operations. So Dedekind's recursion theorem, which can seem so superfluous when applied as above to yield familiar basic operations already known to exist, soon acquires in the general case the theoretical significance of *justifying the circular definition of functions for which no other description is known or even, from a practical standpoint, possible.*

In a letter to Royce, Peirce claimed that Dedekind's "little book . . . is most ingenious and excellent. But it proves no difficult theorem that I had not proved or published years before, and my paper had been sent to him." Now in Peirce's paper [199] on the 'logic of number' he makes essential use of recursive definitions for addition and multiplication, yet without any justification. Did he take it as self-evident that such definitions were sanctioned by the induction principle? There is nothing like Dedekind's theorem (2) anywhere in his published writings, so his claim that Dedekind's essay "proves no difficult theorem that I had not proved or published years before" presumably means that Peirce regarded the theorem as too obviously implicit in the induction principle to bother proving explicitly.[21] Let us examine these views more closely.

At issue is the question whether, and on what basis, under the conditions of the recursion theorem, the existence of a unique function f satisfying the equations (2) is self-evident. An old argument, still common among intuitionists and constructivists, bases an affirmative answer *solely* on mathematical induction. Clearly, the argument begins, f is defined for 0, for we are given $f(0) = a$ outright. Substituting in the second equation then gives us $f(1) = g(f(0)) = g(a)$, then $f(2) = g(f(1)) = g(g(a))$, and so on. By proceeding intuitively in this inductive manner, we have in general that if $f(n)$ is defined, then $f(n') = g(f(n))$ is defined. By induction then, $f(n)$ is defined for all natural numbers, and only these. Induction also shows that f is unique. We can summarize the induction which is supposed to establish the existence of an f satisfying (2) as

(2)′ (i) $f(0) = a$, hence f is defined for 0,
(ii) $f(n)$ *is defined* $\Rightarrow f(n')$ *is defined.*
$\therefore f$ is defined for all n.

This inductive argument is now widely held to be inadequate, ultimately on grounds that go back to Dedekind. Indeed, if we consider a function to be, say, an infinite set of ordered pairs, then (2)′ is clearly inadequate. For it only establishes, at best, the existence, for every k, of the *finite* set

$$f_k = \{\langle 0, f(0)\rangle, \langle 1, f(1)\rangle, \ldots, \langle k, f(k)\rangle\}$$

of ordered pairs: by itself, induction gives no way of collecting together all these finite sets into a single infinite set which would constitute the function f. It is as if one tried in an axiomatic set theory to get an infinite set merely by postulating the empty set and then postulating only the set operations of forming unit sets and pairs, without the axioms of replacement or infinity. The appeal of this kind of reasoning is due to the fact that one has, as Gauss emphasized, already unconsciously assumed that f exists before such an argument begins. But the argument (2)′ is *circular*, for although it is true that, as (ii) says, $f(n')$ is defined if $f(n)$ is, still it can be objected that $f(n)$ *is not really defined unless* f *itself is.* In fact, the difficulty is already apparent when we reflect on (i): when we say that f is defined for 0, we must ask: what *is* f? If the purpose of the argument is to show that there is such an f, we cannot without circularity refer to this f in the beginning of such an argument. The very circularity which Frege and others noted in recursive definitions – the function being defined appears to recur again in the definition – short-circuits any straightforward attempt to justify them by induction alone. Induction can establish uniqueness, but not existence.[22] Still, the argument (2)′ sounds very plausible and we shall do well to examine why.

To begin with, it is clear that the cogency of the criticisms just reviewed of (2)′ depends on what one means by a 'function'. In particular, it depends on using something like the set theoretic definition of f as an infinite object. However, if, like Kronecker and others, one tends to regard a function as a law, a rule, or an algorithm, the situation changes. If one asks what, in general, makes the existence of a rule f for (2) seem so obvious, then the answer would seem to lie generally in our *intuition of effectiveness*: it seems intuitively clear, after reflecting on the equations (2) that there is an algorithm implicit in them. Specifically this intuition establishes that, given the conditions in (2),

(2)″ g is effective $\Rightarrow$ f is effective.

A function is effective if there is an algorithm for it. (2)″ can be regarded as a basic closure property of effective functions. But sound as (2)″ may seem on this intuitive notion, we still must ask whether it can also be proved solely by intuitive induction, i.e. whether recursion and induction rest on the same intuitive foundation.

It was Poincaré, above all, who considered induction as the ultimate principle of mathematical reasoning. As we have seen, Poincaré based infinity and induction on "the mind's power to conceive of the indefinite repetition of the same act, when the act is possible", claiming that we "directly intuit" this power. The essential characteristic of induction itself – Poincaré calls it "reasoning by recurrence" – is that it "contains, condensed, so to speak, in a single formula, an infinite number of syllogisms". The question now is how Poincaré viewed the relation of proof by recurrence to *definition* by recurrence, and the answer seems to be that he viewed these two uses of recurrence as essentially the same. After introducing the recurrence equation for addition he emphasizes, exactly as he does for induction, that it is not a "purely logical" definition; rather

> ... the equality ... in fact, contains an infinite number of distinct definitions each having only one meaning when we know the meaning of its predecessor ([205], p. 7).

Apparently then, Poincaré accepted the argument (2)′ without further ado, so that for him, recursion was implicit in induction. Dedekind, however, showed that recursion must involve in general something more than induction. For discussion call a model $\langle N, a, g\rangle$ which satisfies (P3) an *induction model.* The idea then that recursion can be based solely on induction amounts to claiming that (2) can be proved for induction models. Dedekind, however, showed by counterexamples that this is not the case: if $\langle N, 0, s\rangle$ is only an induction, as opposed to a Peano model, and $\langle M, a, g\rangle$ is any model, then there does not in general exist any function f from N to M satisfying (2). In other words, any valid proof of the recursion theorem must use both (P1) and (P2), as well as (P3). In fact, it can be shown that a necessary *and* sufficient condition for a model to be Peano is just that Dedekind's recursion theorem hold for it, so (2) is actually equivalent to the entire set of 'Peano axioms'. Thus the recursion principle is clearly stronger than the induction principle; in particular, *the induction property does not, like the recursion property, imply the infinity of a model after all.* Induction is certainly a most useful way of dealing with certain infinite sets, but, as we might have expected from our discussion of the inductive argument (2)′, it is powerless to *produce* them. This seems to me to indicate a basic weakness in the point of view typified by Poincaré. Still, it can be shown that every induction model possesses unique operations of addition and multiplication which can be defined recursively, which might help explain the tendency of most writers except Dedekind to base recursion solely on induction: most of them considered only addition and multiplication.

Dedekind's work suggests that the intuition of effectiveness which informally justifies recursion goes beyond the *intuition of finiteness* which justifies induction informally. Some idea of what more may be involved in it may be gleaned by examining Dedekind's proof for his recursion theorem. Although his proof – and indeed, even the statement of the theorem itself – is set theoretic, it has a constructive and informative core. Dedekind first uses his set-theoretic chain concept to define the less-than relation for the numbers of a Peano model $\langle N, 0, '\rangle$: $m < n$ just in case the chain of n is included in the chain of m'. After this, his proof has essentially three parts. First a basic lemma on partial functions satisfying a finite number of initial instances of the equations of (2): for every n, he proves the existence of a partial function f_n defined only for $x \leqslant n$ and such that

$$f_n(0) = a,$$
$$f_n(x') = g(f_n(x)), \text{ for all } x < n.$$

Now this lemma is proved essentially by induction, viz. (P3) – yet even this proof must appeal also to (P1) and (P2) – and solves as much of the existence problem for f itself as can be established by inductive-type reasoning. Next Dedekind shows by induction that for all numbers m, n, x such that $x \leqslant n$, $x \leqslant m$, we have $f_n(x) = f_m(x)$, so that each $f_n(x)$ of the lemma is uniquely determined for all $x \leqslant n$. At this point we have the same problem we encountered in the problematic argument (2)′, with its finite sets f_k which it could generate as approximations of f: how to fuse these functions together into the single total function f. Dedekind's solution is simple: given the infinite sequence $f_1(x), f_2(x), f_3(x), \ldots$ of partial functions, each an extension of its prodecessor, generated by the lemma, he defines explicitly a 'limit' function f for this sequence by the *diagonalization*:

(d) $\qquad f(n) = f_n(n).$

It follows from the lemma and the uniqueness of its partial functions f_n that the explicit definition (d) assigns a unique natural number $f_n(n)$ to every n; f is thus the unique function satisfying the recursion (2).

In defining f here Dedekind employs a variation of a diagonalization technique due to P. du Bois-Reymond [73], which Cantor will use in 1890 to simplify the proof of his theorem on the non-enumerability of the real numbers. Later we shall examine other diagonalization procedures, due to Richard, Ackermann, Gödel, and Kleene. For now let us just say that by 'diagonalization' we mean a process of getting a unary function from a binary function which involves identifying both its variables – and then possibly composing the resulting unary function with further unary functions, e.g. adding one. Thus Dedekind's lemma yields, in effect, a partial binary function $h(n, x) = f_n(x)$ defined only when $n \geqslant x$, whose diagonalization $h(n, n)$ is our solution. Since this intuitively effective process of diagonalization – $h(n, n)$ is surely effective if $h(n, x)$ is – provides the essential construction needed, beyond what is supplied by the inductive reasoning of the lemma, for the existence theorem on recursion, we may ask whether a study of this process might help to clarify how, and to what extent, our intuitions of effectiveness go beyond those underlying induction. For discussion, we can formulate our new closure property for effectiveness as

(d)′ $\qquad h(n, x)$ is effective $\Rightarrow f(n) = h(n, n)$ is effective,

where h is any binary function. Now an intuitive justification of (d)′ would seem to involve essentially our intuitions about the symbolic processes of *substitution* rather than the iterative processes of induction. Still another process involved here which we will have to analyze is that by which $h(n, x)$ itself was formed, and which figured earlier in the applications of the recursion theorem to define binary functions, namely considering the indices in a sequence of unary functions as a variable of a binary function. We shall come back to these points later.

The role played by partial functions in Dedekind's proof suggests a new point of view towards the fallacious argument (2)′: if we could have somehow already proved the existence of at least a *partial* function f satisfying the recursion (2), then the induction of (2)′ would have been a perfectly good way to demonstrate the *totality* of f. However it is not obvious how this point of view could ever be useful, for it is not clear how one would ever be able to prove the existence of a partial function for (2) without already proving the existence of its totalization. It was not then clear how partial functions could play any role at all since it is not obvious how there could be effective partial functions without effective totalizations. Indeed, Frege felt compelled to define numerical functions for everything under the sun! These questions will also come up again.

Thus although Dedekind himself had no special interest in decidability – recall that he published his essay for the express purpose of opposing Kronecker – he nevertheless employed his recursion theorem to prove things whose only alternative justification would appear to be some concept of effectiveness, with the consequence that the theorem and its proof have helped us to articulate and clarify to some extent the structure of that concept. *It even appeared that it might enable us to effectively – and hence mechanically – describe the whole 'creative' process of introducing new functions.* The theorem is also Dedekind's most important and successful application of the credo (D7): not only does it supply proofs for the existence of functions whose existence had seemed to be based on an intuition of effectiveness not accessible to further analysis, it even provides, via the isomorphism theorem, reasonable evidence for the adequacy of the thesis (DT) itself. Indeed, the recursion theorem itself completely characterizes Dedekind's 'simply infinite systems', viz. the Peano models. Of course, all of this assumes that such systems do indeed *exist*, which brings us to the problem of consistency.

4. PROPOSITIONING THE INFINITE

According to Dedekind, the certainty of arithmetic depends entirely on our being able to logically prove the existence of a simply infinite system.

Indeed, Dedekind would have welcomed such a result as a *consistency proof* for this theory, for

> After the essential nature of the simply infinite system, whose abstract type is the number sequence N, had been recognized in my analysis ... the question arose: does such a system *exist* at all in the realm of our ideas? Without a logical proof of existence it would always remain doubtful whether the notion of such a system might not perhaps contain internal contradictions. Hence the need for such proofs ([255], p. 101).

Dedekind's consistency proof is classical: he finds what he thinks is a model for the axioms (P1)-(P3). His proof proceeds in two stages. First he shows, by a construction later used in Skolem's work on denumerable models for axiomatic set theory, that every infinite model has a Peano submodel. Given any infinite model $\langle U, 0, ' \rangle$ Dedekind takes $D \subseteq U$ as the intersection of all subsets of U containing 0 and closed under $'$, and then shows that $\langle D, 0, ' \rangle$ is a Peano model. This reduces his problem to proving the existence of an infinite model, i.e. a model for (P1)-(P2).

Dedekind's program cannot be completed, however, by simply taking a model from some other branch of mathematics, say the real numbers, for he wants to base these other theories on the notion of natural number, and then ground this latter notion on 'logic', which we recall meant for Dedekind *the laws of thought*. Thus he considers the set G of "my own realm of thoughts, i.e. the totality of all things which can be objects of my thought", his *Gedankenwelt*, together with a distinguished element e of G which he calls "my own ego". He then defines a 'thought function' which assigns to any thought s "the thought, that s can be the object of my thought", which we denote by $\phi(s)$. Dedekind claims that the mentalistic system $\langle G, e, \phi \rangle$ is in fact a model, but, unlike the analogous claim for the intuitive number system $\langle N, 0, ' \rangle$, this claim is already problematic.[23] In order to complete his consistency proof, Dedekind must logically show now that $\langle G, e, \phi \rangle$ is an *infinite* model, i.e. he must show that both

(P1)′ $\phi(x) \neq e$, for all thoughts $x \in G$, and
(P2)′ $\phi(x) = \phi(y) \rightarrow x = y$, for all thoughts $x, y \in G$

are laws of thought. Presumably any proof for (P1)′-(P2)′ would have to use the properties of thoughts, and hence would qualify them literally as laws of thought. Dedekind, however, offered no such proofs, saying only that the truth of (P1)′ and (P2)′ was "clear". But apparently to few except Frege who stressed, however, that Dedekind's proof depends essentially on the existence of 'objective thoughts', thoughts that need not actually be thought. Since Dedekind can only actually think finitely many thoughts of the form $\phi(s)$, his proof cannot require such thoughts to be actually thought.

Now if infinitely many thoughts have not yet been thought, the infinitely many thoughts $\phi(s)$ must comprise infinitely many thoughts that are not thought, in which case it cannot be essential to a thought that it should be thought... If there were only thoughts that are thought, the sign '$\phi(s)$' would not always have a meaning; to ensure that it did have, it is not sufficient for 's' to mean something that could be an object of Dedekind's thinking. If this were not the case, then the sign '$\phi(s)$' would have no meaning for the given s ... So the validity of Dedekind's proof rests on the assumption that thoughts obtain independently of our thinking ([90], p. 136).

But Frege does not show that $\phi(s)$ *does* have a meaning for every $s \in G$, nor that Dedekind's proof is really conclusive. There presumably must be objective thoughts *if* that proof is binding, but it remains to be seen whether the assumption of such thoughts could by itself insure its validity, or even that $\phi(s)$ is always defined. Indeed, the assumption that Dedekind needs here, namely

(P0) $\quad x \in G \rightarrow \phi(x) \in G$, for all x,

would be better based on the 'power' which Locke and Poincaré found in the mind for the endless repetition of its own ideas and / or acts – *assuming that Hobbes and Descartes were wrong* about the logical impossibility of one thought being the subject of another. But even the existence of objective thoughts will not insure that the sequence

(S) $\quad e, \phi(e), \phi(\phi(e)), \phi(\phi(\phi(e))), \ldots$

will produce infinitely many thoughts unless both (P1)′ and (P2)′ are also established, no matter how 'endlessly' it be continued. For otherwise it may happen that we could have

(f) $\quad \phi(s) = s$, for some $s \in$ (S),

so that (S) might only generate finitely many *thoughts* after all. The possibility of (f) can be more easily understood by looking at Bolzano's attempted existence proof for infinite sets, a proof which Dedekind in fact tells us is based on a consideration similar to his own proof.

Bolzano claimed to prove the infinity of the "*Menge der Sätze und Wahrheiten an sich*". His idea is that, if we consider, for any truth A, the sentence 'A is true', then the sequence

(S)′ *A*, '*A*' is true, ''*A*' is true' is true, . . .

of propositions (*Sätze an sich*) expressed by sentences of (S)′ will be infinite, because

> . . . der Satz, welchen die Worte '*A* ist wahr' ausdrücken, ein von *A* selbst verschiedener sei; denn dieser hat offenbar ein ganz anderes Subjekt als jener. Sein Subjekt ist nämlich der ganze Satz *A* selbst ([28], p. 13).

But though it is clear that the *sentences* of (S)′, considered as linguistic inscriptions, are distinct – and hence that (S)′ generates infinitely many sentences, if continued endlessly – it is by no means as clear that it generates distinct *propositions*, despite Bolzano's argument.[24] Unless the structure of a proposition both determines its content and is also closely mirrored by the syntax of a sentence expressing it, there is no reason why sentences with distinct subjects could not express the same proposition. Especially if we regard a proposition as the *meaning* of a sentence expressing it. Thus, if we follow Frege and Ramsey in claiming that "it is true that *A*" has the same meaning as *A*, then the sentences '*A*' and ''*A*' is true' can express the same proposition despite having different subjects.[25] Indeed, it also seems to be implied by Tarski's observation that equivalences of the form

$$\text{'}A\text{' is true} \leftrightarrow A$$

follow from the very meaning of the word 'true', that ''*A*' is true' and '*A*' always express the same proposition. How then can Bolzano's sequence (S)′ generate more than *one* proposition?[26] The important point here is that *the more concretely we regard the terms of* (S)′, *the more certain we can be that they stand for distinct entities, whereas the more abstractly we regard them as representing 'contents', the greater the possiblity of identifying them.* Similar considerations arise also in connection with (f) and Dedekind's (S): considering now only thoughts *s* which are expressed by sentences *S* and denoting by $\bar{G}$ the set of such sentences, we can ask whether there may be sentences *S* for which

$$\text{(f)}' \qquad S \in \bar{G} \leftrightarrow S$$

holds in virtue of the thought expressed by *S*. If so, we could say that *S* actually expressed the same thought as $S \in G$. For certain sets $\bar{G}$ of sentences expressing what may be called 'syntactical thoughts' about the sentences of suitable formal systems, Gödel proved in 1934 a general 'fixed-point lemma' to the effect that a sentence *S* of the formalism could always be found for which its formalization of (f)′ could actually be formally proved. This does not by itself refute Dedekind's argument, but does show that his reasoning would not survive one straightforward way of relating thought to language.

Twentieth-century logic, excepting intuitionism, has reacted to the inherent difficulties of Dedekind's argument by gradually replacing considerations of thought by those of language and symbolism.[27] The basic difficulty is that, however saturated with meaning our thoughts or propositions may seem to be, insofar as we have no criterion for their identity we are not sure what they are. Many questions about them can only be settled by a syntactical recourse to symbolism anyway. For example, the inequality

$$\text{(f)}'' \qquad \overbrace{\phi\phi\phi\ \ldots\ \phi}^{1,\,000,\,001}(e) \neq \overbrace{\phi\phi\phi\ \ldots\ \phi}^{1,\,000,\,000}(e)$$

could hardly be established solely by our understanding of Dedekind's thought-function ϕ itself, but simply by counting the occurrences of the *symbol* 'ϕ' on each side of (f)″. This does not, however, show that Dedekind's logical theory of number is circular, for we don't really need numbers as such to establish (f)″: we need only try to put the occurrences of 'ϕ' on each side of (f)″ into 1–1 correspondence. The point is rather Hilbert's that symbolism must play an essential role in our theory of number; in particular, a consistency proof for number theory, if it is to avoid the difficulties of (P1)′ and (P2)′, must focus resolutely on symbols.

As we have said, Dedekind originally regarded the truth of both (P1)′ and (P2)′ as clear. Keferstein objected, however, that neither was clear to him, forcing Dedekind to be more explicit. Concerning (P1)′ he says that

> ... mein Ich kann Gegenstand meines Denkens sein, ist also ein Element von S; aber mein Ich ist doch nicht selbst ein Gedanke (nicht ein in Form eines Satzes ausdrueckbares Urtheil mit bestimmten Subjekt und Praedikat) ([64], p. 4).

Now (P1)′ is presumably true for some object of thought if not for Dedekind's ego – Royce suggested 'my country' – but (P2)′ is more difficult. On this score Dedekind could only reply that

> ... sind a und b verschiedene Elemente von S, so sind auch die Bilder a' und b' verschieden, weil der Gedanke a' von dem Subjekt a, der Gedanke b' aber nicht von dem Subjekt a, sondern von dem Subjekt b handelt ... Was ist nun hierbei misslungen? (*ibid.*).

This is in fact the very principle we found Bolzano using in his attempt to prove the infinity of his sequence (S)′ of truth-attributions: propositions or judgments with different subjects must be different. Indeed, for the special case of judgments of the form $\phi(a)$, this is just what Dedekind's (P2)′ says; his reply does not justify it, but *only repeats it in words*. Dedekind's explanations in fact only served to make Keferstein more doubtful than ever:

> Denn der Beweis stuetzt sich offenbar ganz wesentlich auf die Voraussetzung, dass das benutzte Abbildungsverfahren immer weiter fortgesetzt werden kann, indem der Gedanke s', dass s Gegenstand meines Denkens sein kann, selbst wider Gegenstand meines Denkens sein kann und so unbegrenzt weider. Die Endlosigkeit dieses Processes laesst sich selbstverstaendlich sehr leicht sein logisch begruenden, und insofern waere gegen ihre Benutzung nichts einzuwenden, aber anscheinend beabsichtigt Dedekind ... gerade den Begriff des Endlosen zu vermeiden und das Unendliche im Sinne des 'eigentlich Unendlichen' Cantors ..., also als ein fertige, nicht als ein erst werdendes zu fassen ([143a], p. 5).

Here Keferstein grants too much, for Dedekind was hardly in a position to logically prove that his thought-function ϕ could be endlessly reapplied. In short, (P0) is a *premise* of his proof, the justification of which would seem to rest on something like Locke's power for the repetition of our ideas. But as we have already seen, (P0) by itself could at best produce an infinity of *symbols*.

J. Keyser also emphasized the dependence of Dedekind's proof on (P0), and asked his readers to compare it with the dependence of Euclidean geometry on the axiom of parallels, even if they were not as free to deny (P0) as that axiom:

> Suppose that, in like manner, we say here, 'In the absence of *proof* that an act once found to be mentally performable, we will assume that such is not the case', then, whatever else might result ... one thing is at once absolutely certain: Dedekind's 'argument' would be quite impossible ([144], p. 549).

Keyser misleadingly called this assumption – essentially (P0) – the "axiom of infinity", calling it a "new presupposition of thought", and charged Dedekind with circularity in his proof. Russell, still believing he could logically prove the existence of infinite sets, correctly pointed out that no *one* of Dedekind's 'postulates' presupposed anything infinite, but that together they implied it, remarking that "it is indeed their purpose to do so". He also argued that human mortality made it certain that no act could be literally repeated endlessly, but emphasized that fortunately logic did not depend on mental processes anyway, "and would be equally true if there were no mental processes" ([226], p. 259). But he conceded to Keyser that "other proofs" for the existence of infinite sets

... such as the one from the fact that the idea of a thing is different from the thing, are not appropriate to pure mathematics, since they ... assume premisses not mathematically demonstrable (*ibid.*, p. 258).

A more succinct summary of Dedekind's proof one can hardly imagine! Shortly after writing these words, when it became evident to him that his own more mathematical proof for infinite sets could not pass the test of the theory of types which he had meanwhile adopted to avoid the paradoxes of set theory, Russell himself postulated the existence of infinitely many individuals as a new axiom of logic – later adding that "whether the axiom is true or false, there seems no known method of discovering" ([227], p. 143).[28]

The paradoxes caused Dedekind himself to withdraw his proof, and even to doubt whether human thought was completely rational, for whatever else it was or wasn't, Russell's set of non-self-members was certainly a possible object of *thought*. Cantor described Dedekind's G as an "absolutely inconsistent multitude". But Dedekind did not lose faith in the core of his theory of number, or in the "inner harmony of our logic", or in the creative power of the mind to produce infinite sets:

... ich glaube, dass eine strenge Untersuchung des Schöpferkraft des Geistes, aus bestimmten Elementen ein neues Bestimmtes, ihr System zu erschaffen, das notwendig von jedem dieser Elemente verschieden ist, gewiss dazu führen wird, die Grundlagen meiner Schrift einwand frei zu gestalten ([65], p. 343).

But, Dedekind's proof being the prototype of known attempts to prove the existence of something infinite, should we not wonder whether its dependence on an absolutely inconsistent concept may not also betoken something absurd in the very idea of an existent infinite?

Royce saw Dedekind's work, especially his *Gedankenwelt* proof and recursion theorem, as a significant contribution to the foundations of idealistic philosophy, the latter providing the basis for a theory of "the general nature of any iterative or recurrent process of thinking". On the other hand, the "very profound" *Gedankenwelt* proof established a "parallelism between the structure of the number series and the skeleton of the ideal Self", which previous idealistic philosophers, including Hegel, had been unable to make clear. But just this was needed, Royce argued, to bring Hegel's idea of a "completely self-conscious Thought, none of whose acts failed to be its own intellectual objects" into clear focus. Royce summarized the significance of Dedekind's work for idealism as follows:

The intellect has been studying itself, and, as the abstract and merely formal expression of the orderly aspect of its own ideally conceived complete Self, and of any ideal system that it is to view as its own deed, the intellect finds precisely the Number System ... Their formal order of first, second, and in general, of *next*, is an image of the life of sustained, or, in the last analysis, of complete Reflection. Therefore, this order is the natural expression of any recurrent process of thinking, and above all, is due to the essential nature of the Self when viewed as a totality ([224], p. 538).

Royce thus saw Dedekind's work as a basis for claiming that the human intellect, at least when ideally self-conscious, is infinite, though he did not critically examine the difficulties connected with Dedekind's proof. Peirce, who also believed that the human mind was infinite, applauded Royce's attempt to rehabilitate Hegel's philosophy of mind by applying the theories of infinity of Cantor and Dedekind. Elsewhere he remarked that "the idea of representation involves infinity, since a representation is not really such unless it be interpreted in another representation" [201], possibly influenced by Royce's ingenious interpretation of Dedekind's definition of infinite sets in terms of 'self-representative systems', systems which contain in a proper part of themselves a perfect representation of themselves. Indeed, according to Royce:

An object that contained, as part of itself, a perfect picture of itself – in other words, a self-representative object or system, of the type here in question – would of necessity prove to be an object whose complexity of structure no finite series of details could exhaust; for it would contain a picture of itself, within which there was to be found a picture of this picture, and a picture of this second picture, and so on without end ([225], p. 28).

Royce felt that Dedekind had distilled the mathematical core of this intuitive notion. Later we shall see that the rehabilitation of Dedekind's theory of number by Hilbert and Gödel does indeed lead inevitably to formal systems with very strong, though not perfect, self-referential properties, and which lead to an extension of Dedekind's recursion theorem.

Royce's analysis of Dedekind's concept of an infinite set focusses on the notion of similarity and is hence valuable for reminding us of the relation between Dedekind's logical distillation of this notion and the classical Greek confrontation with infinity necessitated by the Greek discovery of incommeasurability. Von Fritz [94a] has argued that this discovery depended in fact on the construction of an infinite sequence of similar figures, each contained within its predecessor. In particular, investigation of the ratio of the diagonal to the side of a regular pentagon revealed that the numerical evaluation of this ratio leads to an endless series of smaller and smaller similar pentagons inside the original one, as

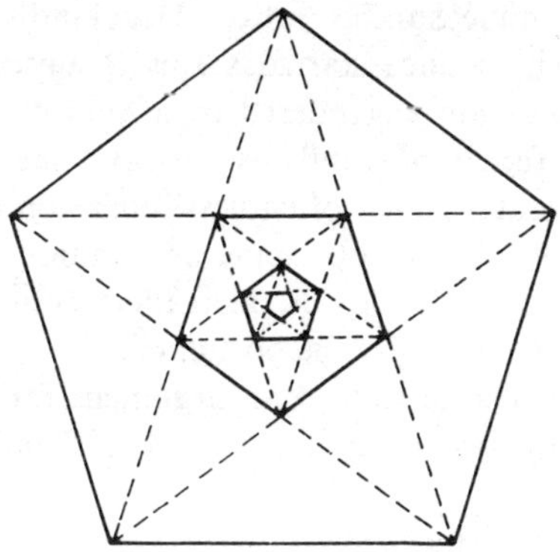

in the figure. One can show that the ratio in question is rational just in case this series is finite, hence it must be irrational, if indeed it can be meaningfully determined at all. The question now is how the Greeks knew that such sequences of similar figures really do go on infinitely. The initial impression that they do so is seen on closer scrutiny to depend on a number of strong geometric axioms, in particular Euclid's axiom of parallels. Indeed, the very existence of regular pentagons and rectangles depends on this axiom. Hence, *insofar as this axiom was not admitted as self-evident, we must also deny that the existence of incommeasurable ratios was self-evidently established.*[29] Indeed, from the standpoint of practical measurement such magnitudes are only a theoretical fantasy, as any carpenter will attest. The Greeks did of course correctly formulate what it would *mean* to say that such magnitudes had a ratio: this is the celebrated Eudoxian theory of proportion elaborated in Bk. V of Euclid. This theory depends essentially on the archimedean axiom and hence on the infinite number sequence, which is presupposed by Euclid. It was precisely the attempt to 'logically' prove this existence assumption which we have been scrutinizing. Dedekind's attempt depended on his abstract logical concept of 'similar' sets, namely those between which a one-to-one mapping exists: infinite sets are those which are similar to a proper subset. This concept appears at first sight to be considerably simpler than the similarity transformations of euclidean geometry, and hence Dedekind's 'infinite' sequence (S) to be freer of dubious assumptions than the Greek sequences of similar figures. But we have seen that this was an illusion, indeed that the assumptions required for Dedekind's proof – such as the very existence of G – lead to paradox. Like the original Greek proof of the existence of incommeasurable ratios, Dedekind's proof produces a sequence, viz. (S), which once started seems to unfold endlessly 'of its own accord', an impression engendered by a kind of mental inertia that Dedekind's 'thought function' ϕ lulls us into. But such sequences require more than Locke's mental power of endless repetition to scale the heights of the infinite: unless other troublesome assumptions are satisfied, they either peter out or go in finite circles. That there even are infinite heights to be scaled must now, we realize, be postulated.

Meanwhile Cantor also advanced an argument for the infinity of the human understanding, opposing it to the traditional argument for its finiteness:

> Man füht so oft die Endlichkeit des menschlichen Verstandes als Grund an, warum nur endliche Zahlen denkbar sind; doch sehe ich in dieser Behauptung wieder den erwähnten Zirkelschluss. Stillschweigend wird nämlich bei der 'Endlichkeit des Verstandes' gemeint, dass sein Vermögen rücksichtlich der Zahlen beschrankt sei. Zeigt es sich aber, dass der Verstand auch in Bestimmte Sinne unendliche, d.i. *überendliche* Zahlen definieren und von einander unterscheiden kann, so muss entweder den Wörtern 'endlicher Verstand' eine erweiterte Bedeutung geben werden, wonach aldann jener schluss aus ihnen nicht mehr gezogen werden kann; oder es muss auch dem menschlichen Verstand das Prädikat 'unendlich' in gewissen Rücksichten zugestanden werden, was meines Erachtens das einzige Richtige ist ([44], p. 176).

Cantor's remarks concern two arguments which may be encapsulated as follows:

(FM) Our mind is finite; therefore only finite numbers are conceivable (consistently definable).

(IM) We can consistently define and distinguish infinite numbers; therefore our mind is in a certain sense infinite.

I have added qualifications concerning consistency to Cantor's formulation, because it is debatable whether we can be said to conceive anything coherently unless we do so consistently, and Cantor presumably meant 'conceive' in the sense of having a coherent theory. This qualification cannot be ignored, as was shown by the paradoxes. Leaving this problem aside for now, we can say that Cantor's argument (IM) is at least as good as the argument (FM) that he criticizes. But how good is this? Dialectically speaking, i.e. against a proponent of (FM), his position is strong: for he had his own set theory to back up his premise, whereupon his conclusion, or something like it, follows from (FM). But considered in and of themselves, both (FM) and (IM) are rather weak, for they are of a piece with the obviously absurd argument:

(Δ) Our mind can define and distinguish triangles; therefore, our mind is in some sense triangular.

We would have to plead special properties of the notions of finite and infinite which carried (FM) or (IM), and enabled us to distinguish them from (Δ). For example, if we knew that a consistent theory of infinite sets was impossible, we would then presumably already have the conclusion of (FM). But considered as an argument, (FM) itself is quite infirm. With (IM) there is even the further problem of the very *meaning* of its conclusion: the mind boggles at literal senses of 'infinite' here. But if we think such arguments are too vague and weak to bother with, consider the following familiar but perfectly analogous argument against mechanism:

(L) Our mind can define and prove the existence of non-recursive sets (which by Turing's thesis are non-mechanical). Therefore our mind is non-mechanical.

Now our examination of Dedekind's work left it doubtful whether we can prove the existence of any infinite set. But even granting such sets, (L) clearly shares the same fallacy in common with (FM), (IM), and even (Δ): by attending to, defining, and having concepts of various (possibly intentional) objects, our mind does not itself thereby acquire the essential properties of these objects. Now (L) is in fact nothing but a version of the familiar argument that by proving Gödel's incompleteness theorem, the mind thereby shows that it cannot be a machine.

Recently another argument against mechanism and mind-body identity has been advanced by R. Abelson [1] that is closely related to Dedekind's argument for infinite sets. He argued that the identity thesis must be false since, whereas our brains are capable of only finitely many states, our minds, being able to think in principle of *any* number, must thereby be capable of infinitely many states. This would also imply that the mind is not a Turing machine. Abelson's argument has been formulated in more detail by S. Nathanson [193] essentially as follows: (i) A person can in principle think of any particular number; (ii) the set of natural numbers is infinite. Therefore, (iii) the set of thoughts or mental states a person is capable of is infinite. But (iv) the set of any person's brain states is finite. Hence, (v) there are more thoughts than brain states and so (vi) thoughts are not identical with such states. The first thing to note is that (vi) does not follow from (v), but rather only that *some* thoughts are not identical to brain states. But the basic fallacy in this argument was found by M.

Lockwood [170] who pointed out that the sub-argument from (i) and (ii) to (iii) already fails. In particular, he argues, quite correctly, that we need not be in *distinct single states* to think of each number, but rather that since 'thinking of a number' in general involves symbolic calculation, it is best analyzed as a process of passing through a '*chain* of states'. But on this analysis the mind can clearly still think of infinitely many numbers while having only a finite number of states. In fact, this is just what a Turing machine does: by passing through finite sequences of its finitely many states, it discriminates an infinite number of distinct input numbers and / or calculates an infinite number of outputs. Indeed, Abelson's analysis of 'thinking of a number' is altogether too static.

However, the interesting thing about this argument and the several replies that it called forth is that no one thought of questioning the premise (ii) that the set of natural numbers is infinite. All parties to the discussion just assumed that infinite sets really existed, in particular, the natural numbers, just waiting for our thoughts to catch hold of them as best they could. It does not seem to have occurred to any of them that the one serious attempt to prove the existence of their infinite set of natural numbers, namely Dedekind's, *failed precisely because it dealt with the world of our thoughts*. Dedekind saw no other way of proving the existence of the infinite set of numbers than by *independently* proving the infinity of our thoughts. How amused he would be to see Abelson and others now trying to prove the infinity of their thoughts by appealing to the supposed infinity of these numbers! Mathematics long ago replaced Dedekind's proof by an axiom, the truth of which, Russell emphasized, we have no means of discovering. But one has tended to somehow forget this whole episode and imagine that somehow or other mathematics really does supply us with infinite sets. But not all mathematicians have forgotten it, as the following pointed words of P. Finsler eloquently testify:

> Die Unendlichkeitsaxiom besagt, dass es unendlich viele Dinge gibt. In der realen Welt ist es sehr wahrscheinlich nicht erfüllt. Man braucht also eine ideale Welt. Gibt es in der idealen Welt unendlich viele Dinge? Wenn man das nicht zeigen kann, dann ist die ganze Unendlichkeitsmathematik hinfällig, man hat nur hypothetisch Sätze ... Es ist wohl die wichtigste Aufgabe einer wirklichen Grundlegung der Mathematik, dieser Punkt abzuklären, also zu zeigen, dass es unendlich viele Zahlen gibt. Das ist nicht leicht, aber auch nicht unmöglich ...
>
> Wenn man einfach behauptet, es gibt unendlich viele Zahlen, ohne zu wissen, ob das stimmt, dann ist das unwürdig und unehrlich. Wenn ein Kind fragt: ist es wahr, dass es zu jeder Zahl immer noch eine grössere gibt, kann man ihm dan *mit gutem Gewissen* antworten: ja, das ist wahr? ....

> *Unwürdig* ist es auch, in der Mathematik so zu tun, als ob es unendlich viele Zahlen gäbe, auch wenn man selbst gar nicht daran glaubt. Viele leugnen das Unendliche und wollen trotzdem Differential und Integralrechnung unterrichten ... ([85], pp. 175–6).

Indeed, many mathematicians have no belief whatever in anything infinite. In the next chapter we shall examine the attempt of one such mathematician, D. Hilbert, to justify our talk of infinity nevertheless. He tried to save Dedekind's theory by freeing it from its dependence on the inconsistent *Gedankenwelt*. In fact, one of the paradoxes which Hilbert tried to avoid could be used to prove Abelson's claim that we can think of infinitely many numbers. For if we could not, then there would have to be a least number, say λ, that we could not think of. But are we not thinking of λ after all? Indeed, doesn't this also prove the infinity of the numbers themselves? Can't we, after all, always 'think up' a number greater than any given one? Apparently not, for as tempting as this thought is, and as often as we pass it on to our children, we have seen that Dedekind's ingenious attempt to incorporate this idea into pure mathematics failed. Indeed, his attempt to deduce the infinity of the numbers from his own thinking fared no better than Descartes' attempt to deduce an infinite God from his. To repeat Locke again, our ideas are not always proofs of the existence of things.

NOTES

1 Hence, for Locke, our ideas need not prove the existence of things even if they are clear and distinct.

2 One could perhaps invoke unconscious mental processes which continue, so to speak, 'on their own', but infinite repetition hardly seems plausible as an unconscious desire of mankind, unless perhaps, we view the desire for immortality as such.

3 [169], p. 292.

4 See Cantor [44], p. 409. See Lodge [171] for modern scepticism about continuity.

5 See Kustaanheimo [160] for an introduction to this program and Kanitscheider [141], for a discussion of how it approximates differential equations of classical physics in finite geometries by recursions. In fact, there are several research programs underway seeking to solve fundamental problems in physics by introducing discrete space and time.

6 These geometries still contain the parallel postulate, but no continuity axioms, and modified order and congruence axioms. See Kustaanheimo [159].

7 These attempts are discussed at the end of Chapter III below. Becker [6] formulated a certain principle of 'transcendental idealism' (which is very close to a basic assumption made by Kaufmann) according to which questions undecidable in principle could not exist. Carnap [46], after arguing for the finite decidability of all questions on the basis of his pheonomenalistic 'construction theory', goes on to express his agreement with both positivism and idealism on such questions:

In the thesis of the decidability of all questions, we agree with positivism as well as idealism; cf. Becker: ... "According to the principle of transcendental idealism, a question which is in principle (in essence) undecidable does not have any meaning at all ... For there are no states of affairs which are in principle inaccessible to consciousness" ([46], p. 292).

[8] Gödel accepted Turing's analysis as providing a definitive clarification of a 'formal system', but not for 'effectiveness' in general. His reservations are discussed in Chapter IV. I have followed the formulation in Kreisel [155] of Gödel's argument.

[9] See Wang [258], pp. 224ff. for a discussion of the lack of any correlation between those domains known to be theoretically decidable or undecidable and those, respectively, for which we have had success in programs for mechanical theorem-proving or those for which we have not. Indeed, he emphasizes that the best such programs have so far been for subdomains of theoretically undecidable domains.

[10] See Smith [236], Vol. I, p. 294.

[11] See Tarski [245], p. 54.

[12] See Frege [87], pp. 8ff.

[13] See Buck [39] for a detailed discussion and proof of this example. The example below of an inconsistent recursion is also taken from Buck.

[14] See Brouwer [36], [38].

[15] I am following the elegant formulation of the Dedekind-Peano theory of numbers given by Henkin [113].

[16] This is equivalent to Dedekind's requirement that N should be the 'chain' of 0, i.e. the intersection of all sets including 0 and closed under′. This is his basis for (D5).

[17] See Steiner [239] for a discussion of the role of isomorphism in the Hilbert-Frege controversy. We discuss it in Section 2 of Chapter III.

[18] See Henkin [113], pp. 337ff.

[19] Veronese [257], p. 45, complained of Dedekind's use of the "elegant theorem" (2) to reduce meaningful acts of calculation to purely symbolic manipulations.

[20] See Buck [39], pp. 133ff., for a discussion of the difficulty of even devising a notation for an explicit description of such functions.

[21] This view, of course, is very common among various kinds of 'constructivists'.

[22] See Henkin [113] for a discussion of this point.

[23] See below for discussion of this difficult point. See also Helm [112].

[24] Bolzano's assumption that propositions with different subjects must themselves be different is also made by Dedekind in defending (P2)′; see below.

[25] Frege's formulation runs as follows:

So the sense of the word 'true' is such that it does not make any essential contribution to the thought. If I assert that 'it is true that sea-water is salty', I assert the same thing as if I assert 'sea-water is salty'. This enables us to recognize that the assertion is not to be found in the word 'true', but in the assertoric force with which the sentence is uttered ([90], p. 251).

[26] Essentially this criticism of Bolzano's argument was made by König, [152], p. 83.

[27] A lucid and direct formulation of this shift can be found in the introduction of Lyndon [178]. He begins by noting that:

Logic is often said to deal with the laws of thought. Here is meant not the historical or psychological principles governing the processes of thought, but rather those formal structural properties of thought which appear to reflect properties of the real world ([178], p. 1).

But for a theoretical development we need suitable idealizations for thought, reality, and their relation:

For thought we substitute *language*, or, more precisely, a formalized version of parts of everyday language ... For reality we substitute something called a *structure*, which is hardly more than a collection of things suitable for being correlated, as meanings, to various expressions in the language. For the connection between thought and reality we substitute an *interpretation*, that is, a function assigning to certain expressions in the language, as their meanings under the interpretation, certain objects in the stucture (*ibid.*, p. 2).

[28] See Moss [189] for the evolution of Russell's thinking on infinity.

[29] Strangely enough many writers willing to bracket the parallel postulate still cling to the incommeasurabilities that depend on it. In elliptic and hyperbolic geometry the diagonal of a square *can* be commeasurable with its side; only in Euclid's *must* they be incommeasurable, for here their ratio must be the *same in all squares*. Many still imagine that incommeasurability in turn implies infinite divisibility, but Berkeley indicated long ago the correct line in a note to himself, "to inquire most diligently concerning the incommeasurability of diagonal and side – whether it does not go on the supposition of units being divisible *ad infinitum* . . . and so the infinite divisibility deduced therefrom is a *petitio principi*" ([15a], p. 79). He saw clearly that "in geometry it is not prov'd that an inch is divisible *ad infinitum*" (*ibid.*, p. 77). This allowed him to entertain a "diagonal of a particular square commeasurable with its side, they both containing a certain number of m. v." (*ibid.*), thereby harmonizing geometry with his theory of vision as comprising a finite number of "minimum visibles" which he would identify with its points. Berkeley's critics have not appreciated the resources of his finitism, some even believing that incommeasurability and the pythagorean theorem refute it. But he met this objection squarely: "One square cannot be the double of another. Hence the Pythagoric theorem is false" (*ibid.*, p. 19). Berkeley was wrestling with what a contemporary geometer still calls the "Pythagorean paradox": although "it was intuitively evident to the Pythagoreans, and as I write it is intuitively evident to me, that a common measure can be found for *any* pair of straight lines" ([251], p. 15), their theorem together with arithmetic reasoning implies that none can exist for the diagonal and side of a square. Arithmetic logic denies what geometric intuition affirms. Non-euclidean geometry lessens the conflict somewhat, but only at the cost of intuitively appealing features of Euclid's. Berkeley's finitism cries out rather for the finite geometries of Järnefelt and Kustaanheimo, constructed from the residue classes of integral numbers modulo a prime number, which have euclidean geometry as their limit. Indeed,

In the Euclidean geometry certain properties of the finite geometry are smoothed out. But then the finer details, i.e. the fine structure of the plane, space, a.s.o. depending on the repartition of the prime numbers, have disappeared. One could also say that the Euclidean geometry draws a veil over the finer details of the plane, space, a.s.o. This fine structure comes into view only in the finite geometrical models ([138], p. 168).

CHAPTER III

THE MENTAL, THE FINITE, AND THE FORMAL

> Wo Begriffe fehlen, da stellt ein Zeichen zu rechter Zeit sich ein. Dies ist das methodische Prinzip der Hilbertschen Theorie.
>
> P. Bernays

1. KRONECKER VERSUS HILBERT VERSUS FREGE ON GEOMETRY

The interest in effective methods stimulated by Sturm's theorem came to a head when Kronecker demanded that such methods accompany *all* definitions in mathematics. For the irreducibility of algebraic functions his demand reads as follows:

> Die aufgestellte Definition der Irreduktibilität entbehrt so lange einer sicheren Grundlage, als nicht eine Methode angegeben ist, mittels deren bei einer bestimmten vorgelegten Funktion entscheiden werden kann, ob dieselbe der aufgestellten Definition gemäss irreduktibel ist oder nicht (quoted in [9], p. 328).

We can formulate the general demand simply as

(K) Only decidable definitions are admissible in mathematics.

Sturm's result encouraged Kronecker's belief in (K): algebra was his principal field of research and Sturm had shown that its fundamental theorem was susceptible of effective analysis. He shared with Gauss and Dedekind a belief in (D0) and (CM) which provided additional philosophical motivation for (K).[1] Kronecker was also confident that he could strictly 'arithmetize' all of mathematics except geometry which, following Gauss, he assumed to deal with a reality outside the mind, and hence not to be part of pure mathematics. Indeed, no discipline qualified as part of pure mathematics unless it was based solely on the integers, and decidably so, and geometry dealt essentially with continuous magnitude.

To better focus Kronecker's view we compare his program with that of his colleague Weierstrass, who also tried to 'arithmetize' analysis. However, to avoid the circularity of Cauchy's theory of limits and convergent sequences (roughly, each was defined in terms of the other), Weierstrass found it necessary – in order to establish the existence of limits, independently of the sequences they were said to be limits of – to postulate the existence of certain infinite sets associated with a given sequence. But Kronecker objected that Weierstrass's definition of irrational number provided no way of telling effectively whether a given sequence actually defined an irrational number. If you have a law which effectively determines every term of an infinite sequence, then the law itself can replace the sequence:

Selbst der allgemeine Begriff einer unendlichen Reihen . . . ist meines Erachtens nur mit dem Vorbehaltze zülassig, dass in jeden speziellen Falle auf Grund des arithmetischen Bildungsgesetzes der Glieder . . . gewisse Voraussetzungen als erfüllt nachgeweisen werden, welche die Reihe wie die endlichen Ausdrücke anzuwenden gestalten (*ibid.*, p. 327).

Kronecker admitted irrationals only if they could be effectively isolated in rational intervals by such methods as Sturm's, but then they could be replaced by those methods.[2]

Kronecker and Weierstrass both tried to base the continuum ultimately on the integers. Both saw, however, that something else was needed: Weierstrass also used infinite *sets* of them, while Kronecker used decision methods for them. But Cantor argued that Kronecker's 'supply' of number symbols could never completely describe the continuum, being of insufficient *power*:

Ich glaube sogar ohne jeden Zweifel vorhersagen zu können, dass es ihm nicht möglich sein wird, mit dem "ideellen Vorrat" seiner "Bezeichnungen" den *aktual unendlichen Punktvorrat* des . . . Kontinuums "vollständig und auf die einfachste Weise zu beschreiben" . . . und zwar hängt diese meine Überzeugung damit zusammen, dass ich . . . den Satz beweisen habe: die Machtigkeit eines Kontinuums ist höher als die Machtigkeit des Inbegriffs aller endlichen, ganzen Zahlen ([44], p. 384).

A great deal of mathematical logic has resulted from subsequent attempts to unravel the ins and outs of the 'cardinality curse' Cantor put on finitism here, and we shall anticipate the problems it raises for (K) by sketching two possibilities.

(i) Suppose L is a language in which Kronecker's algorithms for computing sequences defining irrational numbers can be expressed. Since each algorithm must be formulated in finitely many words, we argue that the 'Kronecker irrationals' are countable, and hence that there are irrationals in Weierstrass's continuum which do not satisfy (K). But here L is only incomplete relative to a *supposedly existing continuum*: unless we can actually give an explicit description of a Weierstrass irrational which is non-Kroneckerian, it will not be clear that anything is really 'lost' with L. Under these circumstances, L might still be complete relative to everything we can 'get our hands on', and Kronecker might see the imagined incompleteness as begging the question. However, this argument is double-edged (see below).

(ii) Perhaps the incompleteness envisaged for L can be sharpened by the diagonal method: if we can go through L effectively picking out its algorithms for Kronecker irrationals, we get a sequence $f_1(n)$, $f_2(n)$, $f_3(n)$, . . . of all functions for computing their expansions, and the diagonal method yields a function $f_n(n) + 1$ which, because it intuitively satisfies (K), describes a Kronecker irrational which, by the diagonal argument, cannot be computed by any algorithm of (L). Under these circumstances, L would be visibly incomplete *relative to its own ideal (K)*. In short, a complete theory of effective processes may itself be impossible, a possibility which haunted several logicians, including Hilbert, between 1888–1936.

Hessenberg [115] argued that (K), though justified in algebra by the existence of a finite criterion for a number to be algebraic, posed nevertheless an 'unresolvable dilemma' beyond algebra since, although there was provably no such criterion for the transcendental numbers due to their uncountability, we could never construct specific numbers whose transcendence was provably undecidable without landing in paradox. In fact, this objection had already been made to Kronecker:

> Wie mir Herr Geheimrat H. A. Schwarz mitteilt, hat er an Kronecker tatsachlich als Erwiderung auf dessen Einwande gegen Weierstrass' Irrationalzahlentheorie die Aufforderung gerichtet, ein nachweislich unentscheidbare Beispiel anzugeben ([115], p. 620).

Without such an example, Kronecker's demand was apparently hollow.

In 1888, as a young *dozent* beginning his mathematical career, Hilbert took a trip through the mathematical circles of Germany to meet the great masters of his science and hear their opinions concerning its most important problems. The deepest impressions on his mind were left by his meetings with Gordan, H. A. Schwarz, Weierstrass, and Kronecker, as well as the circumstance that on arriving at his first stop in Berlin he found "every mathematician, young and old" discussing heatedly the merits of Dedekind's provocative new essay on the number concept: both the essay and the discussions were to leave a lasting impression on Hilbert.[3]

Gordon entranced the young man with his finiteness problem for algebraic invariants, which he soon solved with a revolutionary nonconstructive existence proof whose brevity made a stunning contrast to the tedious calculating methods of the old "*Algorithmiker*" in which even single formulas could run on for pages. In fact, Gordon's work epitomized the kind of *excessive formalism that Hilbert abhorred.*[4] A friend said of Gordon that "formulas were the indispensable supports for the formation of his thought" ([217], p. 32), whereas Hilbert insisted that "*nur der überlegende Geist, nicht der angewandte Zwang der Formel den glucklichen Erfolg bedingt*" ([128], p. 394). Indeed, Hilbert's new proof complied with what he called Riemann's demand that "*man die Beweise nicht durch Rechnung, sondern lediglich durch Gedanken zwingen soll*" ([126], p. 67), but O. Blumenthal emphasized that it also displayed another enduring feature of Hilbert's work which tended to balance his anti-formalism, namely "*eine überlegende Kenntnis und Beherrschung des Formalismus, die ihm die rechnerischen Hilfsmittel mit fast unbewusster Selbstverständlichkeit in die Hände spielt*" ([128], p. 394). If the force of algorithmic formulas could not replace the reflective mind for Hilbert, they still could not be ignored, at least not until one's mastery of their use was so complete that it could be exercised almost unconsciously. Weyl was also struck by Hilbert's "peculiarly *direct* attack on problems, unfettered by algorithms", elsewhere explaining that he "came of a time in which the algorithm had played a more extensive part, and therefore he stressed strongly a conceptual procedure" ([267], p. 549). However, Kronecker had warned Hilbert that only algorithmic construction and not merely consistent concepts could establish mathematical existence, and neither he nor Gordon could at first accept his new proof as mathematics: "*Das ist nicht Mathematik, das ist Theologie*", said the latter. As it happened, Hilbert and other young mathematicians had been making a 'sport' of transforming transfinite proofs into finite ones, and this practice soon enabled him to find new constructive proofs effectively exhibiting the finite bases of various invariants.

Hilbert's descriptions of effective processes were somewhat more explicit than Kronecker's: at issue in his 1892 paper is a number which "*sich . . . mittels Rechung durch eine* endliche Anzahl *von Handlungen wirklich finden lässt*" ([127], p. 275), while a year later another decision procedure is described as a way "*wie man durch endliche und von vornherein übersehbare Prozesse entscheiden kann, ob δ eine ganze algebraische Funktion . . . ist oder nicht*" (*ibid.*, p. 321). As long as one has such a method, of course, such explicitness is only a luxury, for one only need use it. But if the challenges which Cantor and Schwarz had put to Kronecker's finitism were to be scientifically evaluated, one would clearly have to be as explicit as possible about such methods. Hilbert was now stressing that they produced calculations in a *finite* number of steps by processes *fixed in advance*. In his last paper on invariant theory, he even suggests that the satisfaction of Kronecker's demand for its existence theorems may be the final resting place for a theory:

Auf Grund dieses Sätzes findet die fundamentale Aufgabe der Invariant-theorie ihre Erledigung, nämlich die Aufstellung des vollen Invariant-systems, vermöge einer endlichen Rechnung ... (*ibid.*, p. 382).

Hilbert then concludes his work on invariants by distinguishing three stages in the historical development of any mathematical theory:

(I) *Naive*: formation of concepts and their informal application to solve problems.
(II) *Formal*: discovery and perfection of algorithms for symbolic calculation.
(III) *Critical*: general existence theorems, proved constructively if possible.

These stages anticipate the well-known threefold distinction of his later mathematical philosophy between (I)′ an informal theory, (II)′ a strictly formalized theory, and (III)′ a metatheory in which the consistency of (II)′ is proved constructively. Hilbert placed his own work on invariants in the critical third stage and not the formal stage which relies wholly and uncritically on algorithms. His conception of the critical stage of a mathematical theory further crystallized his work on the foundations of geometry and his closely related ideas on the solvability of mathematical problems, whose historical roots we now briefly survey.

In 1872 the biologist E. du Bois-Reymond, in a famous and influential lecture [71] on the limits of scientific knowledge, argued that certain philosophical problems, especially the nature of mind and the origin of consciousness, were scientifically unsolvable. Echoing La Mettrie he claimed that mind and consciousness could not be explained in terms of the cumulative mechanical action of atoms. His general thesis was: *Ignoramus et Ignorabimus*, which Hilbert apparently found just as abhorrent as the formalism of his day, and, as we shall see, some later criticisms of it are based on his own. Meanwhile, the mathematician P. du Bois Reymond [73a] attempted to extend his brother's thesis of eternal ignorance to include the basic foundational and philosophical problem of mathematics, namely the conflict between the 'Idealist' and the 'Empiricist' concerning the nature and existence of the arithmetical continuum. The conflict, presented in dialogue form, bears a striking similarity to that described above between Cantor and Kronecker respectively, and du Bois-Reymond's claim is that it *cannot be completely resolved by mathematical methods.* The idealist continuum invokes all possible binary sequences, whether given by laws or not, while the empiricist continuum is always 'becoming'. Du Bois-Reymond saw Cantor's proof of the non-enumerability of the idealist continuum as showing that the continuous is not merely the limit of the discrete, that the 'abyss' between them could not be constructively bridged. Since any attempt to do so must remain trapped in Cantor's *Abzahlbaren*, it is impossible to regard the idealist continuum as the "*wirkliche Grenze irgendwelcher empiristischen Folge von Vorstellungen*". It follows, according to du Bois-Reymond, that no rational argument can compel the empiricist to recognize the existence of the idealist continuum. Indeed, during his trip Hilbert had kept notes on important points in which we find Kronecker's views on continuity summarized by the entry: "Only the discrete and the singular have significance". For Kronecker, the continuum remained a meaningless fiction, a view which Hilbert found fascinating, even if disturbing.[5]

The foundations of geometry appealed to Hilbert as a field in which he could mathematically clarify such philosophical problems about continuity. In place of the ontological problem of the existence of the contunuum, he posed the metamathematical problem of determining the role played in geometry by continuity axioms, and to what extent they could be *eliminated.* For Klein the importance and novelty of Hilbert's classic [117] consisted indeed in precisely these motives.[6] These questions were the inevitable outgrowth of the 19th century trend towards empiricism in geometry, *for no other geometric axioms go further beyond experience than those of continuity.* Hilbert was also anxious to oppose Kronecker's exclusion of geometry from pure mathematics on the ground that it dealt essentially with continuous magnitudes; in fact, as O. Blumenthal later recalled:

> Der Reiz des Gebietes der Elementargeometrie aber lag für Hilbert darin, dass er hier das einfachste Beispiel sah, an dem er sein aus der Zahlentheorie abstrahiertes Ideal eines vollständigen Beweisgebäudes ausserhalb der Lehre von den ganzen Zahlen durchkonstruieren konnte, als Gegenbeispiel gegen die von Kronecker vertretene und von Hilbert immer Leidenschaftlich bekämpfte Auffassung, dass aller Mathematik, die sich nicht unmittelbar an die ganzen Zahl anknüpfen lasse, ein unreinlicher Erdenrest anhafte ([26], p. 68).

Hilbert wanted to show that pure geometry did not depend on any kind of number, and by showing that any plane geometry could be represented as a commutative field just in case Pascal's theorem holds in it, he established that indeed not even *analytic* geometry presupposed number. Specifically, he defined in a purely geometric way without using the archimedean axiom – which, in its dependence on the whole numbers was not in many eyes a purely geometric axiom – two operations on pairs of segments each yielding unique segments and which *satisfied all the algebraic laws of addition and multiplication.* This is Hilbert's famous '*Streckenrechnung*', and the same trick can also be done directly with the points themselves which in analytic geometry are usually 'assigned numbers'. But then numbers were not so different from geometric objects after all: however disparate their epistemological origins may have seemed to be, the essential properties of both were expressed in the field axioms. It was this kind of result that Hilbert's geometry contributed to 'formalism': since such seemingly different mathematical objects could satisfy the same formal laws, the tendency naturally arose to forget the objects and focus on the laws themselves.

Explanations of Hilbert's new 'formality' in geometry are often beside the point. The logical irrelevance of the meanings of the geometrical primitives already urged by Pasch follows from the idea of logical consequence itself and hence is hardly specific to geometry. The need for their 'implicit definition', on the other hand, followed simply from Hilbert's belief, taken over from Kant, that it was impossible to mathematically define all of them outright. Using the axioms themselves to define them was not an idle exercise in 'disinterpretation', but rather a recognition of the fact that essentially different kinds of interpretation were equally possible for them. Insofar as Hilbert leaned towards the physical interpretation urged by Riemann we may describe his early geometric work as a way-station between classical and modern physics. (As we shall see, however, he did not eschew spatial intuition, but tried rather to clarify and generalize it, and even eventually to base all mathematical knowledge on it.) Some important roots of Hilbert's formalism are to be found in fact in certain late 19th century developments in classical physics, particularly the work of Hertz on electromagnetic theory and mechanics. Since it was directly from the famous introduction to Hertz's mechanics [114a] that Hilbert took his formulation of the basic goal of the axiomatic foundations of geometry, an appreciation of this formalism may profit from a brief sketch of this development.

Boole said of his algebra that "every system of interpretation which does not affect the truth of the relations supposed, is equally admissible" ([26a], p. 278). This formalistic freedom to fish out various interpretations from a democratic pool aimed at widening the physical applications of his algebra and led to the notion of an abstract theory. But as Enriques observed, the value of this notion remained largely unappreciated until it became married to "the positivistic philosophy which aims to free physical doctrines from underlying metaphysical hypotheses and to see in them merely mechanical *models* of reality" ([76], p. 116). He was thinking primarily of Maxwell's electromagnetic theory for which Maxwell himself constructed several mechanical models. But Hertz no longer asked which model was the 'right' one, insisting rather that "Maxwell's theory is nothing else than Maxwell's equations". Poincaré even proved that any physical theory whatever admitted, under suitable conditions, an infinity of experimentally indistinguishable such models, and hence that the choice between them must remain "inaccessible to positive methods" ([205], p. 224). By eschewing in particular such models for Maxwell's 'underlying medium' of aether, Hertz established his electromagnetic field as an independent reality with its own experimentally determined mathematical structure.[7] New applications of Maxwell's theory, however, sometimes depended on surprising interpretations for hitherto obscure terms, and had to be confined moreover to large-scale phenomena in which one could safely ignore the microscopic structure of matter. It was thus clear that applications of even so imposing a theory as Maxwell's would always require, as Hilbert expressed it to Frege, "a certain measure of good will and tact".

Hertz also formulated a general principle for good theory construction which Hilbert would later apply to geometry, namely the rule that "we form for ourselves images or symbols of external objects, and the form we give to them is such that the necessary consequents of the images in thought are always the images of the necessary consequents in nature of the things pictured" ([114a], p. 1). This presupposes, Hertz conceded, "a certain conformity between nature and thought", but above all also the *logical consistency* of our images. The very meanings of the basic concepts of a theory, Hertz insisted, depend on this consistency. That 'force' and 'electricity' had never been adequately defined was shown by the fact that physicists were "continually attacked as to the nature of electricity". And yet, Hertz asked: "Can we, by our conceptions, by our words, completely represent the nature of any thing? Certainly not" (*ibid.*, p. 7). If we think we better understand, say, the nature of 'gold' or 'velocity', this is only because, according to Hertz, "With the terms 'velocity' and 'gold' we connect a large number of relations to other terms; and between all these relations we find no contradictions which offend us. We are therefore satisfied and ask no further questions" (*ibid.*). But around such terms as 'force' and 'electricity' we have "accumulated more relations than can be completely reconciled amongst themselves", and it is just our dim awareness of these inconsistencies that "finds expression in the confused question as to the nature of electricity" (*ibid.*, pp. 7–8). These inconsistencies can only be removed, according to Hertz, by a "logical analysis of the elements" of a science, though he admits that they may not prevent triumphant applications of a theory.

Hertz's views not only influenced Hilbert's belief in the impossibility of completely defining the basic concepts of a science, but also contain the germ of his idea that a suitable consistent set of axioms could go a long way towards establishing some meaning for them nevertheless. Frege objected that Hilbert's axioms were thus "saddled with something that is the function of definitions" ([91], p. 7), and were unable to unambiguously fix the meanings of Hilbert's primitive geometric concepts. One must first define one's terms, adhering to "the tenet that in a definition nothing is asserted, rather something is stipulated" (*ibid.*), and then *assert* one's axioms. Frege also objected to Hilbert's use of a cartesian model to prove the consistency of the euclidean axioms, for since these axioms are *all true*, Frege claims, their consistency follows immediately without any need of proof. Euclid's axioms are true but unprovable, according to Frege, "because our understanding of them derives from that non-logical basis that we call intuition" (*ibid.*, p. 9).

To Frege's first point Hilbert replied, echoing Boole, that "surely it is self-evident that every theory is merely a framework or schema and that the basic elements can be construed as one pleases" (*ibid.*, p. 13). And if they are construed as, say, chimney sweeps and their relations suitably reinterpreted to make the axioms true of these things, the theorems will hold of them as well:

> In other words, each and every theory can always be applied to infinitely many systems of basic elements. For one merely has to apply a univocal and reversible one-to-one transformation and stipulate that the axioms for the transformed things be correspondingly similar (*ibid.*, p. 14).

This is especially true, Hilbert thinks, of the "assertions of a theory of electricity". Hilbert's claim is thus that any theory can only describe its basic domain up to isomorphism: its axioms can never distinguish isomorphic models. To repeat the words of Weyl, "the idea of isomorphism demarcates the self-evident insurmountable boundary of cognition". Nor can this indefiniteness in any theory be improved by looking for definitions of such concepts as 'point' for "one is then looking for something that can never be found, because there is nothing there, and everything gets lost, becomes confused and vague, and degenerates into a game of hide-and-seek" (*ibid.*, p. 12).

Now this unavoidable state of affairs, according to Hilbert, is not at all the shortcoming that Frege takes it to be, but is rather a "tremendous advantage" as far as *applications* of a theory go. But of course "the application of a theory to the world of appearances always requires a certain measure of good-will and tact: that for points one substitutes bodies as small as possible; for straight lines . . . perhaps light rays" (*ibid.*, p. 14). The advantage of eschewing such definitions of point and line as Euclid gave them was that it cleared the way for those "tactful" operational definitions through which alone any theory can be applied at all. Indeed, the points of geometry need not even be taken "as small as possible", but rather can be taken, with a little good will, as large as the sun. But there are *limits* to the applicability of any theory, "and it would require a large measure of bad intentions indeed if one wanted to apply the more precise propositions of plane geometry or Maxwell's theory of electricity to appearances other than the ones for which they were intended" (*ibid.*). But even intended applications may eventually fail, and only bad intentions would insist today on applying euclidean geometry to astronomy or high-energy physics, or for that matter, on applying *any* kind of geometry to sub-atomic phenomena.[8] On the other hand, a theory may inherit totally unexpected applications, Hilbert's favorite example being that of the euclidean axioms of linear order and congruence to the genetic variations in *Drosophilia* flies produced by cross-breeding. On the difficult problem of anticipating the applications of a theory Hilbert could only suggest to Frege that "the more developed a theory is and the more ramified its structure, the more selfevident will be the manner of its application to the world of appearances" (*ibid.*). It is well known that applications of theories are often guided by formal analogies they bear to other theories. In sum then, Hilbert's view was that the 'points' of geometry cannot be defined outright, but only variously interpreted from one application to another, and that this is equally and unavoidably true of the basic geometric relations since the axioms of geometry cannot logically distinguish between different isomorphic interpretations which are nevertheless possible. Frege replied: "I should not shirk from the admission that 'point' cannot be defined" (*ibid.*, p. 17), but "concerning what you say about the applicability of a theory and the univocal and reversible one-to-one transformation, I reserve the right to make a reply" (*ibid.*, p. 21). As far as I know, Frege never replied to these points.[9] He did, however, touch the other side of Hilbert's isomorphism problem with his penetrating comparsion of Hilbert's axioms with "a system of equations with several unknowns, where the solvability and particularly the univocity of the determination of the unknowns remains doubtful" (*ibid.*, p. 18).

This brings us to the question of the consistency ('solvability') and completeness ('univocity') of Hilbert's axioms. What could have been meant by the latter? Hilbert's argument so far concerned only the logical indistinguishability of an infinite subclass of possible interpretations, and implied that the most the geometric axioms could do would be to impose the same logical structure on all possible interpretations, i.e. 'univocity' up to isomorphism. Now in fact Hilbert did find such axioms, but his claim that they determined only one cartesian geometry involved an additional appeal to Leibniz's principle of the identity of indiscernibles.[10] Whether or not this would have satisfied Frege, he may have missed the point altogether, for his example of an axiom system for congruence arithmetic which is supposed to embarrass the implicit definition of numerical congruence by failing to decide whether or not $2 \equiv 8 \pmod 3$ fails precisely because it lacks this 'completeness' property of Hilbert's axioms.[11] And when he further denies that Hilbert's axioms "could decide the question of whether my pocket watch is a point" (*ibid.*), even if they were univocal, he was perhaps shrinking from the undefinability of 'point' after all. Indeed, Hilbert's axioms could no more decide such a question than Newton's laws could decide by themselves whether Frege's watch was moving relative to absolute space. Nevertheless, there clearly are tactful applications of these axioms in which it *is* taken as a point.

Hibert's basic motivation in constructing a 'complete' euclidean system was to secure *non*-euclidean geometry. Various gaps in Euclid's proofs traditionally negotiated by tacit appeals to intuition were gradually filled in the ensuing centuries by the explicit introduction of new axioms, notably those of order, congruence, space, and essentially stronger continuity axioms allowing the proof of theorems beyond the reach of Euclid. The problem arose of deciding when enough axioms had been formulated to cover *all* such appeals. Euclid himself had filled an essential gap in the theory of parallels with his famous axiom, implying his belief that it did not follow from his other axioms. This seemed to be finally justified by the discovery of non-euclidean geometry during the first half of the 19th century. However, most of the aforementioned axioms surfaced only *after* this discovery, and with the appearance of each new axiom the possibility arose that Euclid's postulate might become provable after all. The original classical models did not satisfy all these other axioms. The analytical surfaces of constant negative curvature constructed by Minding and Beltrami all contained singularities in the vicinity of which free-mobility (congruence) failed.[12] They thus realized only a limited region of the hyperbolic plane, and Hilbert proved that *no* such surface could realize all of it. Lobachevski and Bolyai had shown that hyperbolic trigonometry was consistent if the euclidean assumptions needed to derive it for the imaginary euclidean sphere were, but these assumptions neither comprised a complete system nor realized their whole plane. Perhaps Euclid's axiom was derivable from a more complete system. In fact, Bolyai himself died believing that his work only precluded a proof of Euclid's axiom from the "other axioms" of euclidean *plane* geometry, but that it might well be provable from suitable axioms about euclidean *space*, for, "*Der Raum verbirgt in seinen Eingeweiden gar viele Schätze, die dem auf der Oberfläche Wandelnden nie zu Gesichte kommen.*"[13] And Hilbert later verified that the incidence theorem of Desargues behaves fundamentally differently in space than in the plane: whereas in space it is an obvious *a priori* truth following immediately from the space incidence axioms alone, Hilbert showed that in the plane it cannot be derived even when the axioms of order, continuity, and parallels are added to those of incidence. The problem of completing Euclid's geometry thus involved, for Hilbert, also that of *completing the proof for the possibility of non-euclidean space itself.*[14] He thus sought a consistent mutually independent system of axioms for euclidean space which would be so 'closed' that no further axioms of any kind could be consistently added. But he also saw that the key to both this and Frege's univocity problem lie in a suitable formulation of a stronger continuity axiom.

He thus introduced, in addition to the archimedean axiom, another continuity axiom, his '*Vollständigkeitsaxiom*', which "although not of a purely geometrical nature, merits particular attention from the theoretical point of view" ([117b], p. 25). Roughly, it stated the impossibility of extending the geometric universe of points, lines, and planes, while still satisfying the other axioms.[15] He had already used this idea in his axiomatization of the arithmetic of real numbers where he replaced Dedekind's continuity axiom with the archimedean axiom and an axiom of completeness which "expresses the fact that the totality of real numbers contains, in the sense of a one-to-one correspondance between elements, any other set whose elements also satisfy the axioms that precede" ([119], p. 138). This is equivalent to the isomorphism of all such systems, and the importance of this property for Hilbert was that it allowed him to circumvent the inherent difficulties arising when the real numbers are described by reference to 'all possible laws': rather he can simply describe them as the unique (up to isomorphism) system of things satisfying his 'finite and closed system of axioms' – assuming consistency.[16]

But axioms, of course, only exist that we may *deduce theorems* from them, and apparently Hilbert originally saw this closure property as a basis for claiming that any "new statements" beyond the axioms would "have validity only if one can derive them from those axioms by a finite number of logical inferences" ([116], p. 6). At least this would seem to be the case if by 'valid' he meant something like our 'logical consequence'. But it is not clear that he did so when a year later he also required that no further statement beyond the axioms "be held correct unless it can be derived from those axioms by means of a finite sequence of logical steps" ([118], p. 447). In particular, it is not clear whether his 'unless' here marks a definition or something to be proved. If we call a theory 'deductively complete' if, for any sentence, either it or its negation is derivable in a finite number of logical steps, then it seems that Hilbert was flirting with the temptation to infer deductive completeness from closure or categoricity. But since this inference depends on the system of logic used, he also doubted it, for he could see no clear demarcation of 'logic' at that time, much less a complete system of it.[17] Both the temptation and the doubt were explicitly expressed four years later by Huntington:

> In the case of any categorical set of postulates one is tempted to assert the theorem that if any proposition can be stated in terms of the fundamental concepts, either it is itself deducible from the postulates, or else its contradictory is so deducible: it must be admitted, however, that our mastery of the processes of logical deduction is not yet, and possibily never can be, sufficiently complete to justify this assertion ([134], p. 210).

In fact, Hilbert had already raised such doubts specifically about the integers in a 1901 lecture to the mathematical society in Göttingen on the problems of completeness and decidability, on which Husserl took extensive notes. After reviewing how to axiomatically enforce the '*Abgeschlossenheit*' of various theories, he raised the following objection to the inference of deductive completeness in the case of the integers, as they were recounted by Husserl:

> Hilberts Einwand. – Hatte ich recht zu sagen, dass jeder nur die ganzen positive Zahlen enthaltende Satz aufgrund der Axiome für ganze positive Zahlen wahr oder falsch sei? Hier wäre folgendes hinzufügen: Wenn wir behaupten, dass ein Satz aufgrund der Axiome eines Gebietes entscheiden ist, was dürfen wir dabei ausser diesen Axiomen benutzen? Alles Logische. Was ist das? Alle Sätze, die von aller Besonderheit eines Erkenntnisgebietes frei sind, was unabhängig von allen "besonderen axiomen", von aller Materie der Erkenntnis gilt.
>
> Da kommt man aber in ein schönes Gedränge; im Gebiet der algorithmischen Logik, im Gebiet der Anzahl, im Gebiet der Kombinationslehre, im Gebiet der allgemeinen Ordinalzahlen lehre. Und schliesslich ist nicht die allgemeinste Mannigfaltigkeitslehre selbst rein logisch? ([137], p. 445).

We see the *Entschidungsproblem* here as a glint in its father's eye: how to tell whether a sentence of an axiomatic theory is 'logically determined' by its axioms as true or false. But one had first to determine what logic *was*, to single out, from among these widely differing domains with seemingly equal claims to being 'logical', something one could justifiably call Logic. Dedekind's conception of logic as the 'laws of thought' was so broad that even its consistency had to be questioned. Meanwhile one could only use whatever domain seemed sufficient for a given system or context. For example, in connection with his fundamental result (see Section 2) on *Schnittpunktsätze*, it was clear to Hilbert that some simple '*algorithmischen Logik*' would suffice to derive any *Schnittpunktsatz* from Pascal's theorem and the descriptive axioms. On the other hand, any use of the archimedean axiom clearly required a logic that dipped into the 'domain of number'. The work of Frege and Dedekind on the 'logic of number' suggested that the use of this axiom may implicate one in even more. And finally, any use of his *Vollständigkeitsaxiom* obviously required "*die allgemeinste Mannigfaltigkeitslehre*". The basic ambiguity here is whether, in modern terms, logic shall be first- or second-order.[18]

All of this presupposes that Hilbert's axioms are 'solvable'. Frege could accept this, but not the notion of axiom which called for *proof* of this consistency. He simply wanted to say that euclidean geometry was consistent because its axioms were true. Hilbert's principle was just the opposite: if axioms were consistent "then they are true and the things defined by these axioms exist. For me this is the criterion of truth and existence" (*ibid.*, p. 12). In particular, if

'Every equation has a root'.

is consistent with 'the remaining arithmetic axioms', it is true and its putative roots exist. The history of algebra has indeed gravitated to this formalistic principle. The desire to solve all algebraic equations leads to the notion of an algebraically closed field constructed by 'formally adjoining' new elements to a given field to serve as solutions to equations which had none over it. If a polynomial $P(x)$ has no roots in a field F, this does not imply the inconsistency of the claim that $(\exists x)(P(x) = 0)$ generally, hence it should be satisfiable in a suitable extension field F' arising from F by the formal adjunction of new 'numbers' to F.[19]

Frege replied with an example of his own designed to embarrass Hilbert's principle: could we possibly infer, he asks, the existence of an omnipotent, omnipresent, intelligent being from the consistency of

(i) *A* is intelligent.
(ii) *A* is omnipresent.
(iii) *A* is omnipotent.

together with 'all their consequences'? If we think of Frege's three predicates as primitive ones we seem indeed to be confronted with the obvious syntactical consistency of (i)–(iii) alongside all our theological doubts about the existence of such a being. But these doubts are inseparable from doubts about the very *meanings* of these three predicates and their relations. Thus there must be some doubt about what 'all the consequences' of (i)–(iii) are after all. Indeed, it is really doubtful whether (i)-(iii) are consistent at all, for

(iii)′ *A* can make a stone too heavy for any being to lift.

seems already to be a conseqnence of (iii) alone. But from (iii)′ it seems in turn to follow that

(iii)″ *A* is not omnipotent.

The consistency of (i)–(iii) seems as doubtful as the existence of such a being afterall, but Frege did not claim to refute Hilbert's principle by his example: he questioned rather the means available for proving consistency, and hence its *usefulness*:

This principle is not at all evident to me; and if it were true, it would probably be useless.

Are there any means of proving consistency other than that of exhibiting an object that has all of the properties? However, if one has such an object, one would not need to prove that there is one by the roundabout way of proving consistency (*ibid.*, p. 20).

Ironically, Hilbert did not publish his ideas on direct consistency proofs until 1904 after the paradoxes had given him second thoughts about his consistency principle and had obliged Frege and Dedekind yo withdraw their logical foundations for arithmetic as inconsistent. Not having Frege's response to this new proof-theory, we cannot be sure how "sharply opposed" he really was to Hilbert's principle. But Hilbert was quite explicit in his rejection of Frege's truth principle:

It is precisely this procedure of constructing an axiom, appealing to its truth(?), and concluding from this that it is compatible with the defined concepts, that is a primary source of mistakes and misunderstandings in modern physical investigations. A primary purpose of my *Festschrift* was to avoid these mistakes (*ibid.*, p. 13).

Now if truth can be abused in physics, why not also in geometry, which Hilbert described in the lectures for his *Festschrift* as the "*vollkommenste Naturwissenschaft*"? With one eye on non-euclidean geometry he reserved the right to *question the truth* of Euclid's. Frege, who compared non-euclidean geometry to alchemy and astrology, never replied to this directly, but assured Hilbert that he now understood his desire to "divorce geometry completely from our intuition of space and make it a purely logical discipline like arithmetic" (*ibid.*, p. 14). But Hilbert's other eye had already gazed too fondly on Minkowski's striking use of geometric intuition in his *non*-euclidean 'geometry of numbers' to prove deep arithmetic results without numerical calculation to divorce it (see [128], pp. 344ff. where Hilbert also stressed the fruitful use of hyperbolic geometry in function theory). Since according to Hilbert only certain incidence and order axioms formulate "the basic facts of our spatial intuition", virtually any geometry might use such intuition. But however fruitful its mathematical application may prove, such intuition is unable to guarantee the literal *physical* truth of *any* geometry: Hilbert was emphatic that "Der geometrische Sätze gelten in der Natur niemals mit vollen Genauigkeit, weil die Axiome von den Objekten niemals genau erfüllt werden" ([117a], p. 154). Indeed, only in a *mathematical model* could the "more precise propositions" of any geometry, e.g. those on incommeasurability, hold exactly: neither intuition nor measurement could possibly confirm them. Hilbert sought no divorce from spatial intuition but rather a "logical analysis" of it revealing:

Welches sind die notwendige und hinreichenden und unter sich unabhängigen Bedingungen denen ein System von Dingen unterwerfen muss, damit jeder Eigenschaft dieser Dinge ein geometrische Tatsache entspreche und ungekehrt damit also diese Dinge ein vollständiges und einfaches 'Bild' der geometrische Wirklichkeit seinen ([117a], p. 2).

will try to establish this Hertzian picture of elementary geometry as part of pure mathematics despite its independence from his beloved numbers. Its logical analysis even leads to interesting relations between spatial intuition, mechanism, and Kronecker's equally beloved algebra.

2. LOGIC, INTUITION, AND MECHANISM IN HILBERT'S GEOMETRY

It is clear now that Frege and others were quite wrong in seeing Hilbert as fundamentally opposed to Euclid. Ian Mueller, for example, says that Euclid's derivations, being thought experiments, are "quite different from Hilbertian ones, which are usually said to involve no use of spatial intuition" ([189a], p. 292), a difference which he later explains by saying that Hilbert's derivations proceed "according to the principles of [modern] logic" (*ibid.*), which indeed Euclid could not have used. But then neither did Hilbert: no more than Euclid did he explicitly formulate any logic for deriving his theorems, making if anything even more use of figures to this end. Indeed, we have seen that at the time of his *Festschrift* Hilbert was not at all sure what logic really was. If it seems to have fewer 'gaps' than Euclid this is simply because Hilbert had so many more *axioms* to use than Euclid. We have seen that axioms take a long time to surface and even longer for their proper formulations to crystallize. But, as Prenowitz has pointed out, Hilbert's axioms are still every bit as anchored in visual intuition as Euclid's:

It is true Hilbert's treatment is more precise and rigorous than Euclid's. But the basic spirit remains the same. The propositions still tend to be direct transliterations of raw visual data, and are kept in mind mainly through visual intuition. The proofs usually are pictorially motivated and remembered ([210], p. 1).

For a really "contemporary approach" to classical geometry, Prenowitz argued, "the Euclid-Hilbert visual-centered concepts, bound as they are to the line and the plane, must be replaced by dimensionally neutral ideas" (*ibid.*). This kinship is no accident, for Euclid's framework offered itself as the natural foil for the growing use of abstract logical symbolism in geometry that Hilbert deplored there as much as he had in algebra. Indeed, his students were so impressed by the apparent rigor of these symbolic formalisms that they wondered why a man of Hilbert's originality would want to lecture on geometry, for

It was difficult to see how Hilbert could hope to go beyond what had already been done in this area of mathematical thought. But now in his lectures he proceeded to reverse the trend towards absolutely abstract symbolization of geometry in order to reveal its essential nature. He returned to Euclid's points, straight lines, and planes and to the old relations of incidence, order, and congruence of segments and angles, the familiar figures. But this return did not signify a return to the old deception of euclidean geometry as a statement of truths about the physical universe. Instead – within the classical framework – he attempted to present the modern point of view with even greater clarity than either Pasch or Peano ([217], p. 60).

Peano had translated Pasch's geometry into the 'logistics' that inspired Russell and repelled Poincaré. The latter accused Peano and Hilbert of formulating their axioms so that a *machine* could derive their theorems. But as 'artificial and puerile' as logistics seemed, Poincaré conceded that Hilbert's use of it "explains and justifies itself" ([208], p. 5), for his purposes required the reduction of geometric arguments "to a purely mechanical form". Couturat rightly objected that "Hilbert is a complete stranger to logistics and has never used any logical calculus in his research" ([56a], p. 486), so that Poincaré's criticisms of *him* rather "tend to prove the superiority of logistics over verbal reasoning and simple common sense" (*ibid.*). Peano had shown indeed "the necessity of expressing every mathematical proposition *entirely in symbols*" (*ibid.*, p. 491), but logistics was not "a mechanism from which intelligence is nearly excluded": since *we* must translate arguments into it and draw their conclusions it did not "reduce to a blind mechanism" (*ibid.*). But Poincaré meant that mechanisms that were *literally* blind could do geometry now that Hilbert had

. . . sought to put the axioms in such a form that they might be applied by a person who would not understand their meaning *because* he had never *seen* either point or straight line or plane. It should be possible, according to him, to reduce reasoning to purely mechanical rules, and it should suffice, in order to create geometry, to apply these rules slavishly to the axioms without knowing what they mean. We shall thus be able to construct all geometry . . . without *seeing* it at all ([204], pp. 4–5, italics mine).

We just put our axioms into a reasoning machine and "see all geometry come out of it", provided only that we take care to put in a complete set of them, "For our machine cannot work except according to the rules of logic for which it has been constructed; it ignores the vague instinct we call *intuition*" (*ibid.*, p. 5). We see that Poincaré tied both understanding and intuition in geometry to vision, and a machine's lack of these faculties to its inability to *see* anything. A machine, unlike us, reasons only by logic, not by sight. By implication, it could make no intuitive use of figures that characterizes our proofs. But as Kac and Ulam later claimed, echoing

Poincaré, complete formalization like Hilbert's "permits geometry to be taught to a blind man and even a computer" ([139], p. 672).

Hilbert himself had emphasized the importance of geometrical figures, both in proofs and for maintaining a link between mathematics and its indispensable source of new ideas, the 'outside world'. For new ideas always require new signs which, like Hertz, we "choose in such a way that they remind us of the phenomena" ([118], p. 443). Geometric figures, in particular, are "mnemonic symbols of space intuition" (*ibid.*), which are no less dispensable to mathematicians than arithmetical symbols, whose use is often wrongly assumed to be more rigorous. But these symbols are, after all, only "written diagrams", while the geometrical figures are actually "graphic *formulas*" (*ibid.*, my italics), which are no less a part of geometry than are the numerals of arihmetic – a view strikingly confirmed by his *Streckenrechnung*. However:

The use of geometrical signs as a means of strict proof presupposes the exact knowledge and complete mastery of the axioms which underlie those figures; and in order that these geometrical figures may be incorporated into the treasure of mathematical signs, a rigorous axiomatic investigation of their conceptual content is necessary (*ibid.*, p. 443).

This is the meaning of Hilbert's Kant motto: geometrical knowledge begins with mnemonic symbols of space intuitions, then proceeds to an axiomatic analysis of their conceptual content. Such analysis ends with those ideas of reason, infinity and continuity, whose elimination from elementary geometry will greatly reduce the tact needed in taking the geometrical figures themselves as the principal model of its axioms. This anticipates Hilbert's later idea of the numerals as the subject matter of arithmetic, when this typical Hilbertian interplay between the logical and the intuitive will become the '*Wechselspiel*' between a formal system and an intuitive metatheory focussed on its symbols. Neither can nor ever should be eliminated, for as he claimed in his work on '*anschauliche Geometrie*',

In mathematics, as in any scientific research, we find two tendencies present. On the one hand, the tendency towards *abstraction* seeks to crystallize the *logical* relations inherent in the maze of material that is being studied; and to correlate the material in a systematic and orderly manner. On the other hand, the tendency towards *intuitive understanding* fosters a more immediate grasp of the objects one studies, a live *rapport* with them, so to speak, which stresses the concrete meaning of their relations ([130a], p. iii).

Hilbert stressed that an approach to geometry "based on visual intuition" can often illustrate deep geometric theorems without having to enter into "the details of the analytic argument" (*ibid.*), and hence that intuitive understanding is as important now (1932) in geometry as it ever was. This brings us back to Poincaré's blind machines, and the question of how they might ever hope to find their way in geometry.

The machines known to Poincaré, of course, did nothing but compute. The problem of mechanism, following Hobbes and Babbage, was to show, by suitable programming, how various other cognitive capacities could be subsumed under this seemingly narrow one. Until recently, however, all successes have involved problems whose formal representation in machine language was effected by human programmers, thus relieving the machine of the burden of having to perceive its environment directly. Poincaré was thus justified in emphasizing the inability of early machines to see, but in suggesting that this was somehow an essential limitation of machines he went too far. Machines can now be equipped with 'retinas' for registering differing patterns of light intensity, and analogue to digital devices for converting them into numbers. The problem then is whether a machine can learn to 'understand' these numbers, that is, whether it can describe the contents of a picture on its retina in language comparable to humans. In short, can a computer be programmed to "translate confusing arrangements of numbers into simple, meaningful descriptions of visual scenes" ([215a], p. 214). In fact, there are methods enabling a machine to recognize numerical and simple geometrical figures, and to locate objects in two-dimensional drawings.

A more difficult problem for machine vision however is that of 'scene analysis', namely, to convert two-dimensional images or drawings into descriptions of *three*-dimensional scenes. For example, can a machine understand a *perspective* drawing? Indeed, programs have been devised not only enabling machines to correctly identify geometrical objects in such scenes, but even to classify some such drawings as 'impossible', i.e. as corresponding to no physically possible three-dimensional scene. Such programs assign a 'meaning' to each line in the drawing and classify each corner as belonging or not to one of the allowable types. If such an assignment cannot be completed, the drawing is deemed impossible. If some additional constraints are also satisfied, essentially that the space incidence axioms are not violated, the drawing is deemed possible, that is, a drawing of a three-dimensional scene. Figure 1, for example, is impossible because it violates the space axiom that three points determine a unique plane, while Figure 2 violates the axiom that two planes meet in a unique straight line. (Figures from [215a], p. 240.)

Using these and other space incidence principles as well as corner criteria, machines have been programmed to recognize and describe a variety

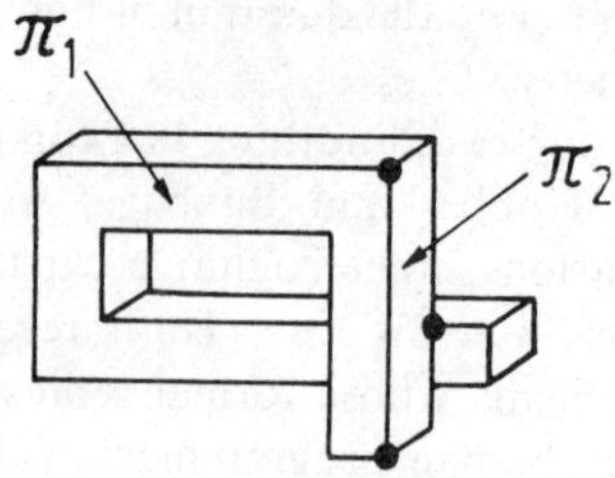

Fig. 1.

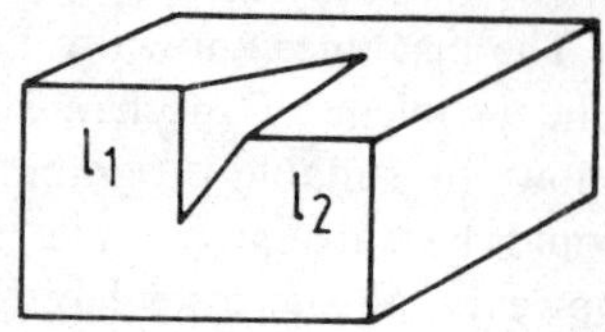

Fig. 2.

of three-dimensional scenes, as well as the impossibility of others. In this sense, they can 'see'. This analysis shows that there must be a cognitive element in machine vision, namely the use of rules and axioms for correctly interpreting what they see. Models extending machine thinking to machine vision also involve a corresponding shift of focus from arithmetical to geometrical principles, thus inviting us to consider anew the basic problem in the philosophy of mathematics of the relation between arithmetic and geometry as part of the more general philosophical problem of the relation between thought and perception.

Meanwhile, machines have also been programmed to *prove geometric theorems* heuristically from given axioms, the most widely studied being the 'geometry-theorem proving machine' of Gelernter [96]. This machine has three main components: a 'syntax computer', a 'diagram computer', and a 'heuristic computer' which coordinates their activities. In particular, the heuristic computer aids the syntax computer in its search for a formal proof of a theorem by making available to it semantic information obtained from a diagram of its premises supplied by the diagram computer. This information may be of the form 'segment *AB* appears to be equal to segment *CD* in the diagram', or 'triangle *ABC* contains a right angle in the diagram'. The diagram computer actually contains only a coordinate representation of the theorem "together with a series of routines that produce a qualitative description of the diagram" ([96], p. 139). Gelernter points out, however, that the behavior of his geometry machine "would not be changed if the diagram computer were replaced by a device that could draw figures on paper and scan them" (*ibid.*). The heuristic computer is thus a simplified model of a geometer like Hilbert who tried to coordinate the interplay between the logical and the intuitive aspects of geometry to facilitate proofs, its major function being "to compare strings generated by the syntax computer . . . with their interpretation in the diagram, rejecting those sequences that are not supported by the model" (*ibid.*, p. 140).

Since the original goal was to get the machine to "produce proofs comparable to those of a high school student", the syntax computer was provided only with a weak *ad hoc* set of axioms rather than Hilbert's complete set. The main point is to simulate the human theorem prover's heuristic use of diagrams, for indeed the semantic interpretation they provide in the case of euclidean geometry "is so useful that virtually no one would attempt the proof of a theorem in that system without first drawing the diagram: if not physically, then in the mind's eye" (*ibid.*, p. 138).

The geometry machine can also employ auxiliary constructions, but these were originally limited to drawing segments connecting points already given in the premises of a theorem: new points not already mentioned in the premises could not be introduced. To obtain more interesting theorems restricted constructions of new points have been proposed. In fact the geometry machine has been belittled for proving only trivial theorems such as that of Thales stating the equality of the base angles of an isosceles triangle.[20] However a criticism made by Peirce long before geometry machines were ever realized is more interesting. Recall that, according to Peirce, no reasoning machine would probably ever be able to wind its way through the labyrinth of von Staudt's long proof of Desargues' theorem. Now the geometry machine of Gelernter cannot indeed prove Desargues' theorem at all, but this need not be true of some feasible extension of it. To understand the nature of the problem Peirce may have had in mind we must turn to his geometrical writings.

Every deduction, according to Peirce, "involves the observation of a Diagram" which in geometric deductions is usually an "optical diagram" ([202a], p. 869). Elsewhere he asserts that this is just what Kant meant by the 'construction' he saw as essential to geometric proof: it is "in general a diagram, or visual array of characters or lines". Pierce distinguishes 'corollarial' deductions "in which any Diagram of a state of things in which the premisses are true represents the conclusion to be true", from 'theorematic' reasoning in which "to the Diagram of the truth of the premisses something else has to be added, which is usually a mere May-be and then the conclusion appears" (*ibid.*, pp. 869–870). Peirce's favorite example of theorematic reasoning occurs precisely in von Staudt's proof for Desargues' theorem which states that

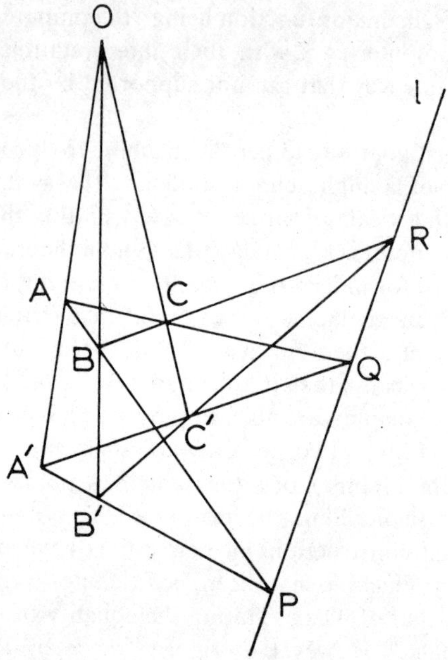

Fig. 3.

(D) If two triangles are in perspective from a point O, then the three points of intersection of their corresponding sides are collinear (and conversely).

Let us note first that (D) is obviously true *in space*, i.e. when the two triangles lie in different planes, for (see Figure 3) since AB and 0 determine a plane containing $A'B'$, these sides must meet in a point P, and similarly for points Q and R. Thus each of P, Q, R lie in *both* of the planes of our triangles and so must be collinear, for otherwise these two planes would not intersect in a line. Indeed, it is obvious by viewing the figure *perspectively*, that the line l must be the intersection of these planes. *The space case of* (*D*) *is thus evident to spatial intuition, resting on a pair of axioms which express Hilbert's geometric facts about space.* This is just what Hilbert sees as implicating us in space intuition in geometry: *the use of the space axioms.* Part of his problem of a "logical analysis of our intuition of space" was just to keep track of the logically necessary and merely convenient uses of the space axioms in proving theorems about 'geometrical reality'. In particular, to keep track of our appeals to properties of *planes*, the simplest elements of space.

An interesting case is the use of space to prove a plane theorem. Thus, suppose we want to prove the plane case of (D). The same figure still applies, though now we view it for what it is, a plane figure. But now (D) is no longer obvious, for as Peirce warned, "for more than two centuries . . . the greatest mathematicians have tried to prove *that* by the diagram required to exhibit it *alone*, and hence have tried in vain" ([202b], p. 871). But if we imagine each of the ten points in the figure connected to a new space point S as in Figure 4 (from [130a], p. 122), then the proof becomes easy, though considerably longer than the space case. The essential idea is to observe that the plane Desargues configuration is the projection from S of the space Desargues configuration we construct and then note that incidence relations are preserved under projection. The advantage, Peirce points out, of using space, i.e. axioms about planes, in either case, is that in space a straight line is the intersection of two planes which "show us exactly where the ray runs, while as long as we confine ourselves to the plane, we have no easy method of discerning precisely what the course of the ray is" ([202], p. 352). For these reasons the plane case was for Peirce "a truly remarkable theorem". He explicitly cited only its extreme length and complex mode of combination of the axioms as the difficulty facing a machine simulation of the above proof of (D), but he could also have invoked the unavailability of diagrams for the machines of

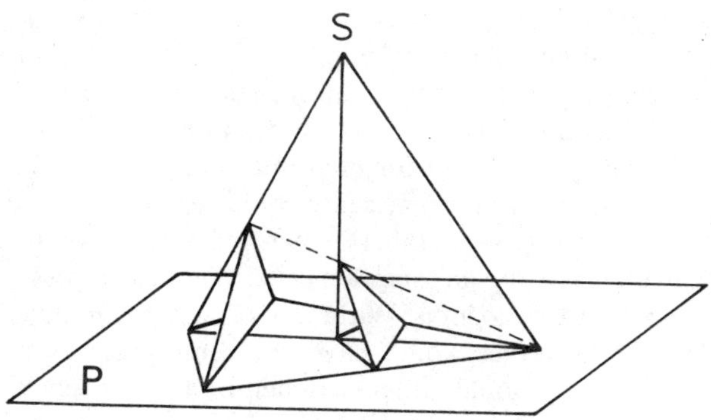

Fig. 4.

his day. Since he saw the difficulty in a machine proof of (D) for the plane as a "lack of initiative", he might also have envisaged the difficulty in getting a machine to add *new* elements to diagrams even if they had them – that is, for them to reason theorematically. In fact, early editions of Gelernter's geometry machine did not do so, but later work showed that it was not an essential limitation.

There is, however, a basic difference between the use of diagrams in proving the space and plane cases of (D). In the space case, though Figure 3 must be viewed in perspective, the proof did not require the addition of new points outside the figure plane. Even this, of course, was far beyond the machines known to Poincaré and Peirce, but the recent advances in scene-analysis by machines show again that this was not a limitation of principle. Our perspectival reasoning with Figure 3 consisted in viewing some of its subconfigurations as determining various planes and then seeing the line of intersection of two of them, i.e. we visually applied our two space axioms. But these are the very axioms that mechanical scene analyzers have applied to Figures 1 and 2 to classify them as impossible. So there is no reason in principle why a machine cannot use space intuition in Hilbert's sense to prove geometric theorems. Indeed, these impossibility classifications *are* such theorems. The use of Figure 4 to prove the plane case is a different matter. Here the machine would be given only a plane Desargues configuration (on plane P of figure 4), and would have to 'take the initiative' in *adding the entire space construction itself*, and then perhaps, view the result in perspective. A machine which feasibly simulated von Staudt's proof would thus have to combine the best of scene analyzers and geometry machines, taking the freest initiative in adding to given diagrams and using its 'space intuition' on the result. Again, I see no limitation in principle, but whatever the prospects for such a machine may be, it is clear from the machines discussed above that Poincaré and Kac and Ulam missed an essential point in claiming that Hilbert had made geometry fit for blind machines, for they make it clear that only machines which in some sense could 'see' would have a ghost of a chance of simulating, say, Hilbert's own geometric activity. At least it would have to make use of diagrams, i.e. concrete representations or models of the geometric axioms. Without its diagram computer the geometry machine of Gelernter would have only empty concepts in Kant's sense. In particular, it would flounder as helplessly in its search for proofs as would have Hilbert's demon. Without its incidence and topological axioms, on the other hand, a scene analyzer would be left with a blind retina. In short, the point *was not to teach geometry to blind machines, but to teach them to see.*

But there is a kernel of truth in what Poincaré said, though it was somewhat misleadingly expressed. He said, we recall, that it was the 'form' in which Hilbert had cast his axioms that made them amenable to mechanical treatment, quoting the following clump of incidence axioms as an example:

On every straight line there are at least two points; on every plane there are at least three points not in a straight line; in space there are at least four points which are not in the same plane ([208], p. 5).

Now as we have seen, this only served to remind one "contemporary geometer" of Hilbert's joint bondage with Euclid to the 'visual-centered' concepts of line, plane, and the third dimension. But Poincaré reacted rather differently:

This statement is characteristic. Anyone who had left any place for intuition, however small it might be, would not have dreamed of saying that on every line there are at least two points, or rather he would have added at once that there are an infinite number of them; for the intuition of the straight line would have revealed to him both facts immediately and simultaneously (*ibid.*, pp. 5–6).

So it was not the form of Hilbert's axioms but their *explicitness* which usurped Poincaré's intuition. For Hilbert, axiomatics was simply a method of making oneself as *conscious* as possible of all one's beliefs for the purpose of isolating the naive dogmatic ones – but it still left one, he claimed, with "all the advantages of belief". The importance of this method to artificial intelligence is clear: for a geometry machine to find von Staudt's proof of the plane case of Desargues' theorem, it would have to be told that 'in space there are at least four points which are not in the same plane'. If it were clever, this hint might be enough. And if its 'visual field' were as Berkeley tried to say our own is originally, namely *flat*, then such a hint would be necessary. The importance of such axioms for Hilbert, as we have said, was precisely to keep a close logical track of spatial intuition: as Bernays put it, "*Der Sinn dieser methode ist, dass an anschaulichen Inhalt nur dasjenige beibehalten wird, was* wesentlich *in den geometrischen Beweise* eingeht" ([17b], p. 203).

But this close tracking of intuition also leads to the discovery that for many important geometric proofs an infinity of points on a line is unnecessary. Despite his acknowledgement that, in considering a plethora of new noneuclidean geometries, Hilbert had "broken down at every point" the barrier of 'absolute geometry' set up by neo-Boetians to confine further geometrical generalization, we see that Poincaré was not yet ready for *finite* geometry. Indeed, what sort of an 'intuition' was it that could have revealed 'immediately' to Poincaré an infinity of points on a line? Obviously a very 'pure' one. But not of the *line* itself, but rather, as we have seen, of the mind's power of endless repetition of certain of its ideas, in this case, of division. On this Lockean power Poincaré rested the basic mathematical method of 'reasoning by recurrence'. Hilbert did, of course, introduce order and continuity axioms having the consequence Poincaré wanted 'immediately', but his ultimate purpose in doing so is to show that the need for the archimedean axiom and the reasoning by recurrence in the form of the least number principle required by Euclid (Book V, Prop. 8) *could be eliminated.* For Hilbert saw that this was indeed the root of that other 'stain' on Euclid's fabric besides his own contested postulate, namely the highly unintuitive and ungeometrical theory of proportion which played for him the role of Descartes' algebra.[21] The feeling that number and recursive processes were foreign to geometry was not uncommon when Hilbert wrote. In his critical review of Grassman's introduction of recursion relations into arithmetic as axioms, Otto Hölder tied this to the old view of Kant and Gauss that geometry dealt only with outer experience, arithmetic with inner experience:

> Axiome möchte diese relationen nicht nennen. Es sheint mir hier anders zu sein als in der Geometrie, deren Grundbegriffe und Grundsätze nicht aus solchen fortschreitenden Parcessen entspringen, wie sie, in der Geometrie so gut als in der Arithmetik, von der Deduktion benutzt werden. Mag man nun die geometrischen Grundsätze als Evidenzen der Anschauung oder, was vielleicht richtiger ist, als aus der äusseren Erfahrung abstrahierte Gesetze ansehen, in beiden Fallen entsammen die Grundsätze sowohl als die Grundbegriffe einem fremden Reich ([132a], p. 591).

Before we pursue these points, it is necessary to consider some general questions about Euclid's own geometry.

Euclid's geometry studies plane figures which, his commentator Proclus tells us, the understanding "projects on the screen of the imagination". And they all lie in *one* plane, which we must thus think of "as projected and lying before our eyes and the understanding as writing everything upon it, the imagination becoming something like a plane mirror to which the ideas of the understanding send down impressions of themselves" ([211], p. 98). What are we to make then of the view that Euclid made essential use of *space* intuition? Does it mean that his study of the figures in his plane depends on its being embedded in space? His geometry was surely discovered by men in space, but in fact Euclid does not use any space constructions like von Staudt's to prove his *plane* theorems.[22] It would seem that he restricted himself to an intuition of a plane 'projected and lying before our eyes' like Berkeley's flat visual field itself. In any case, in his critique of congruent figures as those that 'occupy the same space', Proclus also warns us that it is "altogether unacceptable to shift attention to space, which is far more unknown to us than things in space. At any rate its nature is controversial and difficult to discover" (*ibid.*, p. 153). Would Euclid then have wanted to depend on an intuition of something so controversial? Or perhaps only on an intuition of his *figures themselves and their relations*? Probably, but Proclus also tells us that the "culmination of geometrical inquiry" would be a "more perfect intellectual insight" by which the understanding would "roll up its extensions and figures and view their plurality as a unity without figure." This "superior vision of . . . unextended and essential geometrical ideas" would *emancipate* geometry from "the pictures projected in the imagination" (*ibid.*, pp. 45–6). Descartes' *analytic* geometry was a decisive step in this direction, using Euclid's theorem of the preservation of ratios under parallel projection to define a linear concept of multiplication (see Figure 5). This allowed him to represent arbitrary powers of a segment in the plane by other segments, whereas the Greeks represented the second power by a

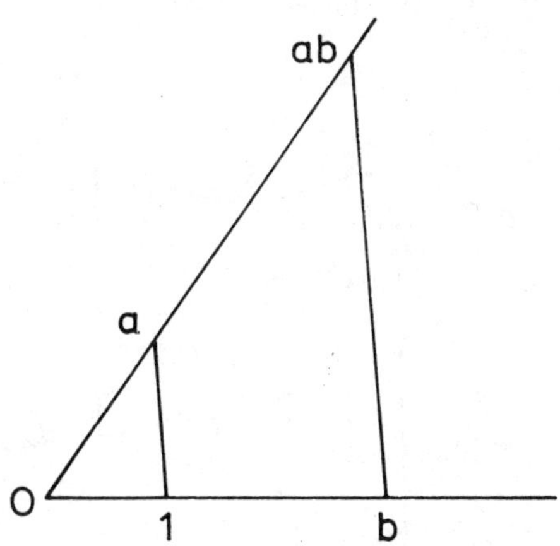

Fig. 5. The uniqueness of the product *ab* of *a* and *b* depends on the uniqueness of the parallel to 1*a* drawn through *b*.

square, the third by a cube, and then stopped for want of further dimensions in space. But two-dimensional figures, according to Descartes, "serve quite as well as surfaces in aiding us to imagine an object which actually has extension" ([69], Vol. I, p. 65). Indeed, since the third dimension is not as 'distinctly conceived' as length and breadth, Descartes lays down as a rule for our understanding that "not more than two different dimensions out of the countless number that can be depicted in our imagination ought to be the object either of our bodily or our mental vision" (*ibid.*).

While Descartes used Euclid's theory of parallels to banish the third dimension from algebra – without asking how the analytic geometry of his plane was possible at all, i.e. on just what geometric laws the algebraic ones depended – Desargues, the father of projective geometry, extended it to a theory of how to *represent* that dimension on the plane. Whereas painters had discovered rules of perspective empirically by trial and error, Desargues formalized them in axioms and theorems. To justify vanishing points he posited infinite points as "goals where parallels tend to meet" and used the parallel postulate to derive a rule for drawing the infinite point of any line. To draw figures, call two triangles 'perspective from a point O' if the lines joining their respective vertices meet at O. Whereas similar triangles have the same form, perspective triangles have the same *appearance* from O. His theorem (D) gives a *geometrical* criterion for a triangle $A'B'C'$ on the picture plane π_1 to be the perspective image from O of a triangle ABC in the object plane π_2 (Figure 3′).

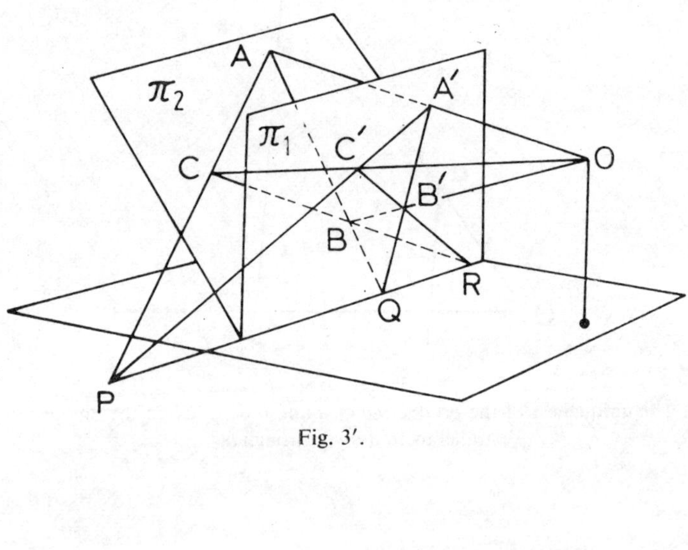

Fig. 3′.

Desargues pioneered the use of space intuition in geometry; the incidence proofs for the plane and space cases of (D) discussed above go back to him. He also found a long proof in the plane avoiding any space construction, but at the cost of using congruence, similarity, and continuity. These proofs raised for Hilbert the question of 'purity of proof' as an important problem of 'critical' mathematics, for:

Der Inhalt der Desargueschen Satz gehört durchaus der ebenen Geometrie an; zu seinem Beweise aber haben wir den Raum gebracht. Wir sind daher hier zum ersten Mal in der Lage eine Kritik der Hülfsmittel eines Beweis zu üben. In der modernen Mathematik [wird] solche Kritik sehr haüfig geübt, woher das Bestreben ist, die Reinheit der Methode zu wahren, d.h. beim beweise eines Satzes womöglich nur solche Hülfsmittel zu benutzen die durch den Inhalt des Satzes nahe gelegt [sind]. Diese Bestreben ist oft erfolgreich und für den Fortschritt der Wissenschaft fruchtbar gewesen ([117a], p. 27).

This is the prototype of problem that occupied Hilbert in his later program: whether and to what extent it was necessary, in proving this finite plane theorem about incidental relations between ten points, to use wholly new concepts like space and congruence, or an infinitistic one like continuity. His answer is:

(I) Without space (D) cannot be proved in the plane without the triangle congruence axiom, even if continuity is used.

which contributes to his logical analysis of space intuition by showing that Desargues' appeal to it through the space axioms can only be avoided by using still other concepts foreign to (D). A geometry machine trying to prove (D) heuristically could thus gain relief from the maze of congruent and similar triangles it would have to navigate in its plane diagram only by taking the initiative in space by adding perspective triangles to it and analyzing the resulting scene. Though we have seen that this is possible in principle, a machine simulation of Desargues is thus far more difficult than one of Thales. But none of the Greeks found (D) either.[23]

Gödel's incompleteness is often compared with the independence of the parallel postulate, but (I) may provide a better comparison. For it shows that, while the plane case of (D) may be evidently true when considered via projection from space (where it is evident), it is no longer provable in a plane contained in space without new axioms. Similarly, we say of Gödel's formal sentence, that while it is evident when considered via the Gödel-numbering as the projection of an evidently true metalinguistic statement, it is no longer provable in the formal language contained in its metalanguage unless wholly new kinds of axioms are added. A Gödel-sentence can be read either as a metalinguistic statement about proofs or a formal statement about numbers, and while it can be evident when read in the former mode, it cannot be seen directly as a literal truth about numbers. Similarly, a Desargues figure can be viewed perspectively as a picture of planes and lines in space, or literally, as a flat configuration of lines. In the former view it can picture an evident truth about the lines of intersection of planes, but when these lines are viewed as a flat configuration, we have no clue as to the meaning of their relations – as Peirce saw. (See Fine [80].)

We recall that Desargues wanted to formalize the rules of perspective, i.e. to find a complete set of rules for representing space on the plane, and that he came up with (D) as a new rule in addition to Euclid's axiom of parallels. Call a plane 'descriptive' if the plane incidence and order axioms as well as Euclid's hold on it. Hilbert, "in order to grasp clearly the meaning of the space axioms" ([117], p. 72), looked for the necessary and sufficient conditions (*geometrische Tatsachen*) for such planes to be embedded in space. His result this time is

(E) A descriptive plane can be embedded in space iff (D) holds in it ('making planes fit for space').

The necessity is immediate from the provability of (D) in descriptive (affine) space. The sufficiency was a remarkable result of the logical analysis of space intuition, showing that Desargues' rules comprised a complete system of linear perspective. To explain this we describe briefly the purely synthetic proof of Schor [228]. We are given only the points and lines of a descriptive affine plane and must construct, using only purely 'logical' means and (D), an affine space with its incidence and order structure containing our given plane. In particular, we must construct all the points, lines, and planes of this space out of the lines of the given plane and then, using only the plane descriptive axioms and (D), prove that the space axioms hold for these space objects.

We begin by fixing a triangle t of our given plane ω and defining as 'space points' all triangles of ω whose sides are parallel to those of t, together with all the points of ω. If ABC and $A'B'C'$ are two space points, their three binding lines a, b, c of ω are, by (D), either parallel or meet at point 0 of ω (see Figure 3″). We take such triples $[abc]$ as our 'space lines'. A space point is 'incident' with a space line $[abc]$ just in case the lines a, b, c pass through the points A, B, C in ω. If $A_1B_1C_1$, $A_2B_2C_2$, $A_3B_3C_3$ are three space points joined by the three space lines $[a_1b_1c_1]$, $[a_2b_2c_2]$, $[a_3b_3c_3]$, we define a 'space plane' to be the set of all space lines

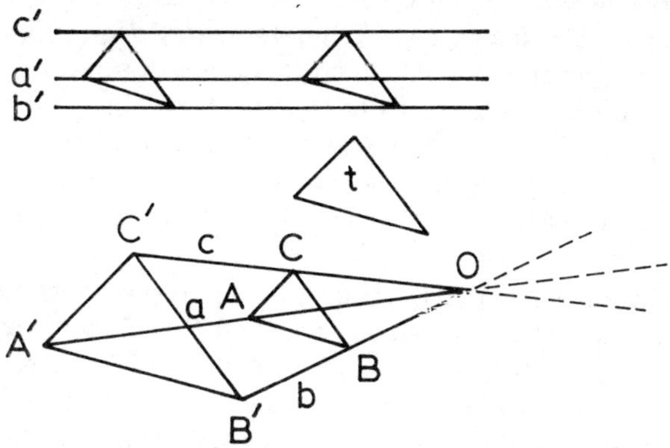

Fig. 3″. Space line [*abc*] passing through space points *ABC* and *A′B′C′* and intersecting ω at 0. Space line [*a′b′c′*], however, passes through two points of a plane parallel to ω.

which intersect at least two of these lines. A space point 'lies in' a space plane α just in case it is incident with a line of α. By repeated use of (D) the space axioms can be proved for these space objects, revealing says Schor the 'logical content' of (D). This illustrates Hilbert's point to Frege that isomorphism pushes science towards abstraction: the geometric axioms cannot distinguish Schor's 'space' from 'real' space. It would be impossible of course if we were saddled with points defined as "that which has no parts".

(E) and (I) allow Hilbert to characterize Desargues' theorem as "the result of eliminating the space axioms": *it represents in the plane the logical content of space intuition.* It would provide 'Flatlanders' or '*Flächenwesen*' and some machines with a logical concept of space they could not intuit, enabling them to prove all the plane theorems we could prove using the most direct expression of our space intuition, viz. the space axioms. The abstract reconstruction of space in (E) is essentially algebraic and indeed Hilbert's own proof of (E) is based on his first answer to Descartes' unasked question of how algebraic representation is geometrically possible:

(A) An affine plane can be represented by a skew field iff (D) holds in it.

Hilbert is thus able to base the algebraic treatment of elementary geometry almost entirely on (D) without need of the axioms of congruence or Archimedes, thereby dispelling the old illusion that in doing analytic geometry one must somehow dip into an antecedently presupposed ontology of numbers and magnitudes. Using Descartes' geometric definition of multiplication, he employs (D) to create an algebra out of the segments themselves, just as Schor used it to create a 'space' out of the points and lines of a plane. Such results reveal a weakness in Frege's view that no meaning can be given to a geometric statement without first fixing the meanings of its constituent terms 'point' and 'line': *they tell us more of the 'meaning' of* (D) *than such definitions could ever hope to.*

One could almost say that (A) had reduced numbers to geometry, at least in their algebraic aspect, much as Frege and Russell had tried to reduce them to logic. But they had focussed on the *recursive* structure of the integers themselves rather than any *field* structure containing them. And in any case, (D) does not quite yield a field as it is unable to enforce the commutativity of Descartes' multiplication. Since the archimedean axiom is needed to prove the commutativity of multiplication from the other axioms for the *real* field, perhaps we cannot eliminate the need for recursive processes from elementary geometry after all. In particular, the concept of integer contained in the archimedean axiom, and its inductive use, may be an ineradicable stain on geometry. But Hilbert found yet another '*geometrische Tatsache*' to rescue him, for "From the accompanying figure it becomes immediately clear that the commutative law of multiplication is none other than Pascal's theorem for the two axes" ([117], p. 94; see Figure 5′). Since Hessenberg showed that Pascal's theorem also implies (D) purely geometrically it gives a field all by itself.

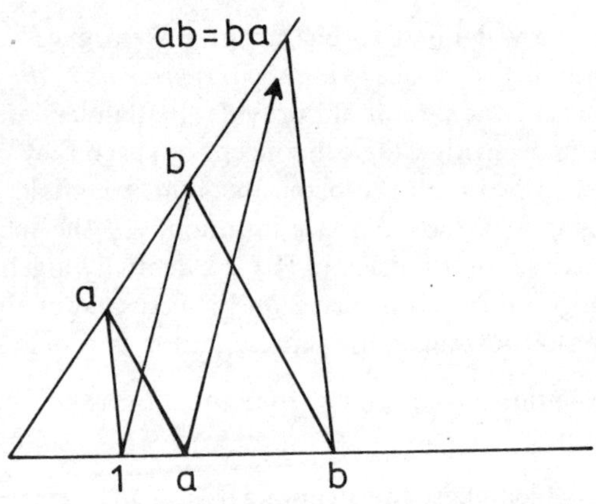

Fig. 5′. Pascal's Theorem and Commutativity of Multiplication.

On the other hand, Hilbert shows that Pascal's theorem does not follow from (D) alone unless continuity is used, but that it can be proved using just congruence. Pascal's theorem thus suggests itself in the context of elementary geometry as a 'finitary' replacement for number, induction, and continuity.

To clarify this Hilbert introduces his 'pure point of intersection theorems' dealing only with finitely many points and lines bearing only relations of incidence and parallelism. These are his so-called '*Schnittpunktsäzte*', of which (D) is an important example.[24] But (D) cannot be derived from plane incidence axioms, even using continuity, while Pascal's theorem, itself another *Schnittpunktsatz*, cannot in turn be derived from (D) unless continuity is used. The question thus arises of whether any purely autonomous theory for these finitary *Schnittpunktsätze* is possible, i.e. whether such *Schnittpunktsätze* could be found from which all others could be proved. Hilbert's result, verifying a conjecture of Wiener [271] which had spellbound him, is that Pascal's theorem alone suffices:

(P) Every *Schnittpunktsätz* follows from Pascal's without using congruence or continuity.

This shows, in Hilbert's later terms, that we can eliminate 'ideal elements' from proofs of these 'real' *Schnittpunktsätze* – though no proof of Pascal's theorem itself could be both 'finite' *and* 'pure': in the absence of continuity a concept of congruence foreign to the theorem itself was needed. Later Hilbert sought, via a finite consistency proof, a result analogous to (P) for arithmetic to the effect that every finite arithmetic theorem could be finitely proved.

Tarski's results [245] crown those above roughly as follows: Descartes observed that all of Euclid's principles are true in a certain algebraic field he inherited, but which later ballooned into the uncountable real field, the real thing. Hilbert showed, on the other hand, that Euclid's elementary geometric principles, since they imply Pascal's theorem, can *produce* a field, a real closed one if not the real uncountable thing. Tarski then showed that if the Euclid-Hilbert system is expressed in *first-order* language, it doesn't really matter which such field it produces, for this language cannot distinguish them logically: a geometric sentence true in one of them will be true in all of them and hence their theory will be complete. Indeed, in this language one can even formulate continuity principles whose consequences are purified of the archimedean stain on geometry.

By applying a close analysis of Sturm's theorem, Tarski even obtains a decidability proof. Thus there exists in principle a 'Tarski machine' which could theoretically prove all theorems of elementary geometry. However, from the point of view of mechanism and artificial intelligence, such a result is of little value. Indeed, it is known that a 'feasible' program for implementing Tarski's result is impossible. But even if it were, it would not throw any light on how humans prove geometric theorems, for they mostly use figures and drawings to that end. The value of Tarski's work lies rather in establishing the importance of first-order logic for generalizing the classical isomorphism of structures to their so called 'elementary equivalence' thereby efacing some distinctions which are not directly intuitive.

Tarski's result completes a tendency found already in Euclid and which dominates Hilbert's work on geometry, namely the elimination of continuity assumptions from elementary geometry. Since none of its theorems deal directly with continuity or infinity, it is natural to ask whether its phenomena could be 'saved' autonomously without reference to them. Hilbert's (P) is a prototype of such an elimination. However, it is a metatheorem about *plane* geometry, and before we can conclude that plane geometry can be developed completely autonomously *we must be sure that we have not presupposed space in any way*. The obvious way of doing this of course is to use the space axioms, as in von Staudt's projection proof of (D). Desargues himself found a proof that stays wholly within the plane, making at least no use of space axioms. Hilbert showed, however, that the

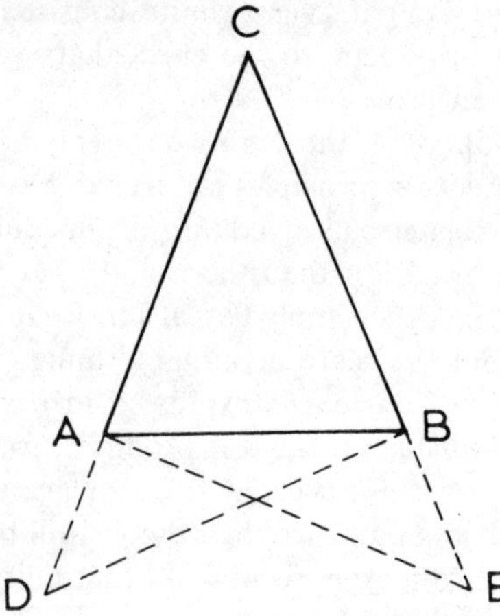

Fig. 6.

side-angle-side triangle congruence axiom is required for this proof and that there are, moreover, simple theorems requiring the application of this axiom to triangles of opposite orientation in the plane. Since such triangles can be superposed only by reflection, the normal interpretation of the congruence relation in terms of rigid motion presupposes space outside the plane even though not explicitly appealing to it by use of the space axioms. For example, Euclid's proof of Thales' theorem on isosceles triangles extends both equal sides by an equal amount and then applies the side-angle-side congruence principle, which he proved by superposition, to the triangles *CDB* and *CEA* (see Figure 6). In this case, however, the superposition would require turning one of the triangles over in space, so that the proof presupposes space even though not explicitly mentioning it. The question is whether it is only Euclid's *proof* which presupposes space or whether perhaps the theorem itself does, i.e. whether *any* proof for it must presuppose space.

This brings us back to Proclus' recommendation that we think of Euclid's plane as "projected and lying before our eyes". Should we see an ambient space for the reflection of its figures? One is tempted here to follow Kant, for whom the third dimension of space was given *a priori* to *Anschauung*, but this faculty was after all only an idealization of a visual field which Berkeley argued to be two-dimensional in and of itself, unconditioned to touch. Even Poincaré admitted that it was only "with a little good will" ([207], p. 34) that visual space can be assigned three dimensions. Berkeley imagined an 'Intelligence' having normal vision but no sense of touch and argued that it would not only have no 'idea' of the third dimension – because distance cannot be immediately perceived – but that in contrast to what Helmholtz claimed for his *Flächenwesen*, it would not even be able to develop *plane* geometry. For since this Intelligence is "without any idea of distance, or outness, or profundity, and consequently of space or body", it would be unable to "conceive the placing of one plane or angle on another, in order to prove their equality; since that presupposes some idea of distance, or external space" ([15a], p. 203). Berkeley's Intelligence would presumably be no more able than Helmholtz's *Flächenwesen* to conceive the superposition of triangles of opposite orientation: neither could prove Thales' theorem unless Euclid's proof can be replaced by one not presupposing space.

These questions and many more are brought to a head by Pappus' proof of Thales' theorem which avoids any auxiliary constructions: given *just* the isosceles triangle *ABC* with $AB \equiv AC$, as in Figure 6′, Pappus

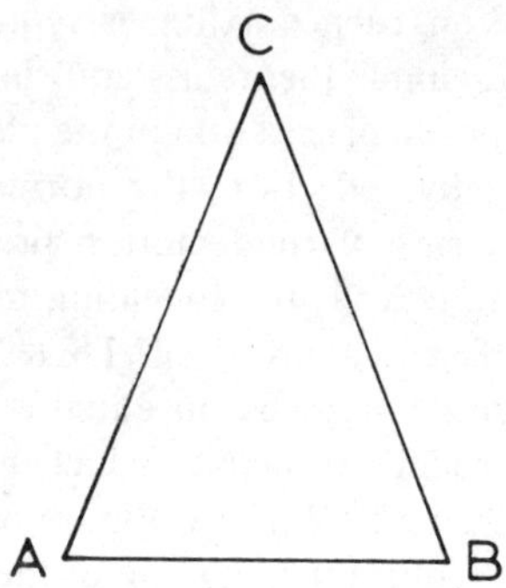

Fig. 6′.

proceeded nevertheless to

... think of this triangle as *two* triangles and reason thus: since *AB* is equal to *AC* and *AC* is equal to *AB*, the two sides *AB* and *AC* are equal to the two sides *AC* and *AB*, and the angle *BAC* is equal to the angle *CAB* (for they are the same); therefore all the corresponding parts are equal, *BC* to *CB*, the triangle *ABC* to *ACB*, the angle *ABC* to the angle *ACB*, and the angle *ACB* to the angle *ABC* ([211], p. 195).

Proclus surmised plausibly that Pappus found this 'proof' by reflecting that in congruence proofs by superposition we end up "uniting the two triangles so that they coincide with each other, thus making them one instead of two" (*ibid.*). Reversing this process, Pappus was able "to see two triangles in this single one and so prove the equality of the angles at the base" (*ibid.*). We see, however, that this proof is incoherent unless reflection in space is presupposed, as was first pointed out by Lewis Carroll [46a], who suggested that perhaps Pappus' triangle could be picked up, turned over and then laid down upon a 'trace' of *itself*! Those bothered by the conjuration of two triangles out of one – and hence by the absence of any criterion of identity for euclidean triangles, were assured by Heath that "we may answer that we keep to one triangle and merely view it in two aspects" ([76b], p. 254). Beginners could even be told, suggested Heath, "that one triangle is the triangle looked at in front and that the other triangle is the *same* triangle looked at from *behind*; but even this is not really necessary" (*ibid.*). He did not say what *is* necessary to make his suggestion coherent, but it raises acutely the status of Euclid's plane universe: could we walk behind 'the screen of our imagination' upon which its figures are projected? Or behind Berkeley's two-dimensional visual field? Or can we avoid these questions by proving the theorem without need of reflection? For Gelernter's geometry machine rediscovered Pappus' proof, but surely did not literally imagine the given triangle as reflected in space. Indeed, all its axioms and diagrams refer only to the plane. What it did was rather to apply a rule of 'syntactic symmetry' allowing it to replace certain sentences about triangles with others arising from it by a permutation of the names of the vertices of triangles. But its proof, which "considered the triangle and its mirror image as two different triangles" ([132], p. 606), is just as questionable as Pappus', the difference being of course that Pappus would have answered our questions by invoking reflection in space. But the problem remained whether his use of it in Figure 6 can be avoided.

Hilbert defined two different orientations for plane triangles and asked under what conditions Thales' theorem or the strong congruence principle for triangles of opposite orientation (reflection case) can be derived from the case restricted to triangles of the same orientation. The result this time was

($\mathscr{R}$) (i) Thales' theorem, and hence also the reflection case of congruence, can be proved if both the archimedean axiom and an axiom of neighborhoods are assumed, but (ii) neither can be proved if either of these continuity assumptions is dropped. However, (iii) both can still be proved if these continuity assumptions are replaced by an 'axiom of embedment', precluding the equidecomposability of any given polygon with any other having boundary points interior to the given polygon but none exterior to it.[25]

It follows that Euclid's plane theory does not logically depend on *Raumanschauung*, *if* we are willing – as Berkeley was not – to assume strong continuity axioms. By $\mathscr{R}$(iii) these can be replaced by what Hilbert called a "very intuitive requirement", but its appeal to arbitrary polygons still depends on the non-elementary concept of number. Given this Berkeley's artificial 'Intelligence' could presumably have developed plane geometry after all, despite having no idea of "external space". But unless the integers are presupposed as *intuitively given*, a *purely elementary geometry of the plane must presuppose the existence of space in which it is embedded.* Perhaps nothing could show better than $\mathscr{R}$ – especially the extraordinary counterexamples he constructed to establish part (ii) – how far Hilbert was determined to follow his analysis of spatial intuition and its relation to continuity. As his last geometry student H. A. Schmidt expressed it:

> Darüber hinaus werden in der Festschrift alle Einzelprobleme nach Möglichkeit unter *Ausschaltung der Stetigkeitsaxiome* behandelt; gegenüberfalls werden Notwendigkeit und Reichweit solcher Axiome einer ausführlichen Untersuchung unterzogen. Dazu gestellt sich als weitere für alle Einzelprobleme charakteristische Zielsetzung das Streben, die *ebene Geometrie selbstständig, unabhängig von räumlichen Voraussetzungen abzubauen* und darüber hinaus die Rolle der räumlichen Daten in der ebenen Geometrie festzuelgen ([127], p. 408).

We may summarize the salient points in our discussion of logic, intuition, and mechanism in geometry as follows.

Thales was perhaps the first man, says Proclus "to notice and assert" the theorem on the base angles of isosceles triangles. Euclid and Pappus discovered proofs of it using reflection which thus presupposed the third dimension without explicitly mentioning it. To prove other earlier theorems, Euclid used a principle of parallelism which Descartes later exploited to introduce linear products into geometry, enabling him not only to replace cumbersome logical deductions in geometry by algebraic mechanical calculation, but also to liberate algebra itself. Earlier algebra, having represented powers by spatial dimensions, had hesitated to go beyond the cubic equation, for "nature does not permit it", said Cardano. Descartes circumvented spatial restrictions on products by representing them in the plane: indeed since length and breadth are the most "distinctly conceived" dimensions, according to Descartes, "we ought not to attend to more than these two simultaneously in the same figure" ([69], Vol. I, p. 64). But simultaneously Desargues, extending Euclid's theory of parallels to a theory of perspective, discovered a theorem on spatial representation – indeed, one in which Hilbert would find the key to Descartes' algebraic representation of the plane – which was not at all perspicuous in the plane, but became obvious if one did 'attend' to the third dimension of its figure and then project back to the plane. Peirce doubted that any machine could produce such proofs, presumably because they could not see the 'optical diagrams' that made it possible. Poincaré, regarding machines as blind mechanisms that could thus only do geometry by pure logic, believed that by developing exclusively the logical side of geometry Hilbert had thus made it accessible to machines. But what Hilbert actually did was make his axioms more explicit in order to determine the necessity of both explicit and implicit uses of space intuition. Though this was indeed useful in the later programming of a geometry machine, and even provided an indication of how it must be extended to prove certain theorems, it contributed nothing per se to its ability to see or attend to figures. An early discovery of the geometry machine was Pappus' proof of Thales' theorem, and it seems fitting for artificial intelligence that a machine should begin its career in geometry with the same theorems that man did, even simulating to some extent the incoherence of one of his proofs. But the ultimate test would be to simulate Hilbert's proofs of his metatheorem $\mathscr{R}$ about Thales' theorem, undertaken to analyze his implicit committment to spatial intuition in plane geometry. Existing machines are presumably light years away from results like this, much less the philosophical motivation for them, but I see no obstacle of principle. Hilbert after all was riding a crest of billions of years of biological evolution, and thousands of years of cultural and intellectual evolution continually enriched geometric science. Geometry machines have as yet no such tradition to draw on, nor even much spatial intuition to analyze, but perhaps they have seen the light.

3. THE PROBLEM OF SOLVABILITY

The goal of establishing the significance of continuity axioms was only a special case of a basic 'ground rule' Hilbert followed in geometry which connects his idea of the 'critical stage' of the development of a mathematical theory with general questions about the solvability or unsolvability of mathematical problems:

> The present treatment is a critical investigation of the principles of geometry ... the *groundrule* was to discuss every question ... in such a way as to find out ... whether it can be answered in a specified way with some limited means ... In fact, if a problem is encountered, or a theorem conjectured, the drive for knowledge is satisfied only if either the complete solution of the problem and the rigorous proof of the theorem are succesfully demonstrated, or the basis for the impossibility of success ... are clearly seen. The *impossibility* of certain solutions and problems thus plays a prominent role in modern mathematics, and the drive to answer questions of this type was often the cause for the discovery of new areas of investigation ([117], p. 106).

Hilbert believed that a science was only alive as long as it had an abundance of well-posed but difficult problems to work on whose solutions would require the introduction of new concepts and the simplification of old methods, and which would unify hitherto unconnected parts of the science. Thus he chose for his 1900 Paris address to present twenty-three research problems in mathematical science. In effect, Hilbert was trying to guess the future development of his science, admitting that "every age has its own problems, which the following age either solves or casts aside as profitless and replaces by new ones" ([118], p. 437). And he guessed well, providing "a chart by which we mathematicians have often measured our progress", as Weyl expressed it some fifty years later. But however difficult some of his problems may have seemed, Hilbert assured his 1900 audience that all of them were "solvable in principle". In conscious opposition to the *Ignorabimus* of du Bois-Reymond, Hilbert claimed that

(S) Every 'definite' mathematical problem has an 'exact' solution, which may well be negative, showing the impossibility of a positive solution.

Hilbert was very impressed by impossibility theorems, such as Abel's on the impossibility of radical solutions for equations of the fifth degree, Hermite and Lindemann's on the "impossibility of constructing the numbers e and π algebraically", and impossibility of proving the parallel postulate from the other geometric axioms. Indeed, as he explained to his Paris audience, it was *just this growing list of such impossibility proofs that persuaded Hilbert and other mathematicians that every definite mathematical problem was solvable*:

Occasionally it happens that we seek the solution under insufficient hypotheses or in an incorrect sense, and for this reason we do not succeed. The problem then arises: to show the impossibility of the solution under the given hypotheses, or in the sense contemplated ... and we perceive that old and difficult problems ... have finally found fully satisfactory and rigorous solutions, although in another sense than that originally intended. It is probably *this* important fact along with other philosophical reasons that gives rise to the conviction ... that every definite mathematical problem must necessarily be susceptible of an exact settlement, either in the form of an actual answer to the question asked, or by a proof of the impossibility of its solution and therewith the necessary failure of all attempts ([118], p. 444).

It is interesting to compare Hilbert's belief in (S) with that of Gödel, who emphasized his belief in (S) in connection with an attempt to refute mechanism: using his own incompleteness theorem, Gödel tried to establish the disjunction: "either the human mind . . . can decide more number theoretical questions than any machine or else there exist number theoretical questions undecidable for the human mind", as he was paraphrased by Wang, who continued his report on Gödel's argument as follows:

Gödel thinks Hilbert was right in rejecting the second alternative. If it were true it would mean that human reason is utterly irrational by asking questions it cannot answer, while asserting emphatically that only reason can answer them. Human reasoning would then be very imperfect, and in some sense, even inconsistent, in glaring contradiction to the fact that those parts of mathematics which have been systematically and completely developed ... show an amazing degree of beauty and perfection ([259], p. 324).

Late I will examine and reject the claim that (S) together with incompleteness entails the denial of mechanism. Now I would like to compare the reasons given by Hilbert and Gödel for their belief in (S). Gödel appealed to (i) the accumulation of positive mathematical results, such as solutions for classes of diophantine equations, the theorem of quadratic reciprocity, and the Euclidean algorithm, and (ii) philosophical assertions about the mind of a piece with (CM) and (DM). Hilbert, on the other hand, while granting that (ii) may engender some belief in (S), and that (i) surely confirms it, was more impressed by impossibility theorems. Such results, however, also occur in theoretical physics and human reason has surely learned that it cannot settle these problems alone. The impossibility of perpetual motion was discovered empirically from repeated failures to produce it. Gödel, interested mainly in applying his incompleteness theorem against mechanism, focussed on the solvability of problems in number theory where his theorem directly applies, whereas Hilbert was concerned with the solvability of problems in all branches of mathematics and science, in particular those whose problems arise from their intimate association with physics. Hilbert is thus more guarded about the philosophical reasons (ii) for solvability:

Is this axiom of the solvability of every problem a peculiarity characteristic of mathematical thought alone, or is it possibly a general law inherent in the nature of the mind, that all questions which it asks must be answerable? For in other sciences also one meets old problems which have been settled in a manner most satisfactory and most useful to science by the proof of their impossibility. I mention the problem of perpetual motion. After seeking in vain for the construction of a perpetual motion machine, the relations were investigated which must subsist between the forces of nature if such a machine is to be impossible; and this inverted question led to the discovery of the law of conservation of energy, which, again, explained the impossibility of perpetual motion in the sense originally intended ([118], p. 445).

It remains to be examined now why Hilbert, and many of his contemporaries, were so struck by the impossibility theorems.[26]

(S) receives some confirmation every time any hard problem is solved at all, yet it was the *originally unexpected* impossibility theorems that Hilbert took as the strongest confirmation for (S), for in these cases we have found a rigorous solution *despite having started out – and then groping, possibly for centuries – in a wholly futile direction.* Such negative solutions clearly involve more than positive ones requiring a proof within a *given* mathematical theory. For the impossibility of a positive solution can generally only be proved relative to given limited means which may initially be described only intuitively or informally, and so require a possibly difficult preliminary analysis of their mathematical content. When the tenth problem on diophantine equations – "to devise a process according to which it can be determined by a finite number of operations whether the equation is solvable" ([118], p. 458) – was raised, no more explicit descriptions of such 'processes' existed than those quoted before of Hilbert himself. It took thirty-six years of logical research before Church could give them their rights in mathematics. Should not more weight be given to such confirming instances of (S)? In any event, those who imagine that Church's negative solution of Hilbert's *Entscheidungsproblem* also refutes his (S) clearly labor under a misunderstanding, especially if they imagine that anyone expected a positive solution. Hilbert certainly did not ([255], p. 384), but it was precisely those problems which were solved "in another sense than that originally intended" which inspired his conviction to begin with. Yet many writers still *confuse (S) with the claim that the Entscheidungsproblem has a positive solution*[27]. But Church's theorem hardly refutes (S): attempts to milk undecidability results for "absolutely undecidable" propositions have not been successful or even clearly conceived. The *Entscheidungsproblem*, though once a "leading problem" of mathematical logic, was also just *another problem falling under (S)* – implying that we should either be able to find a decision method for the predicate calculus, or prove that none exists. Church did the latter, but this does not yield undecidable formulas of the predicate calculus: we just have no *uniform* method for deciding them. Hilbert, however, clearly did not expect all his problems to be settled in the same way! He was well aware that he could easily have formulated "completely inaccessible" problems that would "mock at our efforts" (*ibid.*, p. 438), but they would hardly be 'good' ones providing a "guidepost on the mazy paths to hidden truths" (*ibid.*). The solvability thesis had to be taken with a grain of salt, and later Hilbert would in fact count it among "the great problems of the theory of foundations" ([255], p. 473). Even Church's negative solution depends on a thesis that is *not* universally accepted, so it must be defended if we are to retain some striking confirmations of (S).

On the other hand, a decision procedure for the predicate calculus would hardly solve the likes of Hilbert's fourth problem on "the straight line as the shortest distance between two points", which looked for new geometries which "stand next to" Euclid's as justifiably as those of Lobachevski and Riemann. Geometers had recognized the straight line as the subject of projective geometry and hence that, in the presence of continuity, it was essentially characterized by Desargues theorem. But geodesics satisfy this theorem not only in the classical non-euclidean geometries, but also in Minkowski's where the parallel postulate holds but which is not locally euclidean because the strong triangle congruence axiom fails. Hilbert thus took the weaker triangle inequality, which implies that straight lines are geodesics, as an axiom in its stead and asked for "the construction and systematic treatment of the geometries here possible" ([118], p. 451). Pogorelov [204] shows that the 'construction' of these "Desarguesian spaces" amounts to characterizing up to isomorphism all possible models (metrics) satisfying Hilbert's axioms, a problem he solves for the euclidean and non-euclidean cases as well; but Busemann [43] points out that since Hilbert was unaware of the immense number of such metrics the question of their individual "systematic treatment" was not well-posed.

Hilbert's stress on the 'exactness' of problems and solutions raised a question as to the source of mathematical exactness which, according to Brouwer, "is answered differently by the two sides; the intuitionist says: in the human intellect; the formalist says: on paper" ([13], p. 67). But Hilbert found exactness in *method*:

> The axiomatic method is . . . the instrument suited to the human mind and indispensable for every exact inquiry, whatever its field may be. It is logically unassailable, and at the same time fruitful ... To proceed axiomatically ... is simply to think with consciousness of what one is doing. In earlier times, when they did not use the axiomatic method, men believed in various connections with naive dogmatism. Axiomatics removes the naiveté, but still leaves us with all the advantages of belief ([121], p. 161).

His use of the method to "consciously" examine such connections in geometry also reinforced his belief in the importance of impossibility results. The vague distinction in (S) between 'positive' and 'negative' solutions gives way to that between *provable* and *unprovable* from given axioms by given logical means: impossibility becomes *independence*. The *Entscheidungsproblem* is relevant now, but only after one has found appropriate axioms and logic – and precisely this was at issue in some of Hilbert's problems, the first six of which already run the gamut of axiomatic problems in the foundations of arithmetic, geometry, and physics. The first asked for a proof of the continuum hypothesis from as yet unspecified axioms. The second asked for a 'direct' proof of the consistency of the arithmetic axioms "by means of a careful study and suitable modification of the known methods of reasoning in the theory of irrational numbers" ([118], p. 448) – i.e., of the underivability of a contradiction from them by as yet unspecified logical means. The third problem, to show that the archimedean axiom was essential to the theory of spatial volume, i.e. the impossibility of drawing certain formulas without it, was solved by Dehn.[28] The fourth problem we said sets up new axiom systems and requires a characterization of *all* their models. The fifth problem was to prove that the assumption of differentiability was *not* essential to the theory of Lie groups, especially for their application to the foundations of geometry, for it "contains a postulate which, in the geometric axioms, can be expressed only in a rather forced and complicated manner" (*ibid.*, p. 452). Hilbert's own solution for even the special case of plane geometry, however, required him to *formulate entirely new topological axioms.* And the sixth problem was to axiomatize "those physical sciences in which mathematics plays an important role", producing not only a "complete survey of all conclusions derivable from the system of axioms assumed" (*ibid.*, p. 454), but also a consistency proof for them not relying on either 'experiment' or 'physical intuition'. Such proofs would force us, Hilbert promised, to give "exact formulations of the axioms".

Hilbert's own work in geometry featured independence proofs showing the need for various axioms in proving theorems of Thales, Pythagoras, Desargues, and Pascal. But his approach to such problems never confronted *deduction* directly, using rather what Weyl [268] called the "wonderful trick" of arithmetic models to avoid analyzing "the logical mechanism of deduction" itself. But it must be emphasized in connection with (S) that Hilbert's "rather inharmonic painfully concocted" [38] models required, as Weyl also noted, all his "amazing wealth of invention", not to mention a certain distance from his axioms. As far as problems amenable to the axiomatic method were concerned, the main burden of (S) was essentially that *such models could "in principle" always be found.* And Gödel's completeness theorem provides, for a wide class of such problems, a confirmation of (S): it shows that such independence problems are solvable in principle for first-order theories in the sense that an arithmetical model always exists for showing that a given sentence does not follow from others by means of the "logical mechanism of deduction". The existence of first-order axiomatic set theories shows that this is indeed a wide class of problems. And incompleteness shows that such models will indeed be bizarre and hard to construct insofar as their predicates and relations may be non-recursive. Whereas Boole had tried to widen the applicability of formalisms by finding new equally justified interpretations for them, Hilbert saw the necessity of doing this in a radical way *even to answer questions about the formalism itself.*

Cohen's proof of the independence of the continuum hypothesis (CH) from first-order axiomatic set-theory illustrates a number of these points. His models depend on the introduction of wholly new concepts and techniques which have in turn illuminated other old questions in set theory. And he makes essential use of the completeness theorem, in particular of the existence of a denumerable model for set theory if it is consistent, in order to construct another such model with bizarre properties. But does his result solve Hilbert's first problem? Kreisel has often claimed that it does not because (CH) is decided by second-order axiomatic set theory, but Weston [263] has shown that some of his arguments are faulty, in particular that they beg the question of the existence of a unique intended interpretation for set theory. Kreisel's point of view is based on the categoricity of second-order logic lacking in first-order logic and his belief that the original theories, e.g. of Dedekind, were intended to be categorical. But we have seen that another intention of Hilbert was to realize the old dream of a pure elementary geometry, and that what was needed for this was a language for geometry in which could be formulated an elementary replacement for Dedekind's continuity axiom in such a way that it no longer implied the archimedean axiom. This is precisely first-order logic, which cannot distinguish the real field from any real closed one. Kreisel says that non-standard models are important only for showing the inadequacy of first-order logic, but Weston raises the interesting question of why a language should be regarded as inadequate just because it *can* be interpreted in an unintended way. Kreisel also seems to me to overlook the point that Hilbert insisted on to Frege, namely that not even categoricity can eliminate unintended interpretations: isomorphic interpretations can differ widely in the nature of their objects, perhaps as widely as points and chimney sweeps. It is only in *pure mathematics* that we can get away with applying Leibniz's principle to *identify* such interpretations. Weston even sees no reason why we might not "also be able to develop a model theory of *English* which would allow us to produce, for any given body of consistent English sentences, unintended interpretations under which those sentences come out true" (*ibid.*, p. 291).

Cohen suggests that we may someday regard (CH) as *obviously false*: when we realize that the power-set axiom gives us "an incredibly rich set

. . . which can never be approached by any piecemeal process of construction" ([54], p. 151) such as other axioms like replacement provide. Indeed, Hilbert's own attempt to prove (CH) – by assigning to each number-theoretic function an ordinal measuring the level of a recursive definition he tried to construct for it – failed even to prove its relative consistency, according to Gödel, because he considered only "strictly constructive definitions" ([255], p. 369). Hilbert did not regard (CH) as obviously true, but Cohen's result is still clearly the kind of solution that persuaded him of (S): an impossibility theorem relative to definite limited means settling a problem "in another sense than that originally intended". Gödel even suggested that Hilbert's belief that the continuum problem required entirely new metamathematical methods "would seem to imply (although Hilbert did not say so explicitly) that the Continuum Hypothesis is undecidable from the usual axioms of set-theory" ([217], p. 218).

Like the parallel postulate, (CH) or something equivalent to it must be assumed or denied axiomatically – if, that is, you can find some useful consequences of it or its denial that have some application. Wang has called this comparison "farfetched" because in fact nobody has ever proposed (CH) as an axiom, but this is only because, unlike Euclid's postulate, (CH) has no antecedently-known desirable consequences. But their roles are still sufficiently analogous to merit a comparison of their independence. In geometry we have as a consequence of the order and congruence axioms the infinity of the straight line, and by the continuity axioms the isomorphism of its points with the real numbers. The question is how many points at infinity it has, and the euclidean axiom says exactly one. (The hyperbolic axiom says more than one and the elliptic axiom says none – thereby rejecting the infinity of the line as well.) In (ZF) we have an axiom of infinity postulating (N) and another postulating its power-set $\mathscr{P}(N)$, and the question is how many infinite sets have cardinal numbers between those of (N) and $\mathscr{P}(N)$. So despite differences it is clear that in both cases we have the independence of hypotheses which attempt some fine-tuning of the infinite regions of a structure. Wang says that whereas the independence of Euclid's axiom was followed by doubt about other geometric postulates – later he says "all the postulates" – the independence of (CH) "is not accompanied by doubts about the acceptability of the axioms of *ZF*" ([259], p. 205). But this is only because the doubts of strict finitists about (N) and those of the predicativists like Weyl and Lorenzen about $\mathscr{P}(N)$ set in long *before* the independence of (CH) was shown. In both cases we attempt to fine-tune an infinite structure that was *independently* doubtful. Indeed, many of the Greeks doubted both the infinity of the line as well as its infinite divisibility, but until appropriate axioms for formulating these properties emerged, such doubt could have no clear focus. It is not clear that Wang appreciates these points when he says that doubt about other axioms occurred only after the independence of Euclid's axiom was discovered, or when he says that doubt about that axiom "is often attributed to the difficulty of envisaging the infinite extension of the straight line" (*ibid.*, p. 206).

Finally, in both geometry and set theory, the independence results have left one not only in doubt about the truth of these propositions, but even about the meaningfulness of the question. Hilbert had already realized by 1909 that the question of truth was often less important in foundations than that of provability by limited means:

> Bei gewissen modernen mathematischen Untersuchungen ... ich erinnere an die Untersuchungen über die Grundlagen der Geometrie, der Arithmetik, und der Mengenlehre – handelt es sich nicht sowohl darum, eine bestimmte Tatsache zu beweisen oder die Richtigkeit eines bestimmten Satzes festzustellen, sondern vielmehr darum, den Beweis eines Satzes mit Beschränkung auf gewisse Hilfsmittel zu führen oder den Nachweis für die Unmöglichkeit einer solchen Beweisfuhrung zu erbringen. Die analysis der unendlichvielen Variablen setzt uns in der Stand, solche *Beweiskritischen* Untersuchungen auch im Gebiete der Funtiontheorie und ... der Differentialgleichungen anzustellen ([128], p. 72).

Cohen's result may also only show the unprovability of (CH) by limited means, but no others are yet clearly visible. Until they are it seems that he has justified P. du Bois-Reymond's view that certain questions about the continuum cannot be rationally decided – thereby possibly obstructing some calculations of Laplace's demon – while at the same time confirming Hilbert's view that every definite humanly significant problem is solvable. By 1930, positivism was embracing the spirit of (S), with its emphasis on relative impossibility proofs, as an answer to his brother's more general *Ignorabimus*:

> Die Frage, in der du Bois-Reymond die unübersteignaren Grenzen der Naturerkenntnis erblickte, die Frage wie Seelenvorgänge mechanischatomistisch erklärt werden können, ist uns das typische Scheinproblem. *Fein Ignorabimus hat für uns keine andere Bedeutung als für die Mathematiker die nüchterne Erkenntnis von der Unmöglichkeit der Quadratur der Zirkels und anderer ähnlicher Aufgabestellungen, die dadurch, dass man sie auf die richtige Form bringt, zugleich erledigt und anulliert werden* (von Mises [185], p. 892).

The problem here, however, is that we have not yet succeeded in formulating the problem of physically explaining mental processes in a sufficiently definite way for it to be settled. Nor have attempts to prove this impossibility by applying Gödel's incompleteness theorem been successful.

4. CONSISTENCY, DENUMERABILITY, AND THE PARADOX OF RICHARD

Hilbert's work on the foundations of geometry soon led to the greatest impossibility problem of all; the problem of consistency. He had clarified the *significance* of continuity principles for elementary geometry, even if he had not demonstrated the existence of the geometric continuum. Indeed, Hilbert only proved the consistency of the geometric continuum by assuming it for the arithmetical continuum correlated to it in the manner of Descartes, which left him with the basic problem of the existence of the latter. For Hilbert, this problem had to be approached by directly proving the consistency of the arithmetical axioms and was so formulated in his Paris lecture. His approach essentially follows the very same 'ground rule' that he had applied to the problem of the significance of continuity, namely, to ask whether the 'non-elementary' proof of the consistency of the arithmetical axioms which proceeded by simply describing their intended model, and hence begged the question, could somehow be replaced by a more 'elementary' proof in which nothing *continuous* was assumed or used.[29] Frege, we recall, had insisted in a letter to Hilbert prior to the latter's 1904 paper in which this problem is faced for the first time, "Surely it is impossible to prove consistency in any other way" ([91], p. 15). But Hilbert was beginning to realize that if one were to incorporate the *logic* – which in his investigations on geometry had been left at the level of '*gedanklichen Schlusse*' – into the axiomatic method, then the transcendental problem of the possibility of the continuum could be reduced to a concrete combinatorial impossibility problem in Bernays' words:

> Hier handelt es sich nicht um die Möglichkeit einer stetigen, unendlichen Mannigfaltigkeit . . . sondern um die Unmöglichkeit eines mathematischen Beweises mit bestimmten Eigenschaften. Ein mathematischer Beweis ist aber, in Unterschied von einer stetigen, unendlichen Mannigfaltigkeit, ein konkretes, in allen Teilen überblickbares Objekt; er muss sich, wenigstens grundsätzlich, von Anfang bis Ende vollständig mitteilen lassen ([17], p. 97).

The decisive stimulus for this new approach was, of course, the recent appearance of the paradoxes which in Hilbert's view were having a "downright catastrophic effect" on mathematical science. In particular, he was alarmed that they had caused Frege, Cantor, and Dedekind to give up foundational research, for they had "probed more deeply into the essence of the integer" than anyone else. It is interesting that, as a "major goal" of further research into logic and the integer, Hilbert required not just the avoidance of paradoxes, but their clarification:

(CP) . . . the set-theoretic paradoxes . . . show, it seems to me, that the conceptions and means of investigation prevalent in logic, taken in the traditional sense, do not measure up to the rigorous demands that set theory demands. *Rather, from the very beginning a major goal of the investigations into the notion of number should be to avoid such contradictions and clarify the paradoxes* ([119], p. 130).

We see here another manifestation of the attitude that lay behind (S): the two great problems of traditional logic had always been the nature of implication and logical consequence, and the resolution of the ancient semantic paradoxes – the *insolubilia* as they had been called since medieval times – and Hilbert was beginning to feel that foundational studies could provide solutions for these traditional problems as well as the new paradoxes of set-theory. This would require, however, not only the avoidance of paradox through the construction of adequate consistent systems for the concepts of integer and set, but a clarification of paradoxes which explained this consistency by isolating just which assumptions, required for a given paradox, were not available in such a system.[30] But the decisive influence on Hilbert's 1904 lecture, as Blumenthal pointed out, came from Dedekind himself, who more than anyone had "clearly recognized the mathematical difficulties encountered when a foundation is sought for the notion of number" (*ibid.*, p. 130). Dedekind had gone too far, however, when he introduced his *Gedankenwelt*, for this led inevitably to the paradoxes of Cantor and Russell. Thus a different kind of proof would have to be found to replace Dedekind's (D2), and it would require a new level of rigor in the axiomatic method.

Hilbert's first idea is to replace Dedekind's universal *Gedankenwelt* with individual 'thought-objects' (*Gedankendinge*), where each such object is treated as a primitive *symbol*, or finite combination of such. As Hilbert was later to express the difference between himself and Dedekind: "*Am Anfang ist das Zeichen*". The objects in his theory of the intuitive system (N) will be symbols themselves. And since Hilbert felt that logic presupposes certain arithmetical notions such as 'set' and 'number', he decided that it was necessary to apply the axiomatic method simultaneously to arithmetic and logic. In fact, he even asks, and sketches an idea, for the strict axiomatization of set-theory. Hilbert's axiomatization thus has the effect of replacing Dedekind's *Gedankenwelt G* with a *denumerable* set *G′* of symbolic *Gedankendinge*; it also replaces Dedekind's "existence in the realm of our ideas" with *formal consistency*: whenever a notion is proved consistent it "receives a definite meaning and content" (*ibid.*, p. 134). This leads to three basic 'principles' which according to Hilbert must govern the "further [axiomatic] elaboration of the laws of mathematical thought." First we have the

(a) *Creative Principle:* At any stage of development of a theory we are free to add any newly created concepts and axioms, as long as they have been proved consistent.

This principle, which "in its freest use justifies us in forming ever new notions", is Hibert's generalization of Dedekind's "free creations of the human mind". The restriction which excludes known paradoxes from the axiomatic development is then given by the

(b) *Constructive Principle:* At any stage of development of a theory, substitution of *Gedankendinge* for variables in axioms and proofs is restricted to those that have already been *defined.*[31]

This principle, which is similar in intent to Poincaré's ban on impredicativity, is intended to prevent the definition of Russell's set and the universal set, and hence also the paradoxes of Russell and Cantor. However, in order to actually *prove* that no formal inconsistency has arisen at a given stage, we need the

(c) *Proof-Theory Principle:* Proofs themselves, being finite sets of *Gedankendinge*, can be made the *objects* of mathematical study, provided they are given an *'exact description'*.

The most interesting aspect of this now familiar principle is Hibert's proposal to clarify the meaning of 'exact descriptions' by the use of recursive procedures (*recurrentes Verfahren*), which we formulate as Hilbert's thesis

(HT) 'Exact descriptions' of proofs can always be given by recursive procedures.

We shall discuss this important thesis, which is clearly in the spirit of Church's, in the final chapter. Meanwhile, we elaborate the comparison with Dedekind by formulating six tasks undertaken and / or sketched by Hilbert which play the respective roles of Dedekind's (D1)–(D6):

(H1) Axiom system A for (P1), (P2), and a fragment of elementary logic with equality.

(H2) Proof that A is formally consistent.

(H3) A is extended to A^+ by addition of 'axioms of mathematical induction'.

(H4) Proof that certain extensions of A^+ 'provide a foundation for the notion of finite ordinal'.

(H5) Proof that A^+ is formally consistent ('consistent existence of the smallest infinite').

(H6) A^+ is further extended to A^+_R by addition of a *schema r of recursion,* yielding progressive extensions $A^+_{r_i}$ as ever new recursive definitions for functions are added as axioms.[32]

Details aside, it is qualitatively clear how the principles (a)–(c) are intended to control the formal development of arithmetic from (H1) through (H6). In general, the burden of justification is now borne squarely by formal consistency. Yet it would seem that the increased safety of such a formal development over Dedekind's theory might well come at the price of being a *much less complete* description of (N) than (P1)–(P3). Dedekind's induction principle (P3), after all, excluded the would-be 'alien intruders' into his Peano model of (N) only by means of a quantifier which ranges over the nondenumerable power-set $\mathscr{P}(N)$ of N, and it would seem *a priori* plausible that a theory having only the denumerably many symbolic induction axioms of (H3) in its place might well let them back in with a consequent loss in arithmetical theorems.[33] We have seen that doubts about the uniqueness and completeness of logic itself had raised doubts in Hilbert's mind about the status of Dedekind's deductive conception of arithmetic despite the latter's isomorphism proof for his axioms. Later, we shall see, when the experimental completeness of first-order logic became manifest, he formulated the problem of finitistically transforming Dedekind's isomorphism into a completeness proof for first-order Peano arithmetic. In his 1904 paper, Hilbert formulated only the fragment (H1) of logic and, in place of completeness, decided to gradually build logic and arithmetic together by means of his creative principle (a). In the years that followed before taking up this program again after the war, Hilbert came to feel that "*eine übersichtliche und vollständige Formalisierung der logischen Schlussweisen*" constituted the main obstacle to the successful development of his program (H1)–(H6) for the axiomatization of number theory.[34] Only then could one relate categoricity to the possibilities of finite logical proofs, in order to carry out explicitly Dedekind's logical conception of the science of arithmetic. Meanwhile doubts about the prospects for a complete formalization of logic, and hence also about inferring forms of completeness from categoricity, continued to grow until E. B. Wilson (1908) warned forcefully that a categorical set [P] of postulates should not divert attention from research into logic:

Does it follow that every such proposition must either be deducible from [P] or be in contradiction with [P]? . . . What, however, does the word deducible mean? The meaning is entirely relative to the system of logic which is available. . . Some may consider that the human mind has instinctively at its disposal all valid methods of deduction. This is a tremendous postulate, and one entirely devoid of other than sentimental value. In fact, if it leads to the abandoning of the research for valid methods of deduction, it is dangerous and worse than useless ([276], p. 436).

Meanwhile, it gradually dawned on Hilbert that his more important problem of proving the consistency of axiomatic systems would require not only a complete, but also an '*übersichtliche*' logic, that is, one so formulated that its application to a set of axioms yields a mathematically surveyable set of theorems. In short, that it permit some 'mathematical insight' into consistency.[35] Gentzen's consistency proofs, e.g., depend essentially on the use of certain 'natural' logical rules which allowed him to keep clear track of the possible proof structures by isolating, in so-called introduction and elimination rules, the effects of the individual logical constants. It is interesting that the earlier attempts of Hilbert and Bernays to formulate axiomatic logical systems with this same property of isolating the roles of the different logical concepts and axioms was motivated by and modelled on Hilbert's separation of the geometric concepts and axioms. Their predecessors, Frege and Russell, had tried rather to reduce the logical axioms to as few as possible, but as Bernays observed in 1928:

Den Zweck der logischen Untersuchung wird aber besser gedient, wenn wir, entsprechend wie in der Axiomatik der Geometrie, verschiedene *Axiomgruppen* voneinander sondern, derart, dass jede von ihnen die Rolle einer logischen Operation zum Ausdruck bringt ([24], p. 10).

Recalling Hilbert's view that the meaning of a geometric proposition or concept was best seen in the role it played in geometric systems, it seems fair to see Bernays' application of this to logic as an anticipation of Gentzen's idea that his rules for the individual logical constants provide a kind of operational meaning for them.[36] Bernays studied a subsystem of 'positive logic' comprising axioms for implication, conjunction, and disjunction and showed that new positive theorems become provable when the system is 'closed' by axioms for negation, analogously to Hilbert's addition of continuity axioms to the elementary geometric axioms. Indeed, according to Bernays:

Wir erkennen hier ganz deutlich, dass die Rolle der Negation diejenige einer *ideales Elementes* ist, dessen einführung den Sinn hat, das logische System zu einer Gesamtheit von einfacherer Struktur abzurunden, gerade so wie das System der reelen Zahl durch die Einführung des imaginaren zu einer übersichtlicheren Gesamtheit erweitert wird . . . (*ibid.*, p. 11).

Hilbert (1922) had already compared the quantifiers with ideal elements, and indeed it was precisely the application of negation to the universal quantifier that became the focus of the intuitionistic critique of this 'idealized' classical logic. These considerations, which lead to the recognition of various logical systems of differing strengths, are relevant to the early charges by Poincaré and Brouwer of the circularity of using intuitive induction to prove the consistency of formalized induction. Not only is the later now applied to essentially more complex properties than the intuitive transfinite induction of Gentzen, but it is also used in concert with the full idealized system of classical logic. In fact, Gentzen needed only part of intuitionistic logic and induction up to ε_0 on decidable properties to prove the consistency of applying the idealized classical logic and induction up to any $\alpha < \varepsilon_0$ to undecidable properties – hardly a 'circular' result. But Poincaré's final opinion on Hilbert's original consistency proof claimed only that: "This demonstration is legitimate only if the principle of induction is admitted, and if it is regarded not as a definition but as a synthetic judgment" ([206], p. 184). It is very instructive in fact to examine, against the background of the early worries recounted above over completeness and consistency, Poincaré's critical attitude towards the axiomatic method and logical research, in particular his belief in their sterility.

We recall that he saw Hilbert's use of the method in geometry as a purely mechanistic account of geometric reasoning, which would in principle allow one to "replace geometry by the reasoning piano imagined by Stanley Jevons". But Poincaré also argued that even though we have a complete and mechanical formalization – he accepted this as a matter of course – this formalization would be useless for discovery, for our machine would blindly churn out an infinite number of theorems, most of which would be "devoid of interest". "Discovery is discernment, selection", and Poincaré saw no way a machine could identify the fruitful theorems.[37] For Poincaré, the axiomatic method was only a method for *embalming* theories whose foundations had been unearthed, but not suited for concrete research on the outer frontiers of a theory: a paper on the future of mathematics concludes with the following comments on "the search for postulates":

> Attempts have been made ... to enumerate the axioms and postulates more or less concealed which form the foundation of the different mathematical theories ... It seems at first that this domain must be strictly limited, and that there will be nothing more to do when the inventory has been *completed*, which cannot be long. But when everything has been enumerated, there will be many good ways of classifying it all A good librarian always finds work to do, and each new classification will be instructive for the philosopher (*ibid.*, p. 45).

We see that no small part of Poincaré's contempt for the axiomatic method was based on his belief that axiomatizations would *always be complete*, as they had been in geometry. In particular, he expected those working on the axiomatization of arithmetic to be as successful as Hilbert was in geometry – on the score of completeness – but he warned nevertheless that the axioms of arithmetic would not thereby become mere conventions. In particular, the induction axiom was anything but a convention: it was a synthetic *a priori* proposition that described a basic property of our mind, namely, Locke's power of the endless repeatability of possible mental operations. The geometrical axioms, on the other hand, were pure conventions, as was shown by the possibility of non-euclidean geometry. Indeed, according to Poincaré, there could never be anything like a 'non-euclidean' arithmetic:

> To convince ourselves of this, let us take a *true* synthetic *a priori* intuition ... If a theorem is true for the number 1, and if it has been proved that it is true of $n + 1$, provided it is true of n, it will be true for all positive integers. Let us try to get rid of this, and while rejecting this proposition let us construct a false arithmetic analogous to the non-Euclidean geometry. We shall not be able to do it ([205], p. 49).

But this is literally false: by the very independence of the induction axiom from the others of Peano arithmetic, we can construct just such a system, one in which *ordinary numerical calculation would go on as usual.* It has also occasionally been suggested that this is just the sort of thing that Gödel's incompleteness has enabled us to do, confronting us once again with the double-edged quality of 'arguments from Gödel'.[38] The common tendency to regard incompleteness as vindicating those who, like Poincaré and Brouwer, have emphasized the primacy of intuition, as opposed to those who emphasize with Hilbert the importance of formalisms, proves rather superficial on closer inspection. Thus consider Poincaré's belief that the axiomatic method, precisely because its complete inventory of the foundations would soon be over, leaving nothing more to do – and being incapable of detecting the interesting theorems produced by its complete mechanical enumeration – would be useless for research and discovery in arithmetic. One can argue that just the opposite has turned out to be the case: the incompleteness theorem shows that as soon as we have finished any specification of a formalism for arithmetic we can, by reflecting on that formalism (Hilbert's '*Wechselspiel*'), discover a new truth of arithmetic which not only could not have been discovered working in that formalism, but – and this is the point that is usually overlooked – which presumably *could not have been discovered independently of working with that formalism.* The very meaning of the incompleteness of a formalism is that it can be effectively used to discover new truths inaccessible to its proof-mechanism, but these new truths were presumably undiscoverable by any other method. How else would one discover the 'truth' of a Gödel sentence other than by using a formalism metamathematically? We have here not only the discovery of a new way of using a formalism, but a proof of the eternal indispensability of the formalism for the discovery of new mathematical truths.

The incompleteness theorem also throws doubt on Poincaré's view that mathematical induction is the fundamental method of proof, being "the only instrument which enables us to pass from the finite to the infinite" (*ibid.*, p. 11). For it shows that given any codification of the method of proving universal statements about all integers by induction we can effectively find a true universal statement $(x)A(x)$ which is not provable by our codified inductions, but only by a wholly new and different method, namely, *metamathematical diagonalization.* The system can only prove each specific $A(\bar{n})$: to 'pass to the infinite proof' of $(x)A(x)$ itself we use a self-referential diagonal argument. Of course, we need also the consistency proof of our system to complete our proof of $(x)A(x)$, showing how wrong Poincaré was to think that nothing could be gained by using induction to prove the consistency of a system with induction. This is a metamathematical analogue of the situation we encountered before with Dedekind's proof of his recursion theorem where induction alone could only generate the infinite sequence of finite partial functions and diagonalization was used to define the desired infinite function is terms of this whole sequence.

It is quite probable that the classical categoricity proofs for arithmetic and geometry helped to persuade Poincaré of the sterility of the axiomatic method. He was very impressed with the usefulness and power of the method of isomorphism in group theory to transmit familiar insights into new domains, but the wide applicability of this method here depends precisely on the non-categoricity of group theory as a whole. The categoricity of (second-order) axiomatic arithmetic proved by Dedekind may well have persuaded Poincaré of its limited applicability to other branches of mathematics. For it is probable that Bourbaki had Poincaré, among others, in mind when they cited this 'historical accident' as the principal reason for the widespread resistance to the axiomatic method found among philosophers and mathematicians at the time of its inception:

For a long time many of the latter could see in axiomatics only the empty subtleties of logicians, something incapable of giving birth to any theory whatsoever. This criticism can doubtless be explained in terms of pure historical accident: the first, and also the most celebrated axiomatizations (those of arithmetic by Dedekind and Peano and that of Euclidean geometry by Hilbert) dealt with *univalent* theories, i.e. theories which were uniquely determined by the totality of the axioms, which were thus incapable of application to any theory other than the ones from which they had been drawn (contrary to what we have seen for group theory, for example). If such had been the case for all the structures, then the reproach of sterility made against the axiomatic method would have been fully justified ([166], pp. 34–5).

The main thrust of formalism, namely the view that univalency was the most that one could hope for in the axiomatic treatment of a theory, was never considered by Poincaré. He did, however, clearly recognize an important consequence of Hilbert's program for proving the consistency of such theories, namely the *denumerability* of Hilbert's symbolic replacement of Dedekind's *Gedankenwelt*. And also the paradox this leads to.

A much debated issue in the foundations of mathematics at the turn of the century concerned distinctions like Dedekind's between sets and functions which were 'decidable' and those which were merely 'determined'. Could such a distinction be given any mathematical meaning at all? If so, what would become of various mathematical theories if one were to forbid consideration of the latter? Did there exist determinate functions which were nevertheless undecidable? Such questions were taken up in the five letters on set theory published by Hadamard [108]. Specifically, they try to come to grips with Tannery's 1897 distinction between functions which are 'describable' and those merely 'defined'. Hadamard emphasizes that the former property has the significance of permitting an effective determination' of a function; but in another letter to Borel, he also explains 'describable' as "definable in a finite number of words". Poincaré meanwhile was insisting – à la Kronecker – that all mathematical objects must be defined in a finite number of words; in fact, he went further and advanced the thesis:

(PT) α can be an 'object of thought' $\Leftrightarrow \alpha$ is finitely definable

as a law of thought ([207], pp. 60–1). In addition to this, Poincaré and others asserted the 'theorem'

(D) The set of all finitely definable objects is enumerable.

On the basis of (PT) and (D) Poincaré and Weyl concluded that

(D)′ Our *Gedankenwelt* of all objects of thought is enumerable.[39]

Richard [219] then brought things to a head by applying Cantor's theorem asserting the non-enumerability of the real numbers to infer

(r) There exist real numbers which cannot be defined in a finite number of words, indeed, which cannot be objects of thought at all.

His famous paradox is already implicit in (r), and pushes to the surface as soon as one applies Cantor's diagonal argument to the countable set of all finitely definable real numbers. But paradox aside, Poincaré opposed (r) as too metaphysical: rather, he argued, should one conclude from (D) that

(D)″ The continuum is denumerable.

For surely every real number should be capable of being an object of thought. That Cantor had disproved (D)″ was, according to Poincaré, "merely an illusion", but it did raise the question: "What did Cantor mean and what did he really prove?" (*ibid.*, p. 61). Let $\mathscr{P}_f(N)$ denote the set of all finitely definable subsets of N. According to Poincaré then, what Cantor "really proved" was more like

There is no effective 1–1 function f which maps $\mathscr{P}_f(N)$ onto N, and which is also predicative.[40]

Broadly speaking, the Richard's Paradox is the problem of clarifying the issues and conflicts arising when one confronts Cantor's proof and theory with (PT), (D)′, (D)″. Poincaré's is one such attempt, but it employs two notions, namely 'effective' and 'predicative', which were felt by the majority of his contemporaries to be 'outside of mathematics'. Others tried, however, to clarify the situation without giving up the theorem of the non-denumerability of the continuum. An early attempt to meet this challenge to set theory presented by Richard's paradox as well as Russell's was Zermelo's 1908 axiomatization of set theory. Whereas Hilbert tried to independently axiomatize the elementary arithmetical parts of Dedekind's essay, Zermelo axiomatized its more general set-theoretic parts.

To begin with, Cantor's proof of the non-enumerability of the continuum and $\mathscr{P}(N)$ can indeed be formalized in Zermelo's system, thereby apparently disposing of (D)″.[41] As for (D)′, Zermelo disposes of it by simply denying there *is* any such universal set, such as Dedekind's *Gedankenwelt*, forcing him to replace Dedekind's proof for infinite sets by an *axiom*. In fact, Russell's paradox itself is formalized in the system in order to prove that no set definable in the system can exhaust its universe of sets. Thus, in Zermelo's crucial separation axiom schema for obtaining subsets, one can only pick out those subsets of a given set which satisfy a 'definite condition' on its elements, and Russell's condition $x \notin x$ is used to produce, for any given set, a subset of it which cannot be an element of it. Finally, Zermelo claimed to avoid (r), and Richard's Paradox generally, by the requirement that one have 'definite' conditions on N to separate elements of $\mathscr{P}(N)$, whereas 'finitely definable,' according to Zermelo, is not a definite property. In fact, Zermelo struck at the heart of Poincaré's position, namely (D) and (PT), by arguing that the concept of 'finitely definable' is meaningless in itself *unless relativized to language*:

> 'Finitely definable' is not an absolute notion but only a relative one, and it is always related to the chosen language, or notation. The conclusion that the set of all finitely definable objects must be denumerable holds only if one and the same system of signs is to be used for all of them, and the question whether a single individual can or cannot have a finite designation is in and of itself meaningless, since to any object we could, if necessary, arbitrarily assign any designation whatever ([255], p. 192).

Zermelo's critique of (D) is quite cogent, and also valuable for emphasizing the linguistic dimension of (PT) overlooked by Poincaré. However, Zermelo's axiomatic defense of set theory against Richard's paradox depends essentially on his notion of 'definite': a condition is said to be definite "if the fundamental relations of the domain, by means of the axioms and the universally valid laws of logic, determine without arbitrariness whether or not it holds" (*ibid.*, p. 201). Poincaré replied that Zermelo's notion, thus explained, was impredicative and hence meaningless.[42]

In 1909 Hilbert invited Poincaré to Göttingen to give a series of lectures on mathematics and Poincaré took the opportunity to explain and defend his views on foundations, especially his critique of set theory. Specifically, he defended (PT) and claimed that the real numbers do not form a "definite class", and hence that it was meaningless to speak of all of them. Both Chwistek and Weyl were present at this lecture and it made a strong impression on them.[43] Within a year, Weyl devoted his Habilitation [264] to the impact of the Richard's Paradox on set theory; in particular, he examined Zermelo's solution, and found the notion of 'definite' as explained by him too vague to be admitted into an axiomatic theory. Weyl proposed to explicate Zermelo's notion as follows. First, he lists five "principles of definition" permitting one to define explicitly further relations from given relations; in fact, Weyl's constructions are equivalent to requiring that all such further definitions be expressed in the language of *first-order* logic. He then proposes:

Unter einer definiten Beziehung soll eine solche verstanden werden, welche auf Grund der beiden Beziehungen $=$ und $\in$ durch endmaligen Anwendung unserer in geeigneter Weise modifizierten Definitions-prinzipien erklärt ist (*ibid.*, p. 112).

This leads immediately, he notes, to the "true kernel" of Richard's Paradox:

Kehren wir nunmehr zur Richardschen Antinomie zuruck, so werden wir auch jetzt noch als einem wahren Kern der Beiden in ihr einander gegen-übergestellten Tatsachen anerkennen mussen, dass man es in der Mengenlehre oder in der logisierten Mathematik nur mit abzahlbar vielen Beziehungsbegriffen zu tun hat, keineswegs aber nur mit abzahlbar vielen Dingen oder Mengen (*ibid.*, p. 112).

Weyl's lecture was reviewed by a young Norwegian mathematician, T. Skolem, who later undertook to sharpen Weyl's formulation of Richard's Paradox to the point where one could get an incompleteness theorem for axiomatic set theory. Meanwhile, Weyl [265] himself gave a more explicit reformulation essentially as follows:

(RP) (i) $\mathscr{P}(N)$ is *mathematically* non-denumerable (by Cantor's theorem),
(ii) $\mathscr{P}(N)$ is *metamathematically* denumerable (be enumeration of the defining conditions on N in axiomatic set theory).[44]

In analogy with Weyl's 'global' form of the paradox, we can formulate a 'local' form closer to Richard's own, except for using the distinction between 'mathematical' and 'metamathematical'. Let δ be a new set of numbers obtained by applying Cantor's diagonal method to $\mathscr{P}_f(N)$ – taking this as the denumerable "model" of $\mathscr{P}(N)$ in (RP)(ii). Then we would have

(RP)′ (i) $\delta \notin \mathscr{P}_f(N)$ holds mathematically, by the diagonal argument.
(ii) $\delta \in \mathscr{P}_f(N)$ holds metamathematically, by the finite (diagonal) definition of δ.

Skolem developed the details of (RP) to get his famous non-categoricity result for axiomatic set theory in 1922. He also claimed, in search of an outright incompleteness theorem along the lines of (RP)′, that one could define a subset of N which could not be formally proved in Zermelo's system to belong to its universe. But unless one could *effectively* define such a set, one could hardly see in this the final indictment of axiomatic set theory claimed by Skolem: such was the wait-and-see attitude of Hilbert's proof theory, according to Bernays:

Die näheren Erörterungen dieses Sachverhaltes knüpfen an das Richardsche Paradoxon an, welches in neuerer Zeit durch Skolem eine schärfere Fassung erhalten hat. Diese Überlegungen besitzen insofern keinen abschliessenden Charakter, als sie sich im Rahmen einer nichtfiniten Matamathematik vollziehen. Eine endgültigen Klärung der hier diskutierten Frage würde erst bewirkt sein, wenn es etwa gelänge, in finiter Weise eine Zahlenmenge anzugeben, von der sich zeigen liesse, dass sie in der axiomatischen Mengenlehre nicht vorkommt ([18], p. 366).

In fact, Skolem had claimed that both his countable model and the new diagonal set could be constructed in accordance with "the finite and recursive mode of thought", but it has turned out that he overstated his case.[45] Still, his program of distilling the incompleteness of axiomatic set theory from Richard's Paradox provided yet another motive for clarifying the notion of effectiveness.

5. FREGE, LATER HILBERT, AND THE PROBLEM OF FORMALIZATION

In the history of mathematics, especially since the 17th century, one can trace a repeated, sometimes convulsive alternation between two styles of mathematics, which surface at different times under different lights and whose differences can be related in the most general way to the distinction between 'form' and 'content'.[46] Proponents of the first are generally distinguished by an orientation toward the *formalism*, or the *symbolism* itself of a mathematical theory (for example, in Euler's function theory a function is always an 'analytical expression') while those of the second tend to emphasize what we now call the interpretation of the formalism (for example, Newton, who sees a derivative as a physical magnitude). Insistence on the pre-eminent importance of one or the other kind leads in turn to familiar differences in the philosophy of mathematics. The following historical assessment of Bell will be well worth our reflection:

Each philosophy has greatly enriched mathematics. The first, seeking interpretations, may be called synthetic; the second, beginning and ending in a formalism within its own postulated universe, may be termed analytic. The designations are merely for convenience, and are not intended to recall Kant's terminology, although the parallel may be suggestive. *The development of the number system is the record of a continual interplay between the synthetic and the analytic approaches* ([10], p. 175; italics mine).

More of this 'continual interplay' below, but let us first consider some views of Frege [89] who sought to establish his brand of 'meaningful' (*inhaltliche*) arithmetic as philosophically superior to the systems of his formalist adversaries, notably Thomae. They will be especially helpful in understanding Hilbert, upon whom Frege's problems made a deep impression, and who formulated some of his later program in response to them.

First there was Frege's relentless and uncanny knack of catching the formalists in their inevitable rendezvous with contentual (i.e. referential) idioms. But even on this score he apparently was willing to settle for one's referential *intention*: "in order to justify mention of the reference of a sign it is enough, at first, to point out our intention in speaking or thinking. (We must then add the reservation: provided such reference exists.)" ([88], pp. 61–62). However, he unhesitatingly postulated objective thoughts as the only possible basis for intersubjectively shared concepts. To the pressing question of how this world of thoughts relates to language Frege gives a disarmingly direct answer:

The world of thoughts has a model (*Abbild*) in the world of sentences, expressions, words, signs. To the structure of the thought there corresponds the compounding of words into a sentence (*ibid.*, p. 123).

Later Hilbert ended up making a similar assumption, though with an essential difference which we discuss below. We recall that one of Frege's defenses of his objective sense of '*Gedanken*' was precisely Dedekind's 'proof' of the existence of infinite sets, which Frege virtually assumed to justify such a use.

Another count on which Frege hounded formalists was the question of consistency, arguing that only their neglect thereof allowed them the illusion of 'creating' new mathematical objects by postulating whatever they desired. But although he pressed them for proofs that their creations did not introduce inconsistency, *the only method he permitted for doing this was to prove that such objects exist*: Frege remained steadfastly a 'realist' in the sense that for him existence implied consistency, but not vice versa. Yet as we have seen he was not as dogmatic as some of his formulations would suggest: as he emphasized in [88] and again in a letter to Hilbert, *he simply could not think of any other way of showing consistency than to give examples.*[47] (Moreover, for him the question of the consistency of the axioms of geometry had in any event no significance because he regarded them as *true*.) Frege's preferred formulation appeals to his distinction between 'concept' and 'object': if an object falls under a concept, then that concept is a consistent one, but a consistent concept may still be empty, i.e. no object falls under it. The opening section of his untranslated 'Retrospect' on his critique of formalism is entitled: "Formal and meaningful arithmetic. Neither way has led us to our goal". This included Frege's own way.

Frege's greatest achievement is perhaps his discovery and rigorous formulation of the predicate calculus – and as it turned out, he found the complete predicate calculus all at once. However, for Frege, 'completeness' did not mean what it now means for us, nor could it. Thus, after introducing in the *Begriffschrift* his basic logical laws and inference rules, Frege shows how some logical laws may be derived from others, claming that

> In this way we arrive at a small number of laws in which, if we add those contained in the rules, the content of all the laws is included, albeit in an undeveloped state . . . Since in view of the boundless multitude of laws that can be enunciated, we cannot list them all, we cannot achieve completeness except by searching out those that, *by their power*, contain all of them ([255], p. 29).

And then, by way of admitting the possibility of alternative logical axioms and primitives, he adds that

> There is perhaps another set of judgments from which, when those contained in the rules are added, all the laws of thought could likewise be deduced (*ibid.*).

To a modern reader this reads initially like a straightforward claim of deductive completeness, namely that all valid logical laws are derivable, but Frege offered no other characterization of a 'logical law' besides its derivability, no reason why it *should* be derivable. Indeed, he can hardly have had our *semantic* notion of validity in mind when he spoke of "all laws of thought", for as van Heijenoort [254a] has pointed out, Frege's universal system, like *Principia Mathematica*, was purely syntactic, aiming at a "total logical reconstruction of the world". Its quantifiers range over *all objects* of a fixed unique universe, the Universe. The semantic notions of validity and satisfaction depend, in the spirit of Boole, on variable universes, and hence have no place in Frege's logic.[48] Closely related to this logical absolutism is the question of separate rights for *first-order* logic, whether it leads any autonomous existence of its own. Frege gave it none. Indeed, as van Heijenoort remarks, for Frege and Russell such formulas as '$(x)(Fx \supset Fx)$' are merely "temporary equivalents" of such second-order formulas as '$(F)(x)(Fx \supset Fx)$', for which validity and satisfiability collapse to truth: when Frege passes from first- to second-order logic, "there is hardly a ripple", as van Heijenoort puts it. This is readily understandable, of course, in view of Frege's overriding interest in defining the natural numbers logically, for only that of second-order suffices for this. Nevertheless, in the fullness of time, first-order logic has inevitably floated to the top of our logical investigations, its various recommendations for centrality in our logical scheme of things being crowned by Gödel's completeness theorem for it. Does this result justify the identification of logic with first-order logic? Does the incompleteness of second-order logic disqualify it from the realm of "pure logic"?

The status of the distinction between first- and second-order logic has recently been reviewed by G. Boolos [28a], who has effectively challenged the cogency of Quine's arguments for seeing this distinction as fundamental, and, in particular, for seeing first-order logic as comprising all of logic. On the really decisive point, however, he seems to me to have begged the question. Thus, the crucial property of first-order logic, in fact a corollary of completeness, is its compactness: if some sentence is a logical consequence of an infinite set of sentences, it must already be such a consequence of some finite subset of that set. Consider now the following potentially infinite list $\mathscr{L}$ of sentences:

a_1: a line has at least 1 point
a_2: a line has at least 2 points
a_3: a line has at least 3 points
⋮
a_n: a line has at least n points
⋮

It is to the everlasting glory and misery of Aristotle that, despite being prepared to accept each of these sentences, he nevertheless felt compelled, in order to escape the labyrinth of the continuum, to deny

A: a line actually has an infinity of points.

The question is whether Aristotle's escape is a *consistent* one. Can he get away with boldly denying that A logically follows from $\mathscr{L}$? In short, is the set $\mathscr{A} = \{\text{not } A, a_1, a_2, a_3, \ldots\}$ logically consistent? Any line satisfying all the conditions of $\mathscr{L}$ would seemingly have to contain more than any finite number of points, and hence to contain an infinite number of them. However, Aristotle based his logic on a finiteness principle restricting any proof figure to finitely many premises, and indeed, compactness transfers this finiteness to the logical consequence relation itself.[49] Thus Aristotle's theory $\mathscr{A}$ is consistent if read as a first-order theory, for A is clearly not a logical consequence of any finite subset of $\mathscr{L}$. On the other hand, the vague feeling that $\mathscr{A}$ is inconsistent can be formalized when $\mathscr{A}$ is formulated in second-order logic, where compactness fails. Boolos sees this as evidence that second-order logic is a much-needed extension of first-order logic, needed to "offer a way of regarding as inconsistent certain apparently inconsistent sets of sentences", such as $\mathscr{A}$. For "It seems impossible, on reflection, that all the statements in any of these . . . sets be true" (*ibid.*, p. 522). The fact that such inconsistency cannot be represented in first-order logic "shows not that they are consistent after all, but that not all (logical) inconsistencies are representable by means of that logic" (*ibid.*). The nub of all this is, of course, that only in second-order logic can such ideas as infinity and denumerability really be *expressed*. This does not mean, however, that second-order logic actually provides us with a sound, finite proof of A from $\mathscr{L}$, for it is incomplete. The 'inconsistency' of $\mathscr{A}$ is rather 'represented' in terms of its consequence relation. This means, as Beth has pointed out, that from the point of view of second-order logic, $\mathscr{L}$ is regarded as an *actually* infinite set of premises – and hence, also, such a line as having as actually infinite set of points. But no more than any other logic is second-order logic able to prove the existence of anything infinite: it merely provides a language in which to express our idea of such. If one does not believe that infinite sets actually exist, however, then, following Aristotle, he will regard the set $\mathscr{L}$ of premises as only potentially infinite, i.e. as only the expression of some algorithm, such as the midpoint construction, for generating more and more points on a line. But then A does not follow. In short, the belief that Aristotle's $\mathscr{A}$ is actually inconsistent presupposes that it is actually infinite. But as Dedekind said, nothing in science should be accepted without proof, and in view of the difficulties he encountered in trying to prove the existence of such a set, one may wonder whether such a belief can be justified at all. And if one does not believe in them, much less in nondenumerable sets, then first-order logic offers itself as a truly objective language. This finitism recognizes indeed the existence of our abstract *ideas* of the infinite and the nondenumerable, but the mere availability of a language for expressing them no more compels it to grant them any objectivity or reality than the continued use in everyday language of names for different spatial positions on the earth compromises the relativity of position which follows already from the inability of purely geometrical language to distinguish between two congruent figures.[50] Finitism will still insist with Hilbert that:

> . . . we must make it clear that 'infinite' has no intuitive significance and, in general, no meaning without more detailed investigation. For there are, everywhere, only finite things. There is no infinite velocity and no force or effect which propagates infinitely fast. Moreover, effects themselves are discrete in nature and exist only as quanta. In general, there exists nothing continuous, which could be divided infinitely often . . . endlessness or infinity, precisely because it is the negation of an omni-privalent state of affairs, is a shocking abstraction . . . ([124a], p. 4).

But let us note that the completeness of first-order logic by itself fails to settle Hilbert's dispute with Frege over the relation between consistency and existence in his favor. And not simply because his axiomatization of geometry in question in the dispute was second-order, for we have seen that he formulated the *Vollständigkeitsaxiom* only long enough to remove any lingering doubt about the independence of the parallel postulate and then set about to eliminate all continuity axioms from elementary geometry. We have also seen, however, that the real issue between them concerned the means available for *proving* consistency. The formalist believes that by focussing directly on the structure of formal proofs he can gain a mathematical insight in the consistency of the deductive mechanism itself. Indeed, the 'semi-intuitionist' Poincaré was far more dogmatic than Hilbert in saying that existence was nothing but consistency, so this view by itself hardly distinguished formalism. What the formalist does is rather to subject the very notion of 'model' to a critique by pointing out the delusion involved in seeing a model for a consistent theory guaranteed by the completeness theorem as anything more than a non-constructively completed theory. The very notion 'existence in the model' is itself purely formal, as becomes especially clear from Henkin's proof modelled on the algebraic paradigm of field completion. I have suggested that Hilbert's belief in the solvability of all problems *is* supported by the completeness theorem, but this is not an especially 'formalistic' belief.

It remains to consider Frege's views on formalization and mechanism which are not only interesting in their own right but exerted a considerable influence on Hilbert. Whereas we now tend to think of formalization as peeling away the meanings of the sentences in a proof, leaving only a formal skeleton to justify its validity, Frege rather thought that sentences only become properly analyzed and interpreted upon formalization in his *Begriffschrift*. This is related to Frege's view of the predicate calculus noted before: its only 'model' is *the* one world – Frege's world, inclusive of its objective thoughts. He admitted of course that meaningful reasoning could be cut from its moorings and suffer an '*Inhaltsentleerung*': this happened in Boole's 'calculating logic', Jevons' logical machine, and above all in formal arithmetic. But in explaining how this is possible Frege cannot avoid dragging in *subjective* thought. Thus in criticizing formal arithmetic Frege admits that someone might well derive a theorem of his 'meaningful arithmetic' without any knowledge of the sense or reference of its signs, but he can only insist that "there would be wholly lacking that train of thought which accompanied the affair for us and actually made it interesting" ([88], p. 186). This is debatable, but in any case Frege's "train of thought" here is purely subjective or psychological, not objective. Moreover experience teaches us that *non*derivability proofs are in general best made by setting our normal trains of thought aside. The essential point, however, is the belief Frege shared with Peano and Couturat that our thoughts can be completely formalized in symbols.

Frege claimed in 1880 that Boole's logic represented only part of our thought, the whole of which he felt could never be replaced by any "purely mechanical activity" ([90], p. 39), though following Leibniz, he urged that every conceptual mistake could be exposed by calculation if only the conceptual content of our thought is appropriately represented in symbols. In fact, it was none other than Frege who fifteen years later helped persuade Hilbert of the value of such symbolism after the latter had complained to him about the growing replacement of words by formulas and symbols in mathematics. Frege subsequently conceded in a letter to Hilbert that an adequate symbolism for expressing certain important mathematical methods did not yet exist, but insisted nevertheless that "where a line of thought can be perfectly expressed in symbols, it will appear briefer and more perspicuous in this form than in words" ([90a], p. 33). Even if such use of symbolism increases the danger of lapsing into a "mere mechanism of formulas", Frege warned Hilbert against equating it with a "thoughtless mechanical procedures" (*ibid.*). *For one can think in symbols.* Hilbert, no less than Poincaré, had feared that such a mechanism would stifle thought, but Frege wisely observed that "what was originally saturated with thought hardens in time into a mechanism which partly relieves the scientist from having to think" (*ibid.*).

Despite his mentalistic opposition to such a mechanism Hilbert replied to Frege expressing complete agreement with his analysis of the role of symbols. But when the paradoxes later obliged Frege and Dedekind to withdraw their objective thoughts as a foundation for arithmetic it occurred to Hilbert that the former had not yet gone far enough in recognizing the importance of symbols: "Wo Begriffe fehlen, da stellt ein Zeichen zu rechter Zeit sich ein", as Bernays [16] was later to express Hilbert's reluctant transformation of Goethe's irony into a new methodological principle for studying arithmetic proofs. His later development of this program, while inspired mathematically more by Dedekind, was inspired philosophically by Frege and Kant: it would turn around an attempted synthesis of Frege's meaningful and formal arithmetic, though within a generalized Kantian framework in which symbols play the role of the *a priori*. In order to systematically investigate the newly emerging problems of consistency, completeness, and decidability, which Frege's rigid semantic scheme could not easily accommodate, Hilbert would try to breathe life into Frege's basic distinction between a *theory* and a mere *symbol game*, for as Raggio puts it,

> Für Frege sind die Theorie und das Zeichenspiel zwei für sich bestehende Grossen; das zweite dient bloss als Mittel zur Sicherung und Vereinfachung der Erkenntnis der ersten. Nur eine aussere Erkenntnisbeziehung verbindet beide Instanzen, ohne ihr ontologisches Wesen anzutasten ([215], p. 229).

Hilbert did not find time to pursue his program until 1917, after a brief but intensive career in theoretical physics during which he worked on the axiomatic foundations of statistical mechanics, radiation theory and elementary optics, and general relativity theory. In short, until he had done considerable work on his sixth problem. It is especially in his work on radiation theory and elementary optics that consistency proofs first take on for Hilbert the significance that they characteristically have in his later work, namely as showing not only that some theory is consistent in and of itself, but that *it does not contradict the laws of some more elementary 'neighboring' theory to which it can be applied.* Thus Hilbert [119a] proved not only the consistency of radiation theory (up to Kirchhoff's law), but also that it could only be used to prove correct laws of elementary optics. This is essentially the relation that he later tried to establish between set theory and arithmetic, respectively, by proving the consistency of the former. The difference, of course, is that whereas he proved the consistency of radiation theory by an analytic model for its *axioms* – as he had done for euclidean geometry – the consistencyof arithmetic and set theory are to be proved by direct examination of the *proof*-structure itself.

Hilbert's geometric work does provide, however, the principal background for his new '*finite Einstellung*', as well as for his generalization of the Kantian framework in which it is developed. It is especially Kant's *a priori* that must be generalized, for according to Hilbert:

> In der Kantschen Apriori-Theorie sind noch anthropomorphe Schlacken enthalten, von denen sie befreit werden muss und nach deren Entfernung nur diejenige apriorische Einstellung übrigbleibt, die auch der rein mathematische Erkenntnis zugrunde liege: es ist im wesentlichen die von mir in verschiedenen Abhandlungen charaktersierte finite Einstellung ([128], p. 385).

What is 'anthropomorphic' in Kant is his particular conception of space and time as 'forms of intuition' and his particular categories comprising the 'form of the understanding'. What is essential for Hilbert about these 'forms' is just their 'finiteness'. In particular a 'transcendental aesthetic' which is not anthropomorphic will demand as given *a priori* for logical inference only "certain extralogical concrete objects that are intuitively present as immediate experience prior to all thought" ([255], p. 376), and will focus on their finiteness and discreteness as essential to their 'form' – thereby generalizing the 'spatiality' presupposed by all such objects. According to Bernays, this is the outcome of Hilbert's work on elementary geometry, specifically his success in isolating and minimizing the spatial intuition presupposed in geometric proofs. For,

> Durch die Erfüllung dieser Forderung machen wir uns von dem speziellen Vorstellungsbereich des Sachgebietes der Räumlichkeit los, und was wir an inhaltlicher Vorstellung benutzen, ist nur jene primitiven Art von Anschauung, welche die elementaren Formen der Zusammensetzung diskreter begrenzter Objekte zum Gegenstand hat, und welche die gemeinsame Vorbedingung für jedes exakte wissenschaftliche Denken bildet – wie dies insbesondere Hilbert in seinen neuren Untersuchungen zur Grundlegung der Mathematik hervorgehoben hat ([17b], p. 203).

For Kant geometrical figures were indispensable parts of mathematical symbolism, but existed in space whose essential *a priori* properties were its three-dimensionality, its continuity, and the measurability of its figures. Hilbert's "axiomatic investigation of their conceptual content" aimed precisely to incorporate them "into the treasure of mathematical signs" (see p. 90). But his *Streckenrechnungen* calculate directly with these discrete signs solely on the basis of finitary *Schnittpunktsätze* without any need of axioms of space, continuity, or congruence, thus freeing these "mnemonic symbols of space intuition" from Kant's three presuppositions of their spatiality.

Now we have seen that for Hilbert the arithmetical symbols, though less mnemonic than geometrical ones, had no essentially different status than they; in particular, one could also take the objects of arithmetic to be

> . . . die Zeichen selbst, deren Gestalt unabhangig von Ort und Zeit und von den besonderen Bedingungen der Herstellung des Zeichens sowie von geringfügigen Unterscheiden in der Ausführung sich von uns allgemein und sicher widererkennen lasst ([121], p. 163).

This commits him, of course, to the '*Gestalten*' of his symbols as a primitive presupposition. What about their '*Räumlichkeit*'? This question is raised by Hilbert's desire to interpret '3 > 2' by saying simply that '1 + 1 + 1' extends beyond '1 + 1'. This is Hilbert's intuitive spatial replacement for Dedekind's set-theoretic definition of order for his Peano models. However, when we write down

$$\begin{array}{l} 1 \quad + \quad 1 \\ 1 + 1 + 1 \end{array}$$

we see immediately that not every spatial interpretation will work. Indeed, according to Bernays: "*Hierzu ist zunächst zu sagen, dass die Behauptung 3 > 2 gemäss der Hilbertschen Erklärung zwar einen räumlichen, aber doch darum nicht einen* metrischen *Sinn hat*" ([17a], p. 161). Something rather like the structure of a Turing-machine tape, each cell bearing at most one stroke, seems to be needed here. The 'transcendental aesthetic',

(H_1) The content of intuition, which must include discrete spatial symbols as an extralogical a priori condition of mathematical knowledge, is always finite.

of Hilbert's '*finite Einstellung*' is the philosophical analogue of the cybernetic constraint that a Turing-machine can only scan a finite number of symbols at a time on its tape. The first to explicitly link an 'Einstellung' on spatial symbols in mathematics with the study of a computing machine, however, was Wittgenstein:

> Erinnern wir uns, dass in der Mathematik die Zeichen selbst Mathematik *machen*, nicht Mathematik beschreiben. Die mathematischen Zeichen sind ja wie die Kugeln einer Rechnenmaschine. Und die Kugeln sind im Raum, und eine Untersuchung an der Rechnenmaschine ist eine Untersuchung des Raumes ([277], p. 186).

Hilbert focused rather on the consequence of (H_1) that infinite sets cannot be given in intuition but can only be *thought* of. And insofar as we think scientifically we think axiomatically: since infinite sets cannot be given to us we can only lay down axioms we intend them to satisfy. Frege and Russell had shown how to formulate these axiomatic thoughts in symbolic formalisms which Hilbert now sought to study metamathematically by means of a primitive geometric intuition constrained by (H_1). In particular, the formalism itself can comprise only a *denumerably* infinite sequence of finite objects. According to Zermelo [278] *this is precisely what 'finitism' comes to: the requirement that an acceptable mathematical formalism contain only denumerably many formulas and theorems*. Apart from this we see that Hilbert cannot have literally meant, though he sometimes wrote as if he did, that the formulas of such a formalism are 'meaningless': they represent mathematical thoughts, even if they are only 'finite pictures' of them. What he often meant was only that certain metamathematical investigations, such as a consistency proof, must be executed without appeal to such 'meaning' as the formulas may thereby have. In Husserl's terms, a consistency proof requires an 'epoche' which 'brackets' the meanings of these formulas.[51]

Hilbert's notion of a 'meaningless symbol', however, has led to considerable misunderstanding. As applied to individual formal sentences, he generally has in mind nothing more than the Duhemian point about the impossibility of testing individual theoretical sentences in isolation from their whole theory (cf. [255], pp. 457, 470). On the other hand, he sometimes described the individual *numerals* as "meaningless in themselves", whereas we are generally encouraged to regard them as names for certain abstract entities called "numbers". But what Hilbert meant was rather that the connection between the two, say between '1' and the number one, is purely arbitrary, in the sense that this numeral has no intrinsic property in virtue of which it refers to this object, nor is in any literal sense a *representation* of it. In short, that the numerals are not 'mnemonic'. In Weyl's penetrating study of the various modes of significance of symbolism in mathematics and physics, Hilbert's idea was characterized as follows:

In der Reichen Skala von modi der Bedeutung, and die wir eben erinnerten, sind die extremen Fälle: auf der einen Seite das Zeichen, das eine getraue Reproduktion des Bezeichneten ist (oder anstrebt), auf der anderen das rein konventionelle oder gar das 'leere' Zeichen, von dem *Hilbert* versichert, dass es überhaupt nicht über sich hinausweist ([270], p. 220).

Hilbert admitted, as did Leibniz before him, that numerals name numbers, if at all, only conventionally, as opposed to the mnemonic symbolism of geometry which can 'picture' or 'indicate' geometric objects. A well-drawn circle, for example, may be plausibly said to picture an ideal one. However, in his lecture on mathematical problems, Hilbert had tempered this plausible but vague distinction between geometrical and arithmetical symbolism, we recall, by the admission that all kinds of mathematical symbolism, the mnemonic no less than the unrepresentative, required an axiomatic study of their conceptual content. Hilbert's problem here, lately discussed by Goodman, is that of elucidating how much of an *analogy* the symbol must bear to its object to represent (viz. re-present) it. Hilbert later noticed, on the other hand, that a formalism *as a whole* can be essentially analogous to the intuitive domian it formalizes, indeed, sometimes a sentence of a formal system looks exactly like the one it 'formalizes' (cf. Gentzen [97], pp. 140ff.). Thus von Neumann [195] describes a formalism as "*ein willkürlich defieniertes logische Spiel*", but then immediately adds "*das freilich mit der klassischen Mathematik weitgehend analog sein muss*" (*ibid.*, p. 257). It is these two properties of the formalism which allow one to motivate the study of 'symbols without meaning'. Von Neumann's fuller statement is very helpful if the foregoing points are kept in mind:

Allen diesen Zeichen ist, da sie zum Formalismus gehören, prinzipiell überhaupt kein Sinn zuzuschreiben. Sie bedeuten nichts, sie vertreten nichts, sie haben nicht mehr Sinn, wie etwa die Figuren auf einem Schachbrett. Aber sie stehen dennoch in einer weitgehenden Analogie zu den Begriffsbildungen der klassischen Logik und Mathematik: denn diese sollen ja durch den auf unsere Zeichen aufgebauten Formalismus ersetzt werden. Wir wollen in den nun folgenden Zeilen ... diese Analogien entwickeln, *und dabei kurz von der 'Bedeutung' der Zeichen sprechen.* Man muss sich aber dabei stets vor Augen halten: Mit dieser 'Bedeutung' der Zeichen sind immer bloss diejenigen Begriffe der klassischen Logik und Mathematik gemeint, die die Zeichen auf andere Basis zu ersetzen berufen sind; im Prinzip sind und bleiben alle unsere Zeichen sinnlos (*ibid.*, pp. 258–9, italics mine).

What emerges from formalism here is the line on meaning analogous to Duhem's on testability: the individual terms, e.g. numerals, of the formalism are generally in themselves meaningless, viz. they are not analogous in any essential way to the corresponding objects or concepts of mathematics; the sentences can be formally (viz. in *form*) analogous to their counterparts, but cannot generally be verified in isolation from the theory; the whole formal theory can bear a throughgoing analogy to the informal '*Begriffsbildungen*' and *thereby* has a 'meaning'.[52] Actually, this is not essentially different from the 'interpretation' the formalism gets in the Tarski-Carnap 'semantics', which could be described as a way of formulating, in the language of set theory, the analogies between a formalism and some branch of informal mathematics.

Hilbert realized as early as 1917 that his consistency problem was not an isolated one but was in fact intimately connected with a cluster of epistemological problems for which his proof theory would have to find mathematical formulations:

> Bei näherer Überlegung erkennen wir nämlich bald, dass die Frage der Widerspruchslosigkeit bei den ganzen Zahlen und Mengen nicht eine für sich alleinstehende ist, sondern einem grossen Bereiche schwierigster erkenntnistheoretischer Fragen von spezifische mathematischer Farbung angehört: ich nenne ... das Problem der principieler *Lösbarkeit einer jeden mathematischen Frage*, das Problem der nachtraglichen *Kontrollierbarkeit* des Resultates einer mathematischen Untersuchung, ferner die Frage nach einem *Kriterium für die Einfachheit* von mathematischen Beweisen, die Frage nach dem Verhaltnis zwischen *Inhaltlichkeit und Formalismus* in Mathematik und Logik und endlich das Problem der Entscheidbarkeit einer mathematischen Frage durch eine endliche Anzahl von Operationen ([120], p. 153).

He emphasized that the last-named question of decidability "deeply touches the essence of mathematical thought".[53] So despite his mentalism, he nevertheless became increasingly interested in a mathematical study of mathematical though itself. In fact, as we see from his list of problems, he resolved to subject Frege's fundamental problem itself to mathematical study: the relation between formalism and meaningfulness. Frege [89] had admitted that, as deep as this distinction cuts, later generations may be able to find unclarities on both sides of the argument. Hilbert's basic thesis was that this very distinction *could be studied and clarified using mathematical methods*.

To this end, Hilbert distinguishes between

(I)′ An intuitive informal mathematical theory,
(II)′ A formalization of such a theory (including sufficient logic),
(III)′ The metamathematical study of the formalization,

a distinction foreshdowed, as we have seen, in his early views on the historical stages of development through which a mathematical theory passes. As discussed above, (II)′ represents a model of the thoughts and propositions in (I)′. But some of the purposes of (III)′ can only be served by not appealing to this relation. Hilbert expresses this by saying that meaningful '*eigentliche Denken*' is transferred to the 'higher level' of (III)′, but he takes it as self-evident that this '*Inhaltilichkeit*' can never be dispensed with. These moves are needed to make *proofs themselves* the object of study, for at the level of (I)′ the numbers are the 'objects' of our study; indeed

(K_p) ". . . die Beweise der Sätze über die Zahlen *fielen schon in das gedankliche Gebiet*. Bei unsere Untersuchung ist der Beweis selbst etwas Konkretes und Aufweisbares; die inhaltlichen Überlegungen erfolgen erst an dem Beweise" ([121], p. 170, italics mine).

This is Hilbert's formalization principle which makes proofs, hitherto regarded as mental objects, concrete and surveyable in intuition, and hence finite by (H_1). This is essentially the proof-theory principle (c) formulated on p. 122. He compares his '*Beweiskritik*' to Kant's 'critique of reason'. But now the core of his conception is this: the development of the total mathematical science will take place under the impetus of a continual interplay between (II)′ and (III)′:

Auf diese Weise vollzieht sich die Entwicklung der mathematischen Gesamtwissenschaft in beständigem Wechsel auf zweierlei Art: durch Gewinnung neuer beweisbaren Formeln aus dem Axiomen mittels fomalen Schliessens und anderseits durch Hinzufügung neuerer Axiome nebst dem Nachweis der Widerspruchsfreiheit mittels inhaltlichen Schliessens.

Die Axiome und beweisbaren Sätze, d.h. die Formeln, die in diesem Wechselspiel entstehen, sind die Abbilder des Gedanken, die das übliche Verfahren der bisherigen Mathematik ausmachen, aber sie sind nicht selbst die Wahrheiten in absoluten Sinne. Als die absoluten Wahrheiten sind vielmehr die Einsichten anzusehen, die durch meiner Beweistheorie hinsichtlich der Beweisbarkeit und die Widerspruchsfreiheit jener Formalsysteme gelifert werden ([122], p. 180).

This is an elaboration of what Hilbert called his "creative principle" in 1904. We see that consistency proofs here do not simply play the static role of allaying doubts, but figure integrally in the development of mathematical knowledge. This interplay between formalism and content is just what Bell found operative in the history of mathematics generally, and represented Hilbert's attempt to formulate metamathematically the interplay between thought and experience which he saw all his life as the main impetus for advances of scientific knowledge.[54] Later formulations also speak of the relation between theory and experiment. The metamathematical interpretation of such interplay restricts thought to that which is formalized in the axiomatic theories of (II)′ and experience to the intuitive perceptual experiments on the symbols of these theories required to prove their consistency. The motivation of this for arithmetical theories was Hilbert's realization that insofar as such consistency proofs required only such experiments with symbols as were 'finitary', they would enable one to eliminate any transfinite elements from the formal proofs of finitary theorems – much as he had done for his *Schnittpunktsätze* in geometry.[55] But we must note that (II)′ is *not assumed to be complete*, with respect to (I) or anything else: it is sufficient that its extensions generated by the creative and constructive principles be consistent.

Somit ist das Begriff 'beweisbar' relativ bezüglich des zugrunde liegenden Axiomsystems zu verstehen. Dieser relativismus ist naturgemäss und notwendig; aus ihm entspringt auch keinerlei Schaden, da das Axiomsystems beständig erweitert und der formale Aufbau, unserer konstruktiven Tendenz entsprechend, immer vollständiger wird ([121], p. 169).

Hilbert also points out in the 1923 paper ([122], p. 182) that we cannot assume decidability, nor even ω-consistency, even claiming outright in the 1925 paper that a "general method for solving every mathematical problem . . . does not exist" ([255], p. 384). Thus for Hilbert the '*Kontrollierbarkeit*' he sought to impose through (H_1) and its suite does not necessarily imply mechanical decidability. Nevertheless, it was understood by formalists that a positive solution of their *Entscheidungsproblem* would raise the spectre of mechanism in mathematics, as we see from the following remarkable words of von Neumann (1927) in which he expresses his belief in the undecidability of mathematics but resolves to carry out his consistency proof despite the *de facto* undecidability of his formalism:

Es scheint also, dass es keinem Weg gibt, un das allgemeine Entscheidungskriterium dafür, ob eine gegebene Normalformel a ist beweisbar, aufzufinden. (Nachweisen können wir freilich gegenwärtig nichts. Es ist also gar kein Anhaltspunkt dafür vorhanden, wie ein solcher Unentscheidbarkeitsbeweis zu führen wäre.) Diese Ungewissheit hindert uns aber nicht daran, festzustellen: Heute ist est nicht allgemein zu entscheiden, ob irgendeine gegebene Normalformel a ... beweisbar ist oder nicht. Und die Unentscheidbarkeit ist sogar die Condition sine qua non dafür, dass es überhaupt einem Sinn habe, mit den heutigen heuristischen Methoden Mathematik zu trieben. *An dem Tage, an dem Unentscheidbarkeit aufhörte, würde auch die Mathematik im heutigem Sinne aufhören zu existieren; an ihre stelle würde eine absolut mechanische Vorschrift treten*, mit deren Hilfe jedermann von jeder gegebenen aussage entscheiden konnte, ob diese beweisen werden kann oder nicht.

Wir mussen uns also auf den Standpunkt stellen: Es ist allgemein unentscheidbar, ob eine gegebene Normalformel beweisbar ist oder nicht ([195], p. 10, italics mine).

It should now be clear, at least in a general qualitative way, how it could be that Bernays [22] could come to emphasize Gödel's theorem as a positive contribution to Hilbert's program. Specifically, as Bernays ([128], p. 211) saw early, it constitutes a remarkable contribution to Hilbert's problem of the relation between formalism and '*Inhaltlichkeit*'. In order to clarify this, we need to make a distinction between two brands of '*Inhaltlichkeit*'. First, there is Frege's, which attaches to classical (nonconstructive) mathematics; secondly, there is Brouwer's, which attaches to constructive mathematics. Now Hilbert's basic problem splits into two depending on whether the methods permitted as meaningful in studying a formalism are classical or intuitionist. (According to formalism, this is all that the distinction between 'syntax' and 'semantics' can come to for mathematical theories.) Hilbert was primarily interested in the relation between formalism and constructive content, and it is this which Gödel's discoveries clarify. The more general relation was studied and clarified by Tarski. Now to see how much of a clarification Gödel's result really was, as well as its relation to mechanism, we examine Finsler's 1926 paper ([255], p. 438) on 'formal proof and undecidability' which anticipated Gödel.

Finsler attempted to develop the basic idea in Richard's paradox, called the "paradox of finite definability" by him, into a critique of Hilbert's program.[56] He is especially anxious to short-circuit the possibility that formulas could ever replace 'logical thought', and he apparently has one eye on mechanism, for he later added, as a kind of epithet to his earlier work, that despite the value of symbols and formulas,

... es gibt aber ebensowenig denkende Formeln wie denkende Maschinen. Die Formeln stellen nur gewisse äusserlich gegebene Beziehungen zwischen der Dinge dar; die eigentlichen inneren Zusammenhänge werden von ihnen 'übersehen', und dadurch können schlimme Fehler entstehen. So kann auch ein formal widerspruchsfreies System inhaltlich völlig falsch sein ([85], p. 177).

Now to the argument which Finsler takes to establish these claims. He begins by considering a 'formal system', we will call it F, whose formulas are constructed from some fixed finite set of symbols "that may contain all signs used until now (or even in the future) for writing or printing" ([255], pp. 441–2). Note already that F does not consist merely of 'meaningless' symbols, but is in effect an entire natural language inclusive of all possible mathematical symbols. Finsler then introduces a 'dictionary', call it D, which 'interprets' F, but generally only in the sense that it specifies, for some word of F which may already have had several meanings or usages, one particular such that it shall henceforth uniquely posesss.[57] We write F_D for Finsler's interpreted unambiguous language. He also introduces over and against the formal domain a "purely conceptual" domain (*reine Gedankliche*), which in other papers is described as the domain of "absolute logic" ([82], [83]), wherein "absolute proofs" are carried out by pure "logical thought". This is essentially Dedekind's *Gedankenwelt*. In the paper now under discussion, the introduction of this domain, call it C, is motivated by reference to Cantor's notion of (absolutely) non-denumerable sets. Finsler considers the propositional form 'α is a transcendental number' and claims that "from the axiom system for the real numbers there follow, purely logically [i.e. conceptually] more than denumerably many propositions" of the given form; indeed, even non-denumerably many *truths* of this form. But a proof in a formal system must be "an array that must be given as such to our perceptual intuition", as Finsler quotes Hilbert. By (H_1), then, it must be finite, and hence the set of theorems denumerable. Therefore there must exist true propositions in C undecided by the considered axiom system. Finsler expressly notes that this reasoning leaves open the question whether the true undecidable propositions whose existence is hereby demonstrated *are actually formally representable in the system*: it could be that every proposition for which the formalism has a '*finites Bild*' is decided therein. The foregoing is his heuristic sketch of his two main results, one concerning definability, the other provability.

A crucial point in Finsler's main arguments concerns the relation between F_D, D, and C: the 'definitions' of D need not consist of symbols *of* F: "they may be thought of as given in a purely conceptual way" ([255], p. 442). Ultimately his arguments hinge on the intelligibility of this assumption. He now defines an object α to be 'finitely definable' if there is a finite string of F_D which D interprets as uniquely determining α, and then considers the sequence S of all finitely definable binary sequences. He argues (non-constructively) that S is denumerable, and then considers the 'anti-diagonal sequence' **d** of S obtained by permuting 0's and 1's on the diagonal of S. The diagonal argument now shows that **d** is not in S, hence not finitely definable by F_D. Finsler interprets this as an existence proof for non-finitely definable objects. (Remember: F_D is an all-encompassing language leaving us, so to speak, no linguistic elbow room for a metalanguage in which the 'finite' definition which the proof appears to give of **d** could be stated.) Finsler resolves the apparent paradox that we nevertheless seem to have finitely specified **d** by claiming that the diagonal definition of **d**, considered as a string of F_D, is not a definition at all, because it leads to contradiction. But, he claims:

The given definition, however, becomes unobjectionable as soon as we transfer it from the formal to the purely conceptual realm and leave the formal out of consideration (*und von Formalen abstrahiert*) ([255], p. 443).

The idea that the diagonal method leads, in certain contexts, to paradoxical results whose correct analyses require epistemological distinctions, perhaps involving thought as such, was not new with Finsler.[58] What is new and important is Finsler's transformation of the above argument into a parallel one concerning, not finite definitions, but finite *proofs*: we now consider the denumerable sequence S' of all finite proofs of F_D, determined by D, which establish, for a given binary sequence, either that 0 occurs infinitely often, or that it does not. S' determines a denumerable sequence S'' of binary sequences, namely those of which the proofs of S' hold. Now form the antidiagonal sequence $\mathbf{d}'$ of S'', and consider the proposition

(f) In $\mathbf{d}'$ the number 0 does not occur infinitely many times.

Apparently (f) is false, for there obviously are infinitely many proofs in S' that 0 does not occur infinitely often in that binary sequence all of whose terms are 1, and every such proof determines a 0 in $\mathbf{d}'$. Hence – (f) is true. Now Finsler argues, as for **d**, that the 'proof' we have just given for – (f), call it (p), although written in D, is 'not unobjectionable' as a formal proof of F, for it would lead to contradiction there by the diagonal argument. But, as above,

(C_p) ". . . the proof becomes unobjectionable as soon as we transfer it from the formal to the conceptual realm and leave the formal out of consideration" (*ibid.*, p. 444).

which may be thought of as Finsler's principle of *the conceptualization of p*, where p is any proof. Insofar as (C_p) is intelligible at all, it is best thought of as a principle opposed to Hilbert's formalization principle (K_p) introduced above (p. 145). Ultimately, Finsler's arguments are directed against the consequence of (K_p) and (H_1) that proofs are finite: the above proof (*p*) for $-$(f) was not 'unobjectionable' as a proof in *F*, i.e. as a finite proof, but (C_p), by 'transferring' it to C, gives it, so to speak, room to exist as it 'really' is, i.e. as an *infinite* proof.[59] Just as before we proved the existence of objects which could not be finitely defined, we now show the existence of propositions which cannot be finitely proved. The idea that proofs in general, especially those by mathematical induction, are really infinite objects was not new, nor even uncommon. In fact, modern proof theory contains an extensive mathematical treatment of the subject, which we discuss below. Meanwhile, the question is: How does Finsler's work compare with Gödel's? And what does it mean for mechanism? As a start, consider the following plea of Locher, which appeared in 1938:

> Immer grösser wird die Zahl derjenigen, welche glauben, dass das eigentlich Mathematische darin und nur darin bestehe, dass man aus einem Axiomsystem mit Hilfe gewisser formalisiert angebbare Regeln weiter Aussagen formal 'herleiten' könne...Konsequent weiter gedacht, muss man dann auch zugeben, dass alle Mathematik in *diesem* Sinne sich durch Maschinen verwirklichen lasse ... Allerdings findet man bezeichenderweise den Hinweis auf die Maschine in der betreffenden Literatur selten offen ausgesprachen. Oder man fluchtet sich in die Position, dass das mathematische Genie eben darin sich auslebe, besonders wirksame Automaten ausfindig zu machen.
>
> Wären die Arbeiten von P. Finsler zur Grundlegung der Mathematik einem weiteren Kreise bekannt und würde wirklich auf deren Gedanken ernstlich eingegangen, so könnte eine gesundende Wirkung von diesen auf manchen im Formalismus Steckenbeleiben ausgehen. Mag man noch so sehr das 'inhaltliche' Denken belächen, gerade die diesbezüglichen Finslerschen Abhandlungen können die heute mancherorts als unmodern taxierte Bewusstseinstatsache wieder beleuchen, dass das mathematische eben nur in denkend erlebten Ideen erfasst werden kann. Merkwürdigerweise findet sich ... der Name Finsler kaum erwähnt ([168], pp. 206–7).

So it might be better to ask: why didn't Locher rather use Gödel's work than Finsler's in warding off the mounting advocacy of mechanism in the foundational literature? Have current anti-mechanists simply found in Gödel a rigorous proof of the need for something like Finsler's (C_p) or the ubiquitous 'Mind' of Lucas, or – best of all – for Hilbert's own '*überlegende Geist*'?

A clue to these questions comes from Finsler's own comparison [83] of his work on undecidability with Gödel's. He claims, amazingly, that while he has indeed shown the existence of propositions formally undecided in a 'general sense', which are nevertheless decided by 'inhaltliche Überlegung', Gödel has not done anything of the kind. Rather, he claims, Gödel only shows that in one (very limited) formal system *Z* we can find an undecidable sentence which is nevertheless formally decided in another formal system Z'. To be sure, Finsler's discussion does fall into error. But he does realize something important about Gödel's work, namely, that it does not really depend on the meanings of the symbols of his formal system.[60] If Finsler's was the only proof we had of incompleteness (and we agreed to accept it), this would indeed pose a problem for (M): not only would every Turing machine be (believed to be) incomplete, but, more important, the essential role played in his proof by (C_p) and C would *prima facie* appear to mark just the kinds of steps that could only be negotiated by the '*inhaltliche Überlegung*' of the mind – in short, the knowledge of this incompleteness would seem to be, on the face of it, beyond a machine. Indeed, machines work only with concrete symbols, having no access to '*rein Gedankliche*'. But Gödel's work, as Finsler clearly saw, puts mechanism back in the running: precisely by showing how to carry out the incompleteness argument *effectively*, he showed that a *machine* can carry it out. I shall argue in Chapter IV that the essence of Gödel's work is nothing but the *mechanization of diagonalization*. Much as Aristotle was said to have brought Plato's forms down from heaven and put them back into matter, Gödel brought Finlser's undecidable propositions down from '*rein Gedankliche*' and put them back into formal systems where they exist as finite sequences of symbols. Since a machine can apply Gödel's purely mechanical routine of diagonalizing a formal system or machine, being able to do this to any machine no longer implies that you are not a machine.[61]

Richard's Paradox made urgent the need for some significant distinction between two senses of 'finitely definable' and Finsler was not the first to find a mentalistic formulation necessary for making it. But Gödel's rigorous technique *eliminated* the need for any such notions: one now saw the distinction one really needed in purely mathematical terms, viz. that between recursive and non-recursive functions. If ever a proof eliminated – to use Hilbert's phrase – a "detour through ideal elements" to prove real results, it was Gödel's! It was a beautiful contribution to Hilberts' idea of the interplay between formalism and meaningfulness.

6. HILBERT'S RULE AND THE PHENOMENOLOGY OF INFINITY

Although Gödel's sentences are finite, it is often suggested that we 'see their truth' by apprehending mentally the infinite proof of it which a finite-state machine cannot. In short, that our mind or thought is in some sense infinite. Hilbert took a firm stand against this, without, however, committing himself to decidability:

Unser Denken ist finit; indem wir denken, gescheht ein finiter Prozess. Diese sich von selbst bestatigende Wahrheit wird in meiner Beweistheorie gewissermassen mit benutzt in der Weise, dass, wenn irgendwo sich ein Widerspruch herausstellen würde, mit der Erkenntnis dieses Widerspruches auch zugleich die betreffende Auswahl aus den unendlich vielen Dingen verwirklicht sein musste. In meiner Beweistheorie wird demnach nicht behauptet, dass die Auffindung eines Gegenstandes unter den unendlich vielen Gegenständen stets bewirkt werden kann, wohl aber, dass man ohne Risiko eines Irrtums stets so tun kann, als wäre die Auswahl getroffen ([122], p. 187).

Let us formulate the premise of this explanation as the further thesis:

(H_2) Our mind and its contents are finite; all thinking is a finite process.

Hilbert's intention in (H_2) was presumably that the number of states our mind can assume is finite. Whereas he thought of (H_1) as playing a role analogous to the transcendental aesthetic, we can think of (H_2) as playing for him that of the transcendental analytic: what is essential about the 'form' or 'categories' of the understanding is that they are finite.[62] But even so interpreted in automata language, (H_1) and (H_2) do not entail mechanism: still missing is a 'program' that would impose recursiveness on behavior. In any event, (H_1) and (H_2) do entail that proofs, whether made concrete in intuition by (K_p) or left to languish in the mind, are finite.[63] In his papers of the early 1920's Hilbert didn't even think it necessary to point this out, so obvious did it seem to him. Only "inanities" made explicit stipulation necessary:

If we pay close attention, we find that the literature of mathematics is replete with absurdities and inanities, which can usually be blamed on the infinite. So, for example, some stress the stipulation, as a kind of restrictive condition, that, if mathematics is to be rigorous, only a *finite* number of inferences is admissible in a proof – as if anyone had ever succeeded in carrying out an infinite number of them! ([255], p. 370).

Now, as this and the previously quoted text clearly show, Hilbert's notion of 'finite', as in '*finite Einstellung*', does not directly involve any notion of decidability, but rather only its basic literal meaning.[64] Rather than try to define this directly, which would undoubtedly go in circles – finite is too basic a notion – we had best glean it from his evolving conception of formalized inductive proof.

We have already seen the implicit pressure Finsler tried to apply to (K_p) and (H_1). In the following year, Brouwer, in expounding his theory of proofs as transfinite, well-ordered structures, made the criticism explicit:

These *mental* mathematical proofs that in general contain infinitely many terms must not be confused with their linguistic accompaniments, which are finite and necessarily inadequate, hence do not belong to mathematics.

The preceeding remark contains my main argument against the claims of Hilbert's metamathematics ([255], p. 460).

Three years later Hilbert [124] introduced so-called 'infinite induction' into his proof theory, by adding to his formal system for number theory a new rule of inference which allows one to infer a universal formula $(x)A(x)$ if it has been proved that the formula $A(\bar{q})$ is correct for each particular numeral $\bar{q}$.[65] Such a rule is generally regarded as having, in effect, infinitely many premises, yet Hilbert is emphatic that it is "in any event finite". Carnap [45], who calls it "Hilbert's rule", questioned the reasonableness of calling the rule 'finite', siding with Tarski [246] that the rule "obviously" has an "infinitistic character". But Hilbert, commenting on a still stronger form of the rule, which he introduces in a group entitled 'transfinite' axioms and rules, had other ideas in mind:

... die sämtlichen Axiome und Schlussschemata ... die ich transfinit genannt habe, doch ihrerseits streng finiten Charakter haben: die in ihnen enthaltenen Vorschriften sind in Endlichen ausführbar. Wenn ferner in einer Formel ausser Ziffern das Zeichen η steht, so ist zwar keine Rede davon, dass wir entscheiden können, ob die betreffende Aussage rightig oder falsch ist ... ([125], p. 121).

These words make it quite clear that for Hilbert a 'finite' rule and one whose correct applicability is always mechanically decidable are not in general the same thing: his '*Einstellung*' focusses resolutely on the practical aspects of the rule, namely that any application can only occur via a description of the 'infinitely many' premises and a constructive proof that they satisfy the conditions of the rule, which both are *eo ipso 'in Endlichen aüsfuhrbar'*. His attitude is essentially that of Wittgenstein: "*Wir kennen die Unendlichkeit aus der Beschreibung. Nun, dann gibt es eben nur diese Beschreibung und nichts sonst*" ([277], p. 155).

The introduction of Hilbert's new rule implies that we are no longer dealing with a formal system in the strict sense, i.e. one whose set of theorems is recursively enumerable. However, the finiteness of any individual application of the rule is in no way impugned by the circumstance that each such application cannot be found in 'exactly the same way'. (Remember that Hilbert had already denied that mathematics was decidable.) In short, it does not follow that the '*Kontrollierbarkeit*' of any particular proof suffers with the introduction of this rule. Still there is a tendency to regard the ordinary formalization of induction, where one takes the wff

$$[A(0) \wedge (x)(A(x) \rightarrow A(x+1))] \rightarrow (x)A(x)$$

as an axiom for each formula $A(x)$, as more 'mechanical' than the rule of infinite induction. But this is to some degree an illusion. True, the set of all induction axioms, as specified above, will be recursive: but these axioms are of no *use* to us unless we can prove their antecedents. And the question, given $A(x)$, whether the antecedent of its induction axiom is provable, is recursively undecidable, just as is the question whether it can be proved to hold of each numeral in order to apply 'infinite' induction to it. If one already thinks mathematics is undecidable, as Hilbert did, the apparent advantages of the ordinary formalization dwindle (cf. pp.208–9).

Hilbert's incorporation of infinite induction into the '*finite Einstellung*', marks an intensification of the interplay envisaged by Hilbert between mathematics and metamathematics. Its use, in what is now called a semi-formalism, requires in general a metamathematical proof of the premise of Hilbert's rule. Thus it has been claimed by Rosenbloom that it is precisely the kind of rule which could only be applied by 'brains', but not by machines:

> We may consider intelligence as the capacity for introspection, the faculty of thinking about one's own methods of reasoning . . . In mathematical terms this means the capacity of using a syntax language for reasoning about an object language. It is in this way that a brain can make use of such rules as [Hilbert's rule]. It seems altogether feasible to incorporate this idea into a mathematical definition of a brain and prove that a brain can solve some problems which a machine cannot ([222], p. 208).

Raggio [215] goes a little futher and emphasizes that the "enormous" difference between a formal and semi-formal system requires the intervention of '*einsichtiges Denken*' and '*einsichtige metatheoretische Reflexion*' for the operation of the latter. Freudenthal [92] claims that Hilbert's rule "would be of no use to the machine, for if the number of constant subjects is infinite, the machine can never complete the task of verifying $F(a)$" (p. 101). Now we know that no machine can solve the general decision problem – a fact which we argued in Chapter I supported mechanism.

The question is, does Hilbert's rule really bring anything essentially new into play beyond a machine? No: due to a result of Shoenfield [232], the 'lawfulness' which is required of the metamathematical description of an infinite proof (see below) can be assumed without loss of generality to be always *recursive*, in this sense: if $(x)A(x)$ is the sentence to be proved by infinite induction, then it is sufficient to require that there be a recursive function $f(n)$ such that for each number n, $f(n)$ is the Gödel-number of a formal proof of $A(\bar{n})$.[66] This shows that one would never lose any true sentences of number theory if, whenever their verification of $(x)A(x)$ required an infinite induction, that induction were left to a Turing machine M_A which for an input of $\bar{n}$ would give a proof of $A(\bar{n})$ as output.[67]

In a fundamental paper of proof theory, Schütte argued for the compatibility of Hilbert's '*Einstellung*' (H_2) with his rule as follows:

> Auf Grund der Endlichkeit unseres Denkens enthält jeder mathematische Beweis nur eine endliche Anzahl von Schlüssen. Diese Tatsache kommt in den Kodifikaten, welche nur endliche Herleitungen berücksichtigen, zum Ausdruck.
>
> Es gibt jedoch Gründe, welche auch für die Heranziehung von Kodifikaten mit unendlichen Herleitungen sprechen, und zwar besonders für die Verwendung der 'unendlichen Induktion' ... Aber solche Kodifikate lassen sich mit der Endlichkeit unseres Denkens in Einklang bringen. Es ist nämlich nur zu verlangen dass die unendlichen Herleitungen immer in bestimmter Weise gesetzmässig aufgebaut sind, so dass sie sich auf Grund dieser Gesetzmässigkeit eindeutig beschreiben lassen, Die Endlichkeit unseres Denkens kommt dann zwar nicht in die Herleitung selbst, wohl aber in der metamathematische Erfassung der Herleitung zum Ausdruck ([229], p. 369).

Schütte's words not only prefaced a basic technical breakthrough for Hilbert's program, shedding new light on the nature of proof-theoretic consistency proofs, they also sum up the philosophical position in a peculiarly direct manner. Building on the work of Gentzen, Schütte developed the theory of cut-elimination on infinite proof trees to such a remarkable level of efficiency and elegance that Kreisel was moved to embrace infinite thoughts:

> There is the old and familiar idea, or: idealization, which regards *thought* and, in particular, a *proof* of a general proposition as an infinite object. Could one fail to take the idea seriously, at least in the sense that infinite objects are better *representations* of proofs than the words we use to communicate proofs, after having seen the wonderfully efficient and elegant theory of infinite proof figures? ([155], p. 203).

Kreisel also warns us that:

> When studying an intensional object such as a finitist proof one must be prepared to find that the differences between thoughts and words are essential (*ibid.*, p. 238).

He sketches a finitist proof for the theorem $2^n > n$, and then by way of motivating this essential 'difference', he wants to raise the "empirical question (possibly difficult . . . but quite meaningful)" of whether particular people who execute the finitist proof actually *think* of it as finite, or rather, say, as an infinite object. But it is neither clear that such inquiry could ever isolate 'thoughts' over and against the words which 'merely express them', nor that it is particularly relevant how we 'think of them'. Hilbert certainly agreed that we think of many things *as if* they were infinite. Kreisel exhorts us to "take seriously . . . what is counted as experience in traditional mental philosophy" (*ibid.*, p. 205), but Brentano, who took traditional mental philosophy very seriously, argued that the very notion of the actual infinite was self-contradictory, while Descartes, the very father of our mentalist tradition, took the finitude of thought as self-evident. Moreover, we have seen that Dedekind was unable to prove even the infinity of the whole set of his thoughts, let alone the infinity of a single thought. A thought no more becomes infinite by being *about* something supposedly infinite than it becomes triangular when it is about triangles. Indeed, traditional mental philosophy is summed up by Brentano's thesis of intentionality which tries to say just what it is about thoughts which sets them apart from, say, the physical properties of words expressing them. And the mark of distinction it finds in thoughts is just that they are 'about' objects. If all that we can add here is that they are not 'strictly finite' when they are about inductive proofs, then the claim that this refutes mechanism is a perfect *non sequitur*: Turing machines quite obviously also carry out such proofs. The notion of infinity does seem to have some plausible connections with that of intentionality; but they do not seem to be connections by virtue of which Brentano's thesis would entail the denial of mechanism.

It is very interesting to note that the writers on foundations who expected, and in some cases claimed, the completeness and decidability of arithmetic were not the formalists, but rather the phenomenologists such as Husserl, Becker, and Kaufmann. (And, of course, Wittgenstein.) Yet obviously none of them thereby regarded mechanism as one whit more plausible: they all held firmly to Brentano's thesis, which explicitly intends to contradict physicalism. Assuming that Turing machines can only be realized physically, they undoubtedly took Brentano's thesis to entail the denial of mechanism as well. So the eventual completeness and decidability of mathematical theories were never, in the minds of those who took traditional mental philosophy seriously, the sort of thing which would have *supported* mechanism.

The case of Husserl is especially instructive. Husserl's repentance from his early psychologism is often attributed to criticism by Frege, but the development of his mature logico-mathematical doctrines seems to stem from his having taken Hilbert's side in his controversy with Frege over geometry.[68] We recall that metatheoretic questions had little place in

Frege, whereas Hilbert's formulations were designed expressly to raise such questions. Husserl became especially interested in Hilbert's completeness problem for axiomatic theories, in particular his 'axiom of completeness', and proceeded to develop a whole theory of the subject himself which culminated in his notion of a 'definite manifold' or 'mathematical manifold in the pregnant sense of the term' (cf. [135], p. 204). From his explanations of the significance of this concept there emerges a program that is essentially like the orthodox interpretation of Hilbert: whereas the latter was trying, wherever possible, to eliminate 'detours through ideal elements' to prove 'real' theorems, Husserl sought to show that by 'adding imaginary elements' to the domain of a complete theory one could take a '*Durchgang durch das Imaginare*' to prove theorems about the primary objects of the theory which would signify a great 'technical advantage' but have no ontological significance.[69] He also claimed that Hilbert, while not "being guided by the philosophical considerations that determined my studies", was nevertheless guided by mathematical motives which "tended essentially in the same direction as those which determined the concept of the definite multiplicity [manifold]" ([136], pp. 96ff.). Husserl regarded these 'philosophical considerations' as ultimately grounded in intentionality:

Just what is it that, *purely formally*, characterizes a self-contained system of axioms as '*definite*', as a system by which actually a *'multiplicity' in the pregnant sense* would be defined? For, as I recognized, the intention of this concept includes a hidden intentional sense. Multiplicity meant properly the *form-idea of an infinite object-province for which there exists the unity of a theoretical explanation* or, in other words, the unity of a *nomological science*. The form-ideas 'theoretically explainable province' (province of a deductive science) and 'definite system of axioms' are equivalent.

It should be noted here that, to be sure, *every* formally defined system of axioms has an infinity of deducible consequences. But the idea of a '*nomological science*', or correlatively the idea of an *infinite province* ... governable by an explanatory nomology, includes the idea that there is no truth about such a province that is not deducibly included in the 'fundamental laws' of the corresponding nomological science ... ([136], pp. 95–6).

Again we meet the connection between intentionality and infinity. Elsewhere he emphasizes that the complete system has only a finite number of axioms, which leads to a certain analogy with Brantano's description of intentionalities in terms of 'intentional inexistence' in the mind: all the truths about a definite manifold (infinite object-province) already exist intentionally (are 'deducibly included') in the finite set of axioms. (Recall that for Hilbert axioms formulate precisely what we can clearly think of.)

So intentionality here appears under the aspect of a *reduction* of the infinite to a 'finite nomology'. As to which domains of inquiry were actually definite manifolds, Husserl claimed that they included those of elementary arithmetic, the theory of rational numbers, the theory of real numbers and Euclidean geometry. In [137], pp. 442ff. he offers 'completeness proofs' for these domains, which we will not go into, except to say that they border on the assumption that these domains are, to borrow a phrase of Wang, "complete by intention".

Husserl also became interested in the mathematical structure of consciousness itself, and raised the question of whether it comprised a concrete, mathematically definite manifold for phenomenology, in the way that space did for geometry:

> We can also set our question in these words: Is the stream of consiousness a genuine mathematical manifold? Does it in its actuality ... resemble physical nature, which indeed, if the ideal which leads to the physical in the last resort is valid and is grasped through rigorous concepts, we should have to designate as a concrete definite manifold? ([135], p. 206).

Though Husserl did not answer this completeness question for his phenomenology, he did sketch a theory, based on his idea of *Reflexionstufe*, of the formal structure of intentional consciousness. The essential mark he found in these reflexions is their hierarchical stratification into 'levels', each with its 'characteristic' or index, so that "to every level there inwardly belong certain possible reflections". A noetic reflection may be directed upon indefinitely many reflexions (as noematic objects) of a lower level which may themselves (as noetic acts) be directed upon other acts as objects, and so on. Husserl's theory bears some resemblance to Cantor's theory of transfinite ordinals, enough in fact to tempt O. Becker [7] into seeking a phenomenological foundation for Cantor's theory in Husserl's. He tried specifically to develop the latter's theory of reflexions to the point where it would provide a '*phänomenale Unterbauung*' for Cantor's second number class. This would provide a phenomenology of Cantor's actual infinite in the sense of exhibiting it "*in der Form des Prozesses . . . als ein echte Phänomen des reinen formalen Bewusstseins und auch sogar des konkreten historischen Daseins*" (*ibid.*, p. 2).

First, constructive phenomonological analogues must be found for Cantor's two 'productive principles' (successor and limit) for his transfinite ordinals which, concretely taken, represent "nothing but the transfinite iteration of an intention". Transfinite reflexion structures are identified with *self-consciousness*. Corresponding to Cantor's successor principle we have the iteration of a reflexion, while to Cantor's limit principle there corresponds Husserl's '*Horizontphänomen*', which Becker elucidates thus:

Nach einer grundlegenden Erkenntnis Husserls ist es gerade das Wesen des endlosen 'offen Horizontes' des 'und so weiter', das unendlich viele Glieder 'vorliegen', die aber mit einem Blick überschaut werden. Die Beherrschung des Unendlichen durch einem endlichen 'Gedanken', das ist der Sinn jedes Horizontes ([7], p. 107).

Becker employs a geometrical symbolism for his reflexion structures which he claims to be entirely perceptual, but this raises the problem of how such symbolism, ensnared by Hilbert's (H_1), can serve the *Horizontphänomen*:

In anderer Hinsicht muss aber zugestanden werden, dass nur eine endliche Komplikation in wörtlichen, unsymbolischen Sinn im Bilde 'vorhanden' ist, während die Verwicklung der Intentionalität transfinit ist (*ibid.*, p. 101).

To relieve pressure from the finiteness principle (H_2), Becker distinguishes between "*die Struktur der Betrachtung über die Ineinanderschachtelung der Intentionalitäten*" and "*die Struktur dieser Ineinanderschachtelung selber*". The former is finite, while the latter, in general, is of 'transfinite complication". This is analogous to Schütte's distinction between the *description* of an infinite proof, and that proof itself. The '*Verwicklung der Intentionalität*' is roughly analogous to the transfinite recursive progressions of theories studied by Turing and Feferman, and Becker emphasized that no mechanical laws could ever guide us through such a development, as the modern theory of constructive ordinals tends to confirm. So Becker's answer to Husserl's question is negative: consciousness is not a definite manifold, but is rather indefinitely open to new progressions of reflexion whose laws of development can never be mathematically fixed.

Becker nevertheless thought he could establish the *consistency* of Cantor's theory because constructions (fulfilled intentions) are *eo ipso* consistent. This was important also because of the applications of ordinals in Hilbert's consistency proofs, which required by their very nature that such ordinals possess '*Zugänglichkeit*' and '*phanomenalität*', and he explicitly included ε_0 among the 'konkret' such ordinals.[70] Later, after Gentzen's consistency proof had appeared, he thus had no trouble in interpreting it phenomenologically:

Die Reflexionsstufen wurden im vorigen zumeist von der noetischen Seite, als ineinandergeschachtelte Akte, beschrieben; entsprechend lassen sich aber offenbar auch die zugehörigen noematischen (gegenständlichen) Strukturen nach der Art ihrer Komplikation darstellen.

> Ein solches 'Noema' oder mentales Objekt bezeichnet nun auch jeder symbolische mathematische Ausdruck; auch zum Sinn der rein formal aufgefassten, 'nichts bedeutenden' komplexen Zeichen selbst gehört bereits eine gewisse Art noematischer Intentionalität. Auch die 'metamathematische' Betrachtung der Beweistheorie wird daher ein transfinite Anzahl von Komplikationsstufen, d.h. noematischen Intentionen, an den ins Auge gefassten Symbol-Komplexen zu unterscheiden haben; in der Tat hat *Gentzen* dies auch getan ([9], p. 387).

Gentzen himself had already tried to justify the finiteness of every descent from ε_0 on which his consistency proof rests by an appeal to its *visualizability*, though he admitted that for ordinals approaching ε_0 it became only "remotely" visualizable, and hence that the "*precise finitist sense*" of transfinite induction was not in general "perspicuous". But, he argued, as we progress up through the construction of the ordinals to ε_0, "nothing new is *basically* added" ([97], p. 196). In any case, no one has seriously challenged Gentzen's claim that his proof is finitist, and many even agree with him. Gödel's second theorem, in particular, is powerless to refute this claim by itself.[71] Finally, even if Becker's phenomenological reflections add nothing to the force of Gentzen's proof, they certainly undermine Weyl's claim that Hilbert's consistency proofs would, if successful, consistitute a decisive refutation of mathematical phenomenology.

Meanwhile, Kaufman [143], starting from Husserl's theory of intentionality, developed a completeness proof for number theory in more detail, based on an intentional analysis of the '*Zahlprozesses*' which results in the claim that Peano's postulates "leave nothing open" about them. Mathematically, of course, it was based on Dedekind's classic isomorphism result for Peano models. But he uncritically assumed that this result carries over to *first-order* theories of these models. The essence of the phenomenological position was almost simultaneously getting its most succinct expression from Wittgenstein:

> Die Mathematik kann nicht unvollständig sein; wie ein *Sinn* nicht unvollständig sein kann. Was ich verstehen kann, muss ich ganz verstehen. Das hängt damit zusammen, dass meine Sprache, so wie sie ist, in Ordnung ist und dass die logische Analyse, um zu vollkommener Klarheit zu gelangen, nichts zu dem vorhandenen Sinn meiner Sätze dazufugen muss. So dass der unklarst scheinende Satz nach der Analyse seinen bisherigen Inhalt unberührt behält und nur seine Grammatik geklärt wird ([277], p. 188).

Kaufmann also believed that his completeness result would "in a short time" be extended to a proof of decidability, but already claimed to have shown the impossibility of ever proving the unsolvability of the decision problem! He attributes the "prevailing doubt" among mathematicians (cf. Hilbert [255]. p. 384, von Neumann [195], Herbrand [114], Bernays

[18]) that such completeness results would actually lead to decidability to their taking too seriously oppositions like that developed in medieval theology between the finiteness of the human mind and the infinity of God, which it cannot grasp. Only now the role of the latter is taken by the natural numbers. It is only this lingering picture which could sustain the "nonsense" that relations could exist among numbers that could not be decided by a finite number of steps.

Hilbert had long harbored a lingering doubt that an elementary logic could ever be found to extract all the truths from a categorical set of axioms. But as the experimental completeness of his formulations of first-order logic became more and more evident, his doubt took the form of wondering whether Dedekind's isomorphism result could actually be so 'finitistically transformed', i.e. whether this logic could logically decide every sentence of formal number theory from his axiom system, and he presented the question, along with several other research problems in mathematical logic, to the 1928 mathematical congress:

> Die Vollständigkeit des Axiomsystems für die Zahlentheorie ... wird zwar allgemein behauptet; aber die übliche Überlegung mit der man zeigt, dass je zwei Realisierungen des Axiomsystems der Zahlentheorie isomorph sein müssen, genügt nicht den Anforderungen finiter Strenge.
>
> Es kommt darauf an, für die Zahlentheorie, deren Bereich sich präzise abgrenzen lässt, den üblichen Beweis der Isomorphie finit umgestalten, so dass dadurch folgendes gezeigt wird.
>
> Wenn für einem Satz *S* die Widerspruchsfreiheit mit den Axiomen der Zahlentheorie nachgeweisen werden kann, so kann nicht auch für *S* ... die Widerspruchsfreiheit mit jenen Axiomen nachgeweisen werden. Und damit in engsten Zusammenhang auch: wenn eine Aussage widerspruchsfrei ist, so ist sie auch beweisbar.
>
> Die Aufgabe ist zunächst hier nur für den Bereich der ganzen Zahlen gestellt worden; *sie ist dann sicher keine philosophische, sondern eine rein mathematische und kann nur durch rein mathematische Mittel gelöst werden* ([123], p. 139, italics mine).

The problem of completeness, which had its historical roots in certain philosophical views about the mathematical mind creating its own objects, was thus finally transformed by *Beweistheorie* into a purely mathematical problem. In fact, his metamathematical problems had often been criticized as too philosophical and he took this occasion to emphasize the mathematical clarity and *solvability in principle* of each of his new problems, exactly as he had done twenty-eight years earlier. And within two years Gödel settled Hilbert's problem about transforming classical isomorphism by one of those extremely complex impossibility proofs of the very kind which had convinced Hilbert of (S), namely, one which depended on the introduction of new ideas and techniques. Hilbert had meanwhile introduced his 'infinite rule' to deal with criticisms that (Finsler's?) had been made against his theory on the score of completeness, but he observed as he had before in 1922 that completeness is not as important as consistency and is best approximated by the creative principle:

Es wird meiner Theorie zum Vorwurf gemacht, dass die Sätze zwar Widerspruchsfrei seien, aber damit noch nicht beweisen wären. Freilich sind sie beweisbar, wie ich hier in einfachen Fälle gezeigt habe. Es stellt sich auch allgemein heraus, wie es von Anfang an meine Überzeugung war, dass die Erzielung der Widerspruchsfreiheit das Wesentliche in der Beweistheorie ist und die Frage der Beweisbarkeit bei eventueller sachgemässer Ausdehnung der Festsetzungen unter Wahrung des finiten Charakters sich dann ebenfalls mit erledigt ([124], p. 429).

Hilbert and Bernays later emphasized that the experimental completeness of their formalisms was quite sufficient for the purposes of *Beweistheorie*, namely, the '*Sicherung*' of mathematical theories which play a role in the sciences and that these purposes in no way require their "*vollen deduktiven Geschlossenheit*". Indeed, the incompleteness was *very much* in accord with the philosophy of mathematics which lay behind these purposes:

Auch verträgt sich unsere Auffassung, dass es sich bei dem inhaltlichen Standpunkt, welcher der Ausbildung des Systems der Analysis und Mengenlehre zugrunde liegt, um eine extrapolierende Ideenbildung handelt, sehr wohl mit der deduktiven Unabgeschlossenheit dieses Systems: die schlussweisen in dem System sind orientiert nach der Vorstellung von einer geschlossenen, völlig bestimmten Tatsächlichkeit und bringen diese Vorstellung formal zum Ausdruck; daraus folgt aber nicht, dass die aus den schlussweisen resultierende *deduktive* (metamathematisch zu konstatierende) *Struktur* jene Eigenschaft der völligen Geschlossenheit besitzen muss. Eine gewisse Art von methodischer Geschlossenheit besitzt sie jedenfalls, nämlich diejenige, auf Grund deren wir bei den üblichen Schlussweisen der Analysis und Mengenlehre sozusagen von selbst im Bereich dieses Formalismus verbleiben, und diese Art der Geschlossenheit ist auch für den Zweck, dem der deduktive Formalismus dient, ganz Ausreichend ([130], p. 289).

Indeed, since mathematical *language* has replaced mathematical thought in this scheme, to demand that its deductive apparatus have the same structure imagined for this '*bestimmten Tatsächlichkeit*' would be analogous to the old fallacy that thought should somehow take on the properties of the objects it is about. After all, this '*Tatsächlichkeit*' was the result, according to Hilbert and Bernays, of mathematical thought distinguished precisely by its tendency to *idealize* the world delivered to it by sensation.[72] Moreover, the formalist transfer of mathematical thought per se to its new metamathematical office is shown by incompleteness to be entirely appropriate, for as I have argued above, the '*Tatsächlichkeit*' on which we base the truth of something formally unprovable is really the formalism itself, and not the idealized structure which admittedly inspired it. In fact, this new office was felt by many opponents of this newer formalism to be unworthy of the dignity of mathematical thought precisely because its problems of completeness, decidability, and consistency seemed mathematically trivial. But the work of Gödel and Cohen has shown us that the question of whether even a single sentence is formally derivable in one of these mechanical little formalisms can indeed be as difficult as any mathematical problem can be, and require all the creativity and novel ideas that a mathematician can muster for its solution. This is just what such otherwise radically opposed critics of formalism as Frege and Brouwer could not imagine. In particular, intuitionism, according to Becker, denied the possibility of *scientific knowledge* of anything like incompleteness or undecidability:

Der intuitionistische Standpunkt der Entscheidbarkeitsfrage gegenüber ist nun der, dass, so fern eine arithmetische Frage zur Zeit tatsächlich unentscheiden ist, sich wissenschaftlich über die Möglichkeit oder Unmöglichkeit, jemals zu ihrer Entscheidung zu gelangen, gar nichts aussagen lasst ([7], p. 12).

This was not, of course, an authorized interpretation of intuitionsim, but I believe that it has much to recommend it. The principal such knowledge we now possess, the mass phenomena of undecidability, depends on Church's thesis, which is indeed denied by many intuitionists. As we shall see, they cannot accept the mechanization of effectiveness that it implies.

NOTES

[1] See Kronecker [158], where he quotes approvingly the passage from Gauss's letter to Bessel from which we have taken (D0) and (CM).

[2] See [158], pp. 352ff.

[3] See Hilbert [124] for his reminiscences; also Reid [217].

[4] See Reid [217], pp. 31ff.

[5] Indeed, as Bernays has recollected, Hilbert had internalized this Cantor-Kronecker conflict in his *own* beliefs about mathematics, and created his *Beweistheorie* as a means of clarifying and resolving it, which was apparently part of a larger conflict:

In addition, two other motives were in opposition to each other — both strong tendencies in Hilbert's way of thinking. On one side, he was convinced of the soundness of existing mathematics; on the other side he had — philosophically — a strong scepticism ([217], p. 173).

See Lodge [171] for the sceptical 'fascination' which discontinuity was exercising over "philosophic mathematical thought" as well as physics by 1910.

[6] See Klein [151], p. 185; also Contro [56]. Hilbert was also interested in what one could

accomplish with *just* continuity assumptions, i.e. topologically. In this connection he showed how, in the case of the plane, to eliminate the differentiability assumptions from the classical Helmholtz-Lie *Raumproblem*. Another motive of Hilbert's geometrical work, the determination of the role of spatial intuition and its relation to continuity, is discussed in the next section.

[7] See Stein [238a], pp. 279ff for a lucid discussion of the meaning of Hertz's famous slogan.

[8] The difficulties in applying the concepts of congruence, length, form, and direction at this level pointed out by Schrödinger [228a] have been recognized as all but insurmountable in recent times.

[9] He does not return to them in his two papers on the foundations of geometry published after termination of his correspondence with Hilbert. It is somewhat disappointing that his recently published correspondence [90a] contains no further letters with Hilbert that might throw any light on these matters. Kambartel [140a] contains an interesting criticism of Carnap's [46] attempt to extend the isomorphism thesis to all of language, but he does not, in my opinion, really face the essential issues that divided Hilbert and Frege, namely (i) *Are* the axioms of euclidean geometry really all true? True of *what*? Frege talked half-heartedly of *Raumanschauung*, but as we shall see Hilbert took this intuition *far* more seriously than Frege! (ii) Just what *are* the means available for proving their consistency? (iii) How can we account for their *application* to the physical world? See Resnick [218] for a useful discussion of (i) and (ii).

[10] See Essler [76a] for some interesting remarks on this. It must be remembered though that the first edition of Hilbert's axioms was not univocal. It is possible indeed that those of his second edition [117b] were completed to meet objections like Frege's, but we shall see in any case that Hilbert also had independent motivation for doing so, namely to complete the independence proof for the parallel postulate.

[11] See Steiner [239] for a discussion of this example.

[12] Klein's model did satisfy this, but he did not yet formulate the continuity axioms appropriately.

[13] J. Bolyai [27a], p. 273. See also Stäckel's discussion on p. 187ff.

[14] See Dehn [66], pp. 190ff and Dehn [66b], pp. 77–78.

[15] See Bernays [22a] and Freudenthal [92a] for discussion and modern formulations of this axiom. Freudenthal's paper is a must for understanding Hilbert's geometry.

[16] There is no doubt that Hilbert underestimated at the time the difficulty in proving their consistency.

[17] Veblen [256a] formulates the notion of logical consequence very clearly but contains no explicit and independent notion of a 'finite logical derivation'. It is here that the term 'categorical' first appears, attributed to Dewey.

[18] First-order logic first begins to emerge in Weyl's habilitation [264], written under Hilbert, which tried to clarify the notion of 'definite' in Zermelo's axiomatization of set-theory. See Tharp [247] for a modern discussion of the distinguished role of first-order logic, and Goldfarb [102a] for its emergence in the twenties.

[19] This provides the paradigm for Henkin's completeness proof for first order logic.

[20] By Dreyfus [71]; but as we shall see, there is more to this example than meets the eye.

[21] Both Hilbert [117a] and Dehn [66] cited Henry Saville's remark on these stains. See Hameister [109] for Hilbert's 'purely geometric' theory of proportion, and Hilbert [120]

for doubts about the archimedean axiom in physical geometry. See also Bernays [23].
[22] Indeed, Euclid never states any space axioms at all – even in his *stereometrical* books: they fallaciously purport to prove our two space axioms solely from his plane axioms. So in *these* books he surely did use his space intuition! But did he in the plane?
[23] A porism attributed to Euclid does come close; see [76b], p. 11. It is clear that such a theorem is only likely to be discovered by someone who is *literally* using their spatial intuition, in Hilbert's sense. It is simply too opaque in the plane.
[24] See Rautenberg [216] for a thorough modern discussion of *Schnittpunktsätze* and their relation to Tarski's model theory of elementary geometry. The cornerstone in the formal reconstruction of the algebraic semantics of a formal geometry in the formalism itself, required for proving its completeness, is the formal proof of the commutativity of its geometrically simulated multiplication – and this is essentially Pascal's theorem. See Schwabhäuser [230] for details of the euclidean case.
[25] Part (i) is due to H. A. Schmidt; part (ii) is due to Hilbert, part (iii) to Hilbert and Bernays. ($\mathscr{R}$) leads to many insights into the theory of plane area and the *geometric* meaning of the pythagorean theorem. See [117], Appendix II and Supplement V.
[26] See Reid [217], p. 13 for some interesting background on this point. Meschkowski [183] is one of the few writers who have appreciated this aspect of Hilbert's ideas on solvability.
[27] Mehlberg [182] for example.
[28] Dehn [66a] also proved Hilbert's conjecture that, although one could use the parallel postulate to prove the existence of rectangles and similar figures of different size without continuity, one could *not* recover Euclid's postulate from such existence assumptions without the archimedean axiom. Both saw this as justifying Euclid's choice of his postulate over these and other seemingly simpler alternatives adopted by later geometers, for a purely geometric theory of proportion cannot be based on them without the archimedean axiom. It also showed how uncritical it was to speak of these alternatives as 'equivalent' to Euclid's postulate. See also Hilbert [117b] and Bernays [23].

Dehn also showed that although Legendre's homogeneity theorem could be proved without continuity, the standard correlations between the three classical geometries and the three possibilities for the angle-sum of a triangle all fail without continuity, revealing a rich variety of new geometries. For example, Dehn constructed a 'semi-euclidean' geometry in which rectangles and similar figures exist, but in which infinitely many lines through a point are parallel to a given line and in which all lines are of finite length. These results have led to the realization that continuity had no place, as it had for Bolyai, in 'absolute geometry'. (He and Lobachevski had needed it to introduce hyperbolic parallels, but Hilbert showed how to eliminate it even here.) One very interesting result of modern absolute geometry is that the 'metric euclidean' planes (i.e. those in which rectangles exist) *are the only finite metric planes* (i.e. containing only a finite number of points). This shows how uncritical was the belief of Poincaré and others that no *a priori* purely mathematical case could be made for the uniqueness of euclidean geometry: it is the only metric geometry which can be made to satisfy the strictest finitist scruples.
[29] Indeed, such a replacement seemed utterly necessary in this case:
Die methodischen Schwierigkeiten der Analysis, auf Grund deren man in dieser Wissenschaft genötig ist, über den Rahmen des konkret Vorstellbaren hinauszugehen, rühren davon her, dass die Stetigkeit und das Unendliche hier eine wesentliche Rolle

spielen. Dieser Umstand würde, auch für den Nachweis der Widerspruchsfreiheit der Analysis ein unüberwindliches Hindernis bilden, wenn dieser Nachweis in dem Sinne geführt werden müsste dass man zeigt: ein System von Dingen wie es die Analysis annimmut — etwa das System aller endlichen oder unendlichen Mengen von ganzen Zahlen — ist logisch möglich (Bernays [17], p. 97).

[30] This plan is carried out in great detail in the second volume of Hilbert-Bernays [130], pp. 263–90.
[31] Hilbert does not actually give this principle a name.
[32] (H6) was not explicitly formulated until Hilbert returned to foundational research after the war. See Ackermann [2].
[33] This theme first occurred to Skolem in 1915, who later attempted to work it out in several well-known papers in the 1920's and 1930's. I shall be discussing some problems connected with it later on.
[34] See Hilbert [128], p. 422.
[35] See Bernays [18], pp. 360ff.
[36] Gentzen himself etablished a close connection between the Hilbert-Bernays axioms for logic and one of his systems of logical rules. See [97], pp. 116ff.
[37] See Poincaré [206], pp. 51ff, especially p. 57. One might think that one could cut through all this by appealing to the decision machine which Tarski has shown must theoretically exist, for example, for elementary geometry: pick any geometric proposition which satisfies Poincaré on the score of beauty, usefulness, or whatever, and then our decision machine would tell us whether it was true. But not only must we ourselves select the desired candidates, it is now known that for this particular case, and several others, we are never going to be able to actually write a *feasible* decision program after all. Thus, while Tarski has shown the *existence* of a recursive function $f(x)$ taking the value 1 or 0 according as the sentence A_x of formal elementary geometry with g.n. x is provable or not, A. Meyer has shown that in general $f(x)$ cannot be computed in fewer than 2^{2^x} steps. This, and several similar results show, as we have already emphasized, that both decidability and undecidability results are in general highly theoretical, having little or no bearing on what can actually be accomplished in practice with computer programs in a given domain.

It is interesting to note that Poincaré apparently thought that the completeness of an axiomatization in progress could always be insured by sufficient care: commenting on Hilbert's list of axioms for geometry, he writes:

The first thing to do was to enumerate all the axioms of geometry. This was not so easy as one might suppose; there are the axioms which one sees and those which one does not see, which are introduced unconsciously and without being noticed.
Is the list of Professor Hilbert final? We may take it to be so, for it semes to have been drawn up with care ([208], p. 3).

As for the *criterion* of completeness, Poincaré further formulates this as follows:

For this we have one criterion, and only one. We must find out whether geometry is a logical consequence of the axioms explicitly stated, that is, whether, if we put these axioms into the reasoning machine, we can make the whole sequence of propositions come out (*ibid.*, p. 5).

Poincaré also seems to have assumed that only a little care in the construction of the machine was sufficient to enable it to extract *all* the logical consequences of the axioms.

[38] See Kleene [149], pp. 430, 514, and Kleene [150], p. 328.

[39] See, in addition to the reference to Poincaré just given in the text, Weyl [264], p. 95. It is interesting, and somewhat curious, that Gödel regarded (D)′ as having 'some plausibility' even apart from the question of expressibility in some language. He considers as an objection to using all ordinals to explicate 'definability', their non-denumerability, and admits:

There is certainly some justification in this objection. For it has some plausibility that all things conceivable by us are denumerable even if you disregard the question of expressibility in some language ([101], p. 86).

[40] See Poincaré [207], pp. 60–1, 67.

[41] This is not quite as trivial as it may seem, inasmuch as there are set theories in the literature, namely Quine's two systems ML and NF, in which Cantor's theorem is not provable without restrictions seriously affecting the generality of some classical theorems.

[42] Poincaré [207], p. 59.

[43] See Chwistek [53], pp. 78ff. for some interesting recollections of this lecture and its impact on Weyl and himself.

[44] See Weyl [265], p. 18, where he distinguishes sets which are '*abzahlbar*' in Cantor's mathematical sense from those which are '*abgezählte*' on the basis of the '*Erzeugungsprozess der Urteilsschemata*'.

[45] Rabin [214] has shown that no countable model for axiomatic set theory can be recursively enumerable. Kreisel [153] had already shown this for special set theories.

[46] I am relying here on the account in Bell [10], who gives a number of interesting examples of this distinction in the history of mathematics.

[47] Frege's summing-up of the consistency vs. existence problem emphasizes the need for a totally different method for proving consistency than was then known. Indeed, it reads like a challenge to Hilbert:

Entweder man unterscheidet nicht zwischen Begriff und Gegenstand und glaubt dann, indem man einen Begriff aufzeigt, damit zugleich den Gegenstand von den gewünschten Eigenschaften zu haben; oder man sieht ein, dass ein Begriff auch leer sein kann, und verlangt von ihm, dass er widerspruchsfrei sei. So einleuchtend die Notwendigkeit dieser Bedingung ist, so fraglich ist es, ob sie hinreiche. Dass kein Widerspruch bestehe – meinen nun wohl manche – leuchte unmittelbar ein, da ein solcher, wenn vorhanden, sofort bemerkt werden müsste. Das ist leicht durch den Hinweis auf die indirecten Beweise zu widerlegen. Dass kein Widerspruch bestehe, muss also bewiesen werden. Nun haben wir zwar den Satz, dass ein Begriff widerspruchsfrei ist, wenn ein Gegenstand unter ihn fällt; aber dieser Satz ist grade in unserm Falle unbrauchbar. Bevor also nicht ein ganz neues Prinzip gefunden ist, mit dem wir die Freiheit von Widerspruchen beweisen können, kommen wir auf diesem Wege nicht weiter ([89], Vol. II, pp. 154–5).

[48] In particular, as van Heijenoort [254] explains, the predicate calculus was for Frege an interpreted *language* (interpreted by Frege's philosophical semantics), not a *calculus* (as was Boole's logic) for whose formulas one could consider various 'models'. Van

Heijenoort wryly quotes Gödel ([255], p. 583) on how, with the introduction of any new deductive system, the question of its completeness "at once arises", remarking that it had been in fact fifty years since Frege introduced the predicate calculus until the question was raised by Hilbert [123].

49 Lear [163a] has actually reconstructed a compactness agrument from Aristotle's texts themselves.

50 Scientific advance often turns on the discovery of such relativity. Another example connected with euclidean geometry is the relativity of length which can seem so intuitively objective in everyday life. Newton extended this to include the relativity of uniform rectilinear motion, but thought that acceleration had an absolute significance. Einstein extended the relativity principle to cover all motion, but it is still hard to avoid the intuitive feeling of something objective in acceleration when we drive around a corner. See Weyl [269] for a discussion of the importance of the discovery of relativities and their relation to objectivity.

51 The following remarks of Husserl, which explain, and anticipate objections (on the score of scepticism) to his method of phenomenological *epoche*, could have been used by Hilbert to explain his attitude towards formal mathematics vis-a-vis mathematics:

> Thus all sciences which relate to this natural world, though they stand never so firm to me, though they fill me with wondering admiration, though I am far from any thought of objecting to them in the least degree, *I disconnect them* all, *I make absolutely no use of their standards, I do not appropriate a single one of the propositions that enter into their systems, even though their evidential value is perfect, I take none of them, no one of them serves me for a foundation* – so long, that is, as it is understood, in the way these sciences themselves understand it, as a truth *concerning the realities* of this world. *I may accept it only after I have placed it in the bracket.* That means: only in the modified consciousness of the judgment as it appears in disconnexion, and *not as it figures within the science as its proposition, a proposition which claims to be valid and whose validity I recognize and make use of* ([135], p. 111).

52 See Hofstadter [132], pp. 49ff for an interesting defense of a thesis closely related to this that meanings for formal systems depend on isomorphisms.

53 We see also that by now Hilbert clearly distinguished the question of the solvability of all mathematical problems from that of effective decidability.

54 See Hilbert [118] and [124a]. By way of contrasting Hilbert's explicit incorporation of symbolic formalisms into the development of mathematics with his formulation of Frege's conception of them (p. 139 above), Raggio formulates Hilbert's view of the relations between the 'instances' (I), (II), and (III) in the following very interesting way.

> Dagegen haben wir es bei Hilbert mit einem Wirbel von Beziehungen zu tun, die die drei oben genannten Instanzen miteinander verbinden. Schon der Unterschied von intuitiver Theorie und intuitiver Metatheorie hat das Fregesche Schema wesentlich bereichert. Aber darüber hinaus besitzen bei Hilbert jene drei Instanzen nur den Status von reinen Bezugspunkten ohne eigenen ontologischen Gehalt. Die gegebene intuitive Theorie, erschüttert in ihren ontologischen Grundlagen durch die Entdeckung der Antinomien und die intuitionistische Kritik, ist bloss ein methodischer Leiftaden, ein *terminus ad quem*, dem es sich durch die Entgegensetzung von formalem System und Metatheorie anzunähern gilt. Das Zeichenspiel seinerseits reflektiert die Struktur nicht

nur der gegebenen intuitiven Theorie, sondern klärt sie wesentlich auf ... Auf diese Weise hat Hilbert ein dem Fregeschen Denken ganz fremdes, dialektisches Moment in die Grundlagenforschung eingeführt. Allerdings sah die zeitgenössische, philosophische Kritik im Hilbertschen Programm meistens nur das Wideraufleben irgendeines Ismus (Nominalismus, Relativismus) und verkannte seine tiefligende Motive, die grösstenteils gerade mit alten philosophischen Aspirationen übereinstimmten: Immanenz des Ausdrucks im Denken; dieses als unendliche Aufgabe und offener Prozess; ... ([215], p. 229).

[55] Hilbert offered this general principle as a way of justifying the significance of consistency proofs to constructivists with 'scruples' about mere consistency. The point was that finitary consistency proofs might provide a useful or even necessary tool for transforming some transfinite proofs of finite theorems into constructive proofs. Recall that Hilbert had made a 'sport' of this in his youth which enabled him to so transform his original transfinite proof for the finiteness of Gordan's invariants. But any such ready transformation of his incredibly complex analytic proof of Waring's purely number-theoretic conjecture surely seemed out of the question. Still, Hilbert was unable to give any actual examples to back up his recommendation for consistency proofs, leaving himself open to the constructivist claim that

Hilbert's implied belief that there are a significant number of interesting theorems whose statements (standing alone) are constructive but whose proofs are not constructive (or cannot easily be made constructive) has not been justified. In fact we do not know of even one such theorem ([25a], p. 350).

Kitcher apparently overlooked this 'implied belief' when he claimed that Hilbert's views on consistency proofs prevented him from even raising "the possibility that the use of ideal statements will yield finitary theorems which were unobtain*able* before" ([140a], p. 104). Hilbert quite obviously considered this possibility as one way of justifying the use of ideal statements, namely, by showing them to be at least a practical if not a theoretical *necessity*. Transfinite proofs were typically shorter and more perspicuous than lengthy finitist ones, and hence, Hilbert thought, also easier to discover – as the in case of his original finiteness proof for algebraic invariants.

On the other hand, Hilbert also considered the possibility of establishing the *safety* of ideal statements in certain contexts, i.e. of showing that they can prove no finitary theorem not eventually provable by fintist means. The constructivist criticism of Bishop here is that a constructive consistency proof could hardly make this eliminability of transfinite proofs any more obvious than it already is from ordinary mathematical practice. This is by no means clear to me, but in any case Hilbert was not dogmatic that such an elimination was always feasible to begin with; he considered this thesis rather to have the same status as his solvability thesis which he took with a grain of salt:

Freilich kann vermutlich ein finiter Satz stets auch ohne anwendung einer transfiniten Schlussweise beweisen werden – wie ja im Falle des Satzes von der endlichkeit des wollen Invariantsystems mein zweiter Beweis zeigt — aber diese Behauptung ist von der Art der Behauptung, dass jeder mathematische Satz überhaupt entweder sich als richtig nachweisen oder widerlegen lassen muss ([121], pp. 187–188).

Kitcher also argues that Hilbert cannot "scotch the sceptic" with a consistency proof because he cannot really *know* the finite decomposability of all his stroke-numerals on which he based induction: according to Kitcher he cannot know this intuitively because of the need to consider the 'indefinite' case '||. . .|' in which "we can no longer survey the array and observe the detail we need to observe if we are to exhibit the inductive property" (*ibid.*, p. 112). But this symbol is just as definite as ' | | | | | ': the dots are merely part of our intuitive symbolism, playing the role of variables, i.e. symbols which can be replaced by others. Induction is based simply on the rule that they can only be replaced by strokes. For Hilbert the finite decomposability of all symbols is guaranteed by his belief that "there are everywhere only finite things". This finitism may indeed beg the question but his moral certainty about it merely measures the depth of *his* scepticism – about the existence of anything infinite (cf. fn. 64).

56 I have also made use of Finsler ([81], [82], [85]) in interpreting this interesting paper. It appears from Bernays [19] and the discussion following Bernays [21] that it was Finsler's work and not Gödel's work that had persuaded many that human mathematical thought transcends every (denumerable) formalism.

57 Presumably it is intended that it assigns meanings to symbols which have not yet even been used.

58 E.g. Broden [35]; Frey [93] even relates diagonalization directly to intentionality.

59 See Finsler [81]. Later he worried about proving the existence of infinite sets.

60 Finsler errs when he claims that the reflection principle $P(\ulcorner A \urcorner) \rightarrow A$ cannot be added to a formal system without making it inconsistent. All that can be shown is that a contradiction could be inferred from any derivation of this principle *from the other axioms*. But his remarks on Gödel's work are interesting:

Nun muss aber bei Gödel ein formaler Beweis nur gewissen formalen Bedingungen genügen ohne Rücksicht auf seinen wirklichen Sinn. Diese abgeänderte, in den formalen Theorien allerdings übliche Behandlungsweise hat jedoch zur Folge, dass jetzt prinzipiell kein Hindernis mehr besteht, den oben angegeben, sehr einfachen und, solange er gedanklich geführt wird, korrekten Beweis für den angeblich formal unbeweisbaren Satz doch formal darzustellen. Dies ergibt aber einen Widerspruch und es folgt, dass Herr Gödel keineswags die Existenz von Sätzen nachgewiesen hat, die in einem allgemeinen Sinne formal nicht entscheidbar wären, sondern er hat gezeigt, dass die von ihm betrachteten Systeme, sofern sie nur gewisse einfache und inhaltlich korrekte Schlüsse enthalten, formal widerspruchsvoll sind ([83], pp. 310–11).

61 This point is elaborated in the last chapter.

62 Heidegger [111] has formulated and elaborated in considerable detail the theme (interpretation) that Kant's first critique is essentially about the finitude of human beings and knowledge. In particular, he argues that, for Kant, "the essence of sensibility lies in the finitude of intuition" (*ibid.*, p. 31). A related argument concludes, on the basis of the "bond" (for Kant) between pure intuition and pure thought, that the categories suffer an "essential delimitation" which implies the "finitude of pure thought and knowledge". But see also Parsons [197] for a penetrating discussion of some difficulties that would stand in the way of carrying through this interpretation of Kant.

63 The point is simply that, if a proof is to be regarded as a 'mental object', then it is one of the 'contents' of the mind, and hence finite by (H_2). Of course, those who regard proofs as infinite mental objects will say that this argument begs the question. Specifically, they hold that an individual proof by mathematical induction is really a sort of infinite

process (perhaps this is why Ackermann [2] says that the formalisms studied in metamathematics provide, in general, a *'finites Bild'* of intuitive proofs).

64 Gödel's theorem has occasionally been interpreted as showing that we can not completely formalize this intention, but one should be cautious about this. It is true, as Trachtenbrot [249] shows, that the incompleteness of formal set theory can be linked directly to the definition of 'finiteness'. Thus he shows that for any formal set theory there are two open formulas $A(x)$, $B(x)$, each satisfied by exactly the finite sets, but such that the sentence $(x)[A(x) \leftrightarrow B(x)]$ is not provable. But inspection shows that the underlying theorem here is just his familiar result that the set of wffs of the predicate calculus which are satisfied by exactly the finite domains is not r.e. And note that even in stating and proving these results — in fact, *in specifying the formal system of proofs itself — we already must presuppose we know what 'finite' means.* The formalist presupposition about 'finiteness' has been admirably summed up by Bernays:

Die wesentliche Gebundenheit der formalen Abstraktion an das Moment der Endlichkeit macht sich insbesondere dadurch geltend, dass bei den Betrachtungen von Gesamtheiten und von Figuren die Eigenschaft der Endlichkeit für den Standpunkt der anschaulichen Evidenz gar kein besonderes beschränkendes Merkmal bildet. Die Beschränkung auf das Endliche wird von diesem Standpunkt aus ganz ohne weiteres, sozusagen stillschweigend vollzogen. Wir brauchen hier kein besondere Definition der Endlichkeit, denn die Endlichkeit der Objekte versteht sich für die formale Abstraktion ganz von selbst. So passt z.B. die anschaulichstrukturelle Einführung von Zahlen nur für die endlichen Zahlen. 'Wiederholung' ist eben vom Standpunkt der anschaulich-formalen Betrachtung *eo ipso* endliche Wiederholung ([18], p. 348).

65 It is sometimes said, e.g. by Carnap [45] and Müller [190], and is generally believed, that Hilbert introduced this rule, more or less *ad hocly*, to get around Gödel's discovery of the ω-incompleteness of the formalisms he was studying. But Bernays [128], p. 215, said explicitly that Hilbert already introduced the rule before he had heard about Gödel's result. Since Hilbert's lecture [124] was delivered in November of 1930, only one month after Gödel published the *abstract* of his paper, it is hardly possible that he could have been motivated by Gödel's result.

66 By the "ordinary ω-rule" we mean the rule which infers $(x)A(x)$ from all of the premises A(1), A(2), A(3), . . . For every ordinal α define a set T_α of sentences as follows: T_0 is the set of theorems of the standard first order formal system P of Peano arithmetic; $T_{\alpha+1}$ is the set of sentences provable from T_α by the ordinary ω-rule, along with their first order consequences; for limit α, $T_\alpha = U_{\beta<\alpha} T_\beta$. It is essentially trivial that T_ω is the set of true sentences of number tneory. By the 'effective ω-rule' we mean the rule which infers $(x)A(x)$ from the existence of a recursive function $f(n)$ such that for each n, $f(n)$ is the Gödel-number of a proof of $A(n)$ in P. Now define, for every ordinal α, a set T'_α of sentences as follows: $T'_0 = T_0$; $T'_{\alpha+1}$ is defined like $T_{\alpha+1}$ except that the effective ω-rule replaces the ordinary ω-rule; similarly for T'_α with α a limit ordinal. The by no means trivial result of Shoenfield is that $T'_\omega \omega = T_\omega$ = the set of true sentences of number theory. This result makes essential use of Kleene's fixed-point recursion theorem which we will discuss in the final chapter. Hilbert's original formulation of the ω-rule was effective in intent, but did not require a particular kind of function which would enumerate proofs of the premises.

[67] Feferman [79], building on Turing's [61] fundamental paper on 'ordinal logics' has established and clarified some interesting completeness results for these logics, called "recursive progressions of axiomatic theories" by Feferman. We sketch his basic ideas and results. Let A_0 be an r.e. axiom system, and R a recursive function of (Gödel-numbers of) wffs defining the set A_0, called a "generating function", which extends the system A_0 of axioms by r.e. many new theorems to a new system $R(A_0)$ of (Gödel-numbers of) wffs defining the set A_0. We can then consider transfinite sequences A_α of axiomatic theories defined as follows:

(1) A_0 is given,
(2) $A_{\alpha+1} = R(A_\alpha)$,
(3) $A_\alpha = U_{\beta<\alpha} A_\beta$, for limit α.

Turing and Feferman have studied such questions as (i) for what ordinals α do such sequences of r.e. axiom systems exist? (ii) are the unions of their theorems complete for various classes of sentences? Of course, the ordinals themselves must be constructively given, and it turns out that a certain maximal set 0 of constructive ordinal notations (integers), whose existence was proved by Kleene [148] (again, as a corollary of his fixed-point recursion theorem proved in the same paper), proves to be suitable for many results. Among the possible generating functions R, attention focusses on certain ones, called "reflection principles" by Feferman, which, in a certain sense, "reflect trust" in the axiom systems they extend. Then, where $\varphi_1, \varphi_2, \varphi_3 \ldots$ is a complete effective enumeration of all unary partial recursive functions and where $t_0(x)$ is an RE-wff (see Feferman [78a]) defining, say, A_0, and R any generating function, an *ordinal logic function* is a partial recursive function h such that:

(1) If $n \in 0$, then $h(n)$ is defined and yields an RE-wff $(h(n))(x)$
(2) $h(1) = t_0(x)$
(3) If $n \in 0$, then $h(2^n) = R(h(n))$
If $\varphi_n(k) \in 0$, $\varphi_n(k) \prec_0 \varphi_n(k+1)$, for each k, and A_m is defined by $(h(m))(x)$ for each $m \in 0$, then $(h(3 \cdot 5^n))(x)$ defines $\bigcup(A_m \mid m \prec_0 3 \cdot 5^n)$

The 'ordinal logic' determined by h is taken as the union $\bigcup\, (A_m \mid m \in 0)$. Each A_m will be an r.e. theory, although this union of the A_m will in general not be r.e. Using Kleene's recursion theorem, Feferman establishes the basic existence theorem that there exists an ordinal logic function h for every RE-wff $t_0(x)$ and generating function R. For the particular generating function R corresponding to the formalization of the effective ω-rule, it turns out the ordinal logic determined, call it $(T_d \mid d \in 0)$, proves every true sentence of number theory, i.e. it proves every sentence in the Shoenfield theory $T_\infty' \omega$ of the previous note. Feferman expresses the meaning of this completeness theorem by remarking that "the set of all true sentences of number theory is a constructive limit of recursively axiomatizable theories proceeding by the given succession principle" ([78], p. 312). His comment about the overall situation in ordinal logics is also interesting:

A general incompleteness theorem for recursive progressions based on arbitrary succession principles (or even just arbitrary reflection principles) would have been dramatic proof of the farreaching extent of incompleteness phenomena. However, the situation has not turned out in this way ([79], p. 261).

Perhaps Smart [235] had some result like this in mind when he rebuts (he credits the objection to Quine) what he calls the "Rosenbloom-Kemeny-Nagel and Newman argument" against mechanism by claiming that a machine could be programmed to add its own Gödel-sentences as additional axioms, and then the process could be continued "through the constructive ordinals" (*ibid.*, p. 110). This is misleading, and, strictly speaking, false. No one machine could find the 'path' through 0 along which the ordinal logic based on R^{ω} collects all the true sentences of number theory. On the other hand, there is not the slightest reason to suppose that a human mind could do this either, or that a machine could not model the 'ingenuity' displayed by a mind in getting as far as it can. Lucas [174] and Good [103] engage in a game of leap-frog through some vaguely specified transfinite extensions of some initial machine, whose *dramatis personae* are Lucas's 'mind' and Good's 'machine'. But the essential point, is simply that a Turing machine could just as well do what Lucas does in this game, so that the ability to play it, and play it ever so well, does *not* imply that the player couldn't be a machine. And the essential point about the ordinal logic results is that they display the remarkable amount of 'self-reflection' a machine could be capable of. The basis for this is the 'self-referential' recursion theorem on which these results depend, which will be discussed in Chapter IV.

68 See Husserl's [137] notes on the Frege-Hilbert correspondence, as well as his posthumously published studies on the completness problem. Husserl [136] later emphasized, with Hilbert, that a great advance of modern mathematics consisted in the circumstance that now mathematical languages ("system-forms") themselves became the objects of mathematical study. See Mahnke [79] for more on this.

69 Husserl's motivation for such results strikingly recalls Kreisel's ([13], pp. 157 ff.) formulation of Hilbert's motivation for consistency proofs:

The concept of the definite multiplicity served me originally to a different purpose, namely to clarify the logical sense of the computational transition through the 'imaginary' and, in connexion with that, to bring out the sound core of Hermann Hankel's renowned, but logically unsubstantiated and unclear, "principle of the permanence of formal laws". My questions were: Under what conditions can one operate freely, in a formally defined deductive system (a formally defined 'multiplicity'), with concepts that, according to the definition of the system, are imaginary? When can one be sure that deductions that involve such an operating, but yield propositions free from the imaginary, are indeed 'correct' – that is to say, *correct consequences of the defining forms of axioms*? How far does the possibility extend of 'enlarging' a 'multiplicity', a well-defined deductive system, to make a new one that contains the old one as a 'part'? The answer is as follows: If the systems are 'definite', then calculating with imaginary concepts can never lead to contradictions ([136], p. 97).

70 See Becker [7], p. 127, and also his discussion on pp. 359ff, of Veblen's principle for giving 'constructive notations' to ordinals of the second number class. His '*Bild-intentionalitäten*' however did not yet explicitly reach ε_0; see p. 353.

71 Gödel suggested in his original paper that there might be finitary proofs not formalizable in his system, but later in [102] he claimed that, unlike induction up to ω^2, Gentzen's ε_0-induction was not "unmittelbar anschaulich" since one can no longer 'survey' the structural possibilities for its descending sequences — and hence that it was not finitistic. But he did not say just when this occurs in Gentzen's "shrittweises Übergehen" from

smaller to larger ordinals and Church has denied that it does at all: "That induction up to ω^2 or up to ω^ω should be allowed as *finit*, but induction up to ε_0 not, is a boundary line so artificial and arbitrary as hardly to be accepted" ([52a], p. 359). He took Gentzen's proof as a "remarkable confirmation" of Gödel's original suggestion, stating flatly that it "shows that there are *finit* proofs which cannot be stated in first-order arithmetic" (*ibid.*). It is indeed very hard to get an epistemological foothold on Gentzen's 'slippery slope'. Gödel even admitted that the alleged abstract element of ε_0-induction may be no more than "die praktische Unmöglichkeit, kombinatorisch allzu komplizierte Verhältnisse anschaulich vorzustellen" (*ibid.*, p. 72), which is compatible with Gentzen's claim that "nothing new is basically added". In any case, Gödel's second theorem is irrelevant to such epistemological questions, and M. Detlefsen [70] has in fact argued convincingly that it cannot refute Hilbert's program. Resnick [118a] claims that "since the consistency of Γ cannot be proved via the methods formalizable in Γ, it surely cannot be proved via (weaker) finitistic methods alone" (p. 136). But we have no reason for supposing that all finitist methods are formalizable in any Γ, nor any for supposing that all proofs formalizable in number theory are finitist. Indeed, Gentzen realized that Gödel's result even allowed the possibility of proving the consistency of number theory by means lying outside it "which can nevertheless be considered to be *more reliable* than the doubtful components of elementary number theory itself" ([97], p. 139). Its doubtful components are the application of classical logic to an infinite domain and of induction to arbitrarily complex predicates. Takeuti [244] has even formalized classical analysis in a conservative extension of first-order Peano arithmetic, which is thus a very strong experimentally complete system whose undecidables hardly occur in mathematical practice, a fact for which Takeuti suggests the following explanation:

When we learned to formalize mathematics, the formalization itself was an important but difficult task. Naturally we chose a very strong system so that it was easy to see that everything could be formalized in the system. It is likely that we do not need such a strong system and hence our identification of mathematical practice with a certain strong formal system could be an illusion (*ibid.*, p. 73).

See Benda [14] for the possibility of a finist consistency proof of Peano arithmetic using the infinite Ramsey theorem.

72 The central example of this tendency is the geometrical continuum idealized from what Hilbert increasingly came to believe to be a world itself finite and discrete. My sketch of his mathematical philosophy has focussed on issues which are more or less directly related to mechanism and logic and has not tried to reconstruct his account of just how this idealization occurs, nor his answer to the question why we nevertheless find the continuum so widely applicable in physical theory, or have such difficulty in finding, in Einstein's words, "anything organic" to take its place. Thus I have largely avoided the decisive problem of both the genetic and systematic relation in Hilbert's philosophy between mathematical geometry, which never in his eyes lost intuition as its handmaiden, and his work in physical geometry which dominated his later thinking. In some fascinating unpublished reminiscences of his 1925–26 stay in Göttingen. Dirk Struik reminds us that "Hilbert, despite his turning to formalism when he had to obtain the utmost clarity through abstraction, so that mathematics looked like a game, a whim of the mind, was clearly aware of the intimate and ultimate relationship of nature and mathematics".

CHAPTER IV

EFFECTIVENESS MECHANIZED

The conclusion that man is not a machine is invalid.

E. Post

1. EFFECTIVENESS, DIAGONALIZATION, AND THE PROBLEM OF RECURSION

The problem of effectiveness tended early, as we have seen, to be conflated with the paradoxical one of 'finite definablility'. That the two were really distinct problems emerges from an observation due essentially to Hilbert, that

$$\alpha \text{ is finitely definable} \Leftrightarrow \alpha \text{ is definable (period!).}$$

The qualification 'finitely' is superfluous: how could any definition be infinite? This underscores a confusion in the formulations of Poincaré and other 'semi-intuitionists'. One might say they committed a 'fallacy of misplaced finiteness'; they were primarily interested in effective definition, but the essential element of finiteness they rightly saw in them was not simply the number of words they used, but rather that the processes initiated when they are applied to appropriate arguments will *terminate after finitely many steps*. We have seen that in order to develop number theory intuitively and constructively 'from below', as opposed to Dedekind's impredicative approach 'from above', they actually needed some notion of effectiveness – beyond that of mere finiteness – to deal with recursion. Of course, Dedekind's recursion theory for the Peano model can be developed in axiomatic set theory, thereby avoiding this problem altogether.[1] Thus, if we base number theory on axiomatic set-theory we face the problem of the definable elements of $\mathscr{P}(N)$, whereas the intuitive constructive approach must give an account of its effectively definable elements, or equivalently, the effectively computable functions on N.

Hilbert's metamathematical approach, however, sought to combine the constructive and the axiomatic approaches in tandem, with the result that the problem of effectiveness eventually emerged as a problem concerning certain axiomatic systems after all, namely, the *elementary* (first-order) formalizations growing out of his program (H1)-(H6) to axiomatize Dedekind's theory. But apparently in these systems one can no longer even state, let alone prove, Dedekind's recursion theorem, hence the focus of the problem necessarily shifts to the possibilities of *formal calculation* of functions whose recursions are added as axioms: one no longer proves their existence, but rather formalizes their calculation. Thus for Hilbert the problem of effectiveness gravitates to this: what functions on N are *formally calculable on the basis of his axiomatization of Dedekind's description of N as a Peano model.* Before Gödel, this problem was beginning to seem hopelessly ambiguous since there was no end in sight to new recursions that had to be continually added as new axioms. Before going into this, let us review and synthesize now some of our history of the effectiveness concept, with particular emphasis on various diagonalization and substitution processes associated with it.

We first note that historically there had been two distinct kinds of suggestion concerning what is essential in effective definitions or algorithms for functions. First we had Hilbert's two 'finiteness' requirements on algorithms formulated in his work on algebra (pp. 75–6 above):

(F) (i) Each computation must terminate in a *finite* number of steps.
(ii) The processes employed in each computation must be *fixed in advance*.

The key question left open here is: what 'processes' are permitted? We may think of Turing's theory as one answer to this question. On the other hand, we had the suggestions of Cantor and Hilbert's (HT) that a theory of effective calculation and 'exact description' focus on *recursive* definitions and procedures. We can formulate this as

(R) Each computation must be effected by repeated use of some *recursive* procedure.

An attractive feature of (R) was that, for simple paradigmatic cases of recursion, it was clear that it satisfied (F) (i), which any approach must surely do. Thus, when $f(n')$ is defined in terms of $f(n)$, it was obvious that one could always work one's way back to the given value of $f(0)$ in a *known* finite number of steps. In fact, it was perhaps *too* obvious, for the task of giving an explicit description of the simple substitution processes by which one 'works one's way back' – the analogue of (F) (ii) needed for the (R) conception – was long ignored as too easy to be profitable. Still, the real question left open by (R) was: what kinds of 'recursion' are allowed? This more difficult question was also underestimated for years, largely because of a tendency of early writers, notably Poincaré, to assimilate the 'recurrence' property of these definitions too closely to the simple iterative pattern of ordinary induction. Thus, for many years recursion was described as the 'process of going from n to $n + 1$', until Hilbert emphasized in 1925, on the basis of Ackermann's discovery of a non-primitive recursive function that (R) would have to encompass 'transfinite recursions'.[2] The essential point about definitions by 'recurrence' has turned out to be not so much their 'inductiveness', but simply that the function (symbol) being defined recurs again in the definiens; in short, they employ a wide measure of (apparent) *circularity* and *self-reference.*

Let us recall that some important closure properties on the class of intuitively computable functions already became evident in connection with Dedekind's recursion theory as soon as we asked how his constructivist opponents could justify it within the framework of their ideas. Thus, his recursion theorem itself was justified intuitively by the closure of effectiveness under simple recursion itself ((2)″ on p. 54 above), whereas a key step in his proof of it was so justified by the closure of effectiveness under diagonalization ((d)′ on p. 56). If we now ask what further will intuitively justify his applications of the theorem to define functions of *two* variables (see pp. 51–2), we are led to assert the following closure principle of *iteration:*

(It) $f_1(x), f_2(x), f_3(x), \ldots$ is any sequence of effective functions $\Rightarrow$ $h(n, x) = f_n(x)$ is an effective function of two variables.

Iteration used together with diagonalization provides a powerful method of getting new effective functions of one variable from whole infinite sequences of them. As we saw in our discussion of Kronecker's prospects for meeting Cantor's challenge to his finitism, it might even seem that the method is strong enough to preclude any complete theory of effective functions. Let us examine this in more detail now in connection with an early 1908 global hypothesis of Borel:

(BH) The class of all effective functions is denumerable.[3]

This hypothesis seems rather plausible, but it leads, in connection with Cantor's argument, almost immediately to a kind of 'Richard's Paradox for effectiveness'. If $f_1(x), f_2(x), \ldots$ is any enumeration hypothesized by

(BH) to contain all effective functions of one variable, then $h(n, x) = f_n(x)$ is effective by (It); but then $h(x, x)$ is effective by closure under diagonalization and $h(x, x) + 1$ is a new effective function not contained in our enumeration. Hence we seem to have an equally good reason for claiming, in contradistinction to (BH), that the class of effective functions is *not* denumerable. Thus we have yet another form of Richard's Paradox:

(RP)″ (i) The effective functions on N are mathematically non-denumerable, by diagonalization.
(ii) The effective functions on N are metamathematically denumerable, by the 'finiteness' of their definitions, or algorithms.

Hilbert especially worried about the possibility of (RP)″(i), which has been nicely described by Rogers:

> It is evident that the diagonalization method has wide scope, for it is applicable to any case where the sets of instructions . . . can be effectively (i.e. algorithmically) listed. At first glance, it is difficult to see how a formal characterization can avoid such effective listing and still be useful. The diagonal method would appear to throw our whole search for a formal characterization into doubt. It suggests the possibility that no single formally characterizable class of algorithmic functions can correspond exactly to the informal notion of algorithmic function. Perhaps, no matter what . . . symbolism and . . . specifications we choose, that symbolism and those specifications can be augmented by stronger symbolism and more complex 'effective' operations to yield new functions . . . Perhaps, indeed, the algorithmic functions form a non-denumerable class, and perhaps there is a spectrum of algorithmic computability upon which all functions fall ([221], p. 11).

Let us isolate a diagonal lemma implicit in Cantor's proof of non-enumerability. Suppose $h(n, x)$ is any binary function. Then by performing successive substitutions of integers for the variable n, we generate an infinite sequence $h(1, x), h(2, x), \ldots$ of functions of one variable, which we describe as *enumerated by simple substitution* from $h(n, x)$. This process is the converse of iteration which forms $h(n, x)$ from a given infinite sequence. Another, more special substitution process yielding unary functions from $h(n, x)$ is to successively substitute each integer for both variables, yielding the values $h(1, 1), h(2, 2), \ldots$ of a function $h(n, n)$ which we call the *diagonalization* of h. We can obtain further functions of one variable by composing this diagonalization with given functions of one variable. The diagonal lemma then is

(DL) From a given binary function $h(n, x)$ we can obtain unary functions by diagonalization which cannot be enumerated from it by simple substitution.

$h(n, n) + 1$, of course, will always be such a function, but $h(n, n)$ itself will satisfy the lemma for a wide class of given $h(n, x)$, typically because it will grow faster than any specific $f_k(n) = h(k, n)$.[4] But however much faster $h(n, n)$ may grow than any $f_k(n)$, it seems intuitively clear that it must always be just as *effective* as any of them, and hence that diagonalization will impose very strong demands on any mathematical theory of effectiveness. In particular, as Kleene was the first to see, the only way of holding onto (BH) is to make specific allowance for *partial functions*, for (DL) depends on the assumption that all functions considered are *total.* Thus, if $U(n, x)$ were any 'universal' function enumerating all effective functions of one variable by substitution, the lemma implies that either $U(n, x)$ cannot be effective or that not all the effective functions enumerated by it could be total. Before Gödel's work it was hard to see how to construct a specific comprehensive class of arithmetical functions which satisfied these requirements, especially one containing a universal function for all effective unary functions, yet (BH) and (It) jointly imply that a comprehensive theory of effectiveness would have to have such functions. These requirements become even more imposing if we consider other kinds of diagonalization. Thus, suppose $H(z, n, x)$ is a tertiary function enumerating $h_1(n, x), h_2(n, x), \ldots$ by substitution as z runs through the integers. Then the diagonalization process of substituting, for the first argument in each $h_z(n, x)$, *its own index* z, forming the binary function $H(z, z, x) = h_z(z, x)$, leads to $H(z, z, x) + 1$ which cannot be enumerated from h – if it is total. Imagine now a hypothetical 'system' of effective functions comprising two enumerations:

(E) $f_1(x), f_2(x), f_3(x), \ldots,$
$h_1(n, x), h_2(n, x), h_3(n, x), \ldots.$

Our requirements on a theory of effectiveness can now be expressed in terms of the 'interaction' between these two sequences. If (E) is to contain *all* intuitively effective functions of one and two variables it would have to possess at least the following properties:

(e) $h_u(n, x) = f_n(x)$, for some u.
(s) Closure under substitution.
(d) Closure under diagonalization.

The universal element h_u of (e) would be required by (It). By (s) we mean that any substitution instance enumerated by any binary function of (E) is again a unary function of (E), i.e.,

$$\text{(s)}' \qquad (z)(k)(\exists j)(x)[h_z(k, x) = f_j(x)]$$

Now (s)′ already implies our newer form of diagonalization, namely,

$$\text{(d)}' \qquad (z)(\exists j)(x)[h_z(z, x) = f_j(x)].$$

For (d) we require also our older form of diagonalization, namely,

$$\text{(d)}'' \qquad (z)(\exists j)(x)[h_z(x, x) = f_j(x)].$$

It follows from (e) and (d)″ that the unary sequence of (E) is itself closed under diagonalization, bringing us very close to the pattern of several paradoxes.[5] But there is still another joint requirement on substitution and diagonalization clearly imposed by our intuition of effectiveness. Consider (s)′: it says, in effect, that if we fix k as a parameter in $h_z(k, x)$, an algorithm *exists* which will compute the resulting function of one variable x, namely, $f_j(x)$. But intuition demands more: if we have an algorithm for computing $h_z(n, x)$ as a binary function, then not only should the unary $h_z(k, x)$ be effective, for each integer k, but we can surely *effectively find its algorithm from z and k*. Hence there ought to be a 'substitution algorithm' for computing the j in (s)′ uniformly from z and k, i.e. one of our effective binary functions h_s of (E) must satisfy

$$\text{(s)}'' \qquad (z)(k)(x)[h_z(k, x) = f_{h_s(z,k)}(x).$$

But diagonalization is unrelenting, and now by (d)″ the diagonalization $h_s(z, z)$ of our substitution function must be identical with one of our unary functions, say f_s. This f_s then will effectively or uniformly satisfy our other form (d)′ of diagonalization, yielding

$$\text{(d)}''' \qquad (z)(x)[h_z(z, x) = f_{f_s(z)}(x)].$$

That (d)‴ holds in current formal theories of effectiveness will later lead to the strongest arguments known for Church's thesis. We have considered (e) – (d)‴ at this time to emphasize the very strong and specific demands which our intuition of effectiveness imposes on any system (E) of functions which might be offered to explicate it. They also suggest that a theory of effectiveness has to be *metamathematical* in the sense that we will have to consider, via our enumerations, the action of our functions on the algorithms of other functions. In particular, we see that such a theory must involve far more than the kind of 'arithmetization' of mathematics envisaged by Kronecker: it will require the explicit arithmetization of the very intuitive processes of enumeration and substitution – though now formalized as explicit symbolic processes – that early writers took for granted. In short, arithmetization in the sense of Gödel. We can also appreciate from the above why Rogers calls our current theory of effectiveness "virtually a theory of diagonalization" ([220], p.31). But it is also clear that the closely related processes of substitution are important to any theory of effectiveness, and I would now like to consider their role in the history of logic.[6]

The first logician to attempt a theory of substitution processes as such was J. König in his 1914 posthumously published attempt [152] to carry out Hilbert's 1904 program for the foundations of logic, arithmetic, and set theory. He introduced a clear metalogical symbolism for describing the results of various substitutions which has remained in wide use. He emphasized that the processes of substitution have nothing to do with the 'meaning' of the primary logical symbols upon which they operate. He also tried to prove that the mental processes of substitution are effective – '*endlichen Denkprozesses*' which are 'completely describable'. This takes the form of setting up recursive definitions for various substitution operators, but the theory is left as a hypothesis:

(KH) All formal substitution processes are completely describable by recursion.

This is a companion piece to Hilbert's (HT) concerning the recursive description of proofs, covering formal processes not explicitly treated by Hilbert in 1904.

The decisive studies of substitution were, however, inaugurated by the combinatory logic of Schönfinkel and Curry, again, both inspired by Hilbert.[7] The three principal motives of combinatory logic were (i) the analysis of substitution, (ii) the analysis of the concept of a constructively definable function, and (iii) the clarification of the logical paradoxes. Only the first motive was present in Schönfinkel's work, while (ii) and (iii) – which offers a specific attack on Hilbert's 1904 problem (CP) of clarifying the paradoxes – have come to the foreground in Curry's work.

Schönfinkel observed that the use of variables to communicate logical truths is to some extent beside the point: although the proposition

$$\vdash \neg P \vee (P \vee Q)$$

may appear to be about the variables 'P' and 'Q', or objects they denote, a little reflection shows that it expresses a relation between two functions, namely, the truth functions denoted by '$\vee$' and '$\neg$'. He also establishes an analogous point about the individual variables and quantifiers of the predicate calculus, leading to the construction of a function calculus for the expression of logical laws, the primitive elements of which, since Curry, are called 'combinators', and whose primitive operation is that of 'application' of combinators (and their combinations) to their arguments. The principal interpretation is that combinators are functions whose arguments are other functions, including themselves: self-application is freely permitted.[8]

Curry [57] argued that combinatory logic might enable logic to fulfill a longstanding promise to provide an 'analysis of thought'. Previous logics, notably those of Frege and Russell, had concentrated on reducing the "postulates and primitive ideas to their lowest terms", but had employed *rules* which were highly complex. Now rules, Curry argued, were the real "port of entry of intelligence" into any symbolic logic: the only judgment actually required in working with such a system is whether the rules have been correctly applied. And since, further, no reasoning at all can be done in the system without rules, it follows that rules "represent the atoms of thought, so to speak, into which the reasoning can be decomposed" (*ibid.*, pp. 367–8). Hence, for the analysis of thought the simplicity of the postulates is not enough: "it is even more important to so choose the rules that they involve, in their application, only the simplest actions of the human mind" (*ibid.*, p. 368). It was an open question whether a system was possible which could be both "adequate for the whole of logic" and yet employ simple rules, but Schönfinkel had taken a major step in his analysis and simplification of a fundamental class of complex rules too long ignored, namely, those controlling the processes of substitution (see below). But a price must be paid for the increasing simplicity of the primitive framework: the deductions in the system become much longer and more complex than informal ones. Hence, if we are to use logic to analyze thought, we must adopt the metatheoretic point of view of Hilbert:

Indeed if we are to dissect the reasoning into microscopic pieces it is but natural that more of them should be necessary to bring about a given result. Consequently we must adopt a point of view sugested by Hilbert. With each theory there is associated a metatheory in which we reason intuitively about the theory. In this metatheory we can derive more and more complicated rules by showing, in general terms, how a particular consequence of the derived rules can actually be deduced from the primitive ones. The aim of mathematical logic is, in fact, not to reduce mathematics to a formalism, à la *Principia*, in which all steps explicitly appear; but rather to analyze logic with a view to obtaining a greater command over its use, and a more profound understanding of its nature (*ibid.*, p. 368).

One apparently simple and therefore neglected process had been the substitution of constants for one or more variables of a propositional or numerical function. But from the point of view of combinatory logic this attitude was not justified. Its painstaking attention to the details of substitution, however, was largely motivated by the possibility of viewing it as an operation on functions:

A notion closely related to substitution is that of transformation of functions. Suppose we regard a function as having inherent in its definition a certain order of its variables. Then permuting these variables in any way, or making two or more of them alike, will produce new functions related to the old; let us call them transforms of the original function, and the operations by which they are produced transformations . . . The study of substitution is to a certain degree equivalent to the study of these transformations (*ibid.*, p. 369).

Schönfinkel had shown that "neither the notions of propositional function nor that of substitution need be assumed as primitive". The main result implicit in Schönfinkel was that his function calculus was closed under explicit definitions, a property now called 'combinatorial completeness'. To explain this, suppose that the function calculus had two constants A and M and we define F by using variables:

$$Fxyz = A(Mxy)(Mxz).$$

Schönfinkel asked now whether this F could also be constructed in his function calculus from A and M without using variables, i.e. could a combinator Q be found such that F could be defined using only Q, A, M, and application.[9] He showed that the presence in the function calculus of his two primitive combinators S and K was sufficient to carry out every explicit definition of this kind in the calculus from given function constants. Suppose then that $\mathscr{F}[x]$ is any expression built up from given function constants, combinators, and the variable x. Then there exists a combinator Q constructed from just S, K, and the given function constants by application only defined by this expression,[10] i.e.,

$$(\exists Q)(a)[(Qa) = \mathscr{F}[a]].$$

This has the force of a comprehension principle: any condition that can be explicitly written in the calculus defines a combinator of the calculus (with S and K as the only primitive combinators). A close connection was soon discovered between the function calculus of Schönfinkel and Curry and the λ-calculus of Church which did not try to eliminate variables, but merely postulated that every expression $\mathscr{F}(x)$ defined the function $\lambda x \mathscr{F}(x)$ formed by λ-abstraction: the combinator Q above plays the role of this abstraction.[11]

Curry's work went beyond Schönfinkel not only in the formulation of explicit rules governing the function calculus, but in the definition and study of new combinators. In particular, he defined the diagonal combinator W which, when applied to an argument representing a function of two variables, transforms it into a function of one variable by identifying its variables.[12] Further study of this diagonal combinator revealed that, combined with a certain other combinator, it could be used to state a very general form of Russell's paradox in combinatory logic, which Curry communicated to Hilbert in a letter of 1929.[13] Implicit in Curry's construction is a species of combinators now known as 'fixed point combinators' which play a fundamental role in combinatory logic, especially when used as a foundation for arithmetic. However, at this time, Hilbert was still struggling with his own axiomatic foundation for arithmetic: his joint work with Bernays, announced for publication in 1923, had encountered unforeseen complications, making consistency proofs more difficult than anticipated. The real difficulty, however, concerned recursion.[14]

As we have seen, recursions only entered explicitly into Hilbert's 1904 program on the metatheoretic level as a means of giving an 'exact description' of formal proofs, viz. (HT). He had yet to face the problem of formalizing these recursions: the effect of Dedekind's recursion theorem had somehow to be incorporated into his elementary axiomatization despite not being provable therein. Hilbert and Bernays pursued two different approaches to this problem. At first they just added what Hilbert called the 'ordinary' recursion schema as axioms in addition to an axiom schema for induction.[15] Secondly they also investigated the possibilities for defining some of these recursively introduced functions explicitly in terms of functions and relations already available, and elementary logic. Thus, beginning with the simplest elementary axiom system with only the successor function, they showed that the less-than relation could not be defined in it; and by a simple diagonal argument that even if suitable axioms for the less-than relation were added, one could still not define addition explicitly. Then it was shown by an analogous diagonal argument that after addition of the recursion axioms defining addition, it was not possible to explicitly define multiplication. After adding a recursion defining multiplication, the case was examined for exponentiation, but they were neither able to define it explicitly nor prove that it could not be so defined in terms of 'successor', 'plus', 'times', and logic. Nevertheless, because of the traditional algebraic view of arithmetic as the study of the field of integers with the operations of plus and times, they studied this axiomatic system in detail, which has come to be known as Z. Despite not having the exponential function, one could still use it to express, e.g., all the specific cases of Fermat's thereom for each exponent $n > 2$, so that it was by no means a trivial system. Moreover, since Ackermann's attempted consistency proof for a system allowing all ordinary recursions as axioms had become bogged down in difficulties concerning the assignment of ordinals to recursive definitions to measure their complexity, systems like Z appeared to present easier targets for consistency proofs.

Skolem, meanwhile, in his fundamental paper of 1923 basing arithmetic on the 'recursive mode of thought', showed that a considerable amount of ordinary number theory could be done on this basis, essentially a free-variable formulation of primitive recursive arithmetic. In fact, Skolem felt that he had finally provided a "foundation for mathematics" that satisfied Kronecker's (K), for it

> . . . is a consistently finitist one; it is built upon Kronecker's principle that a mathematical definition [*Bestimmung*] is a genuine definition if and only if it leads to the goal by means of a *finite* number of trials ([255], p. 333).

It is quite possible that both Skolem and Hilbert believed in 1923 that *all* effective or finitary functions could be defined by primitive recursion.[16] Nevertheless, Ackermann refuted this hypothesis in 1925 by using a nested double recursion to define a clearly effective function of one variable which grows faster than any function definable by an ordinary primitive recursion. The essential clause in Ackermann's recursion was in effect

$$(3)' \qquad F(z + 1, m, n + 1) = F(z, m, F(z + 1, m, n))$$

which, along with suitable boundary clauses, summarizes in a single double recursion the infinite sequence (3) of ordinary single recursions on p. 52 which arose from Dedekind's theorem. By an obvious generalization of the iteration principle this F already had to be admitted as effective, but Ackermann showed how to actually set up recursion equations for it, and also that its diagonalization $F(z, z, z)$ grew faster than any function definable by primitive recursion. (3)′ was called 'nested' to bring attention to its greater degree of 'self-involvement' than double recursions that had been hitherto studied: previously F had indeed occurred on the right side of its defining equations, but it could only be substituted there for a variable of a *known* function of z and n, whereas on the right side of (3)′ values of F occur *nested in F itself*.[17]

Hilbert's 1925 lecture on the infinite contains a general discussion of the significance of Ackermann's discovery, as well as recursion in general, for the foundations of arithmetic. Hilbert points out that, given the "elementary means" of substitution (composition) and recursion ("schema of the derivation of the function value for $n + 1$ from that for n"), all known methods of explicitly defining an arithmetic function by "elementary processes obtained from arithmetic operations" can be reduced to these given means.[18] The significance of this reduction is that

> The method of search for the recursions required is in essence equivalent to that reflection by which one recognizes that the procedure used for the given definition is finitary ([255], p. 388).

Let us formulate this as a thesis about definitions formulated in elementary arithmetic:

(HT)′ Arithmetic definitions can be recognized as 'finitary' $\Leftrightarrow$ they can be reduced to 'recursions'.

Hilbert then emphasizes, however, that in view of Ackermann's result, any attempt to 'survey' the permissible recursions can restrict attention to the 'standard form' (of ordinary recursion) only at the cost of permitting the recursion variables to range over functions as well as numbers. Thus he introduces in effect a functional Ψ of one number and one function variable which is implied by Dedekind's recursion theorem (2) on p. 51, i.e.,

$$f = \Psi(a, g) \Leftrightarrow f(0) = a$$
$$f(n + 1) = g(f(n)).$$

Ψ can now be used to define an infinite sequence of numerical functions by the scheme

$$(3)'' \qquad F_0 = \alpha,$$
$$F_{z+1} = \Psi(\mathscr{B}(z), F_z),$$

where α is some initial function, $F_{z+1} = f$, $F_z = g$, and $\mathscr{B}(z)$ a 'boundary number' for F_{z+1}. Thus, for the general term F_{z+1} we have, by the above definition of Ψ, the recursion

$$(3)''' \qquad F_{z+1}(0) = \mathscr{B}(z),$$
$$F_{z+1}(n + 1) = F_z(F_{z+1}(n)).)$$

If we now write $F(z, n)$ for $F_z(n)$ we recover essentially the nested double recursion (3)′ without the parameter m. This shows that more complex recursions like (3)′ can be reduced to the standard *form*, viz. (3)″, if we allow functionals taking functions as arguments. Hilbert refers to the likes of (3)″ as 'schemas of general recursion', and sometimes also as 'higher recursion', and emphasizes that they must enter into (HT)′. But these still do not exhaust the recursions needed for (HT)′, for there is a generalization of ordinary recursion, according to Hilbert, which must be considered, namely 'transfinite recursion':

The generalization of ordinary recursion is transfinite recursion: it rests upon the general principle that the value of a function for a value of the variable is determined by the preceeding values of the function (*ibid.*, p. 386).

These 'preceeding values' are to be given by a well-ordering having an arbitrary ordinal of Cantor's second number class as its measure. We might formulate the transfinite generalization of ordinary recursion as

$$f(0) = a$$
$$f(n + 1) = g(n, f(T(n + 1)))$$

where g and T are given functions and $T(n + 1) \prec_\alpha n + 1$, for some ordinal α, called the 'level' of f. At the time, Hilbert thought that by finding suitable recursions of ever greater complexity he might be able to settle Cantor's continuum hypothesis affirmatively by correlating, with each number-theoretic function, the ordinal which measured the level of its recursive definition. Together with his claim that Cantor's diagonal method could be used to construct recursive definitions of ever higher level, this seems to indicate that his overall picture of effectiveness was essentially that described on p. 178 above by Rogers; in particular that the class of effective functions might form an uncountable spectrum.[19] In any event, Hilbert does sketch a formalism in which these successively higher types of recursion could be formalized, but at the cost, it seemed, of forever having to extend the formalism *itself* into the transfinite: he was struggling with Cantor's curse on finitism. One idea he had for easing the difficulty was to *arithmetize* the process of formalizing these recursions, but this proved to be "too complicated":

Der Gedanke, die symbolischen Formeln in solcher Weise durch Aufeinanderfolgen von Ziffern darzustellen und dadurch die Regeln für die Bildung der symbolischen Formeln und für die Ausführung der formalen Beweise in arithmetische Regeln zu übersetzen, was von Hilbert bereits im Zusammenhang mit seinen Überlegungen zum Cantorschen Kontinuumsproblem gefasst worden. Es shien aber damals, dass die Durchführung dieses Gedankens sich allzu kompliziert gestalten wurde ([107], p. 216).

Presumably the use of this arithmetization, already implicit in (HT), was envisaged for the proof of a basic 'lemma' asserted by Hilbert to the effect that higher and transfinite recursions could always be eliminated from the "formation of functions of a number-theoretic variable."[20]

Let us now try to summarize the basic problem about recursion raised for Hilbert by Ackermann's discovery. He wanted to formalize the possible recursions, but in view of the unexpected "variety of ways of passing from n to $n + 1$" revealed by Ackermann, he faced a certain dilemma. Either (i) in order to preserve a fixed standard form of recursion whose application can be studied without loss of generality, we allow higher recursions such as (3)″, in which case the treatment is not elementary and hence its consistency neither obvious nor easy to prove, or (ii) we allow only the transparently consistent recursions on numerical variables, in which case Ackermann's result for (3)′ raises the acute problem of specifying the most general form of such recursions. Ackermann had seemed to open the floodgates of recursion and diagonalization, and it hardly seemed possible to find any *single* new recursion principle for defining new functions from given ones which could yield all cases not covered by Dedekind's principle of simple or primitive recursion.

We now conclude this section by indicating a general 'thesis' implicit in much of the foregoing, and to which (HT), (R), (KH), and (HT)′ all gravitate, namely,

(R)′ The effectively computable functions are those definable by the 'recursions of the most general kind'.

Though implicit in much of Hilbert's writing in the 1920's, (R)′ does not explicitly emerge until Herbrand and Gödel. But this is still a far cry from the thesis (CT) of Church which identifies effectiveness with the precisely defined mathematical concept of recursiveness: (R)′ is only a kind of intuitive 'meaning postulate' which analyzes one intuitive concept in terms of *another intuitive concept*. The idea of the 'most general recursions' appealed to by (R)′ is really no less informal or intuitive or vague than that of effectiveness itself. It would presumably make no sense to ask for a refutation of (R)′ itself, for it just seems to express what some mathematicians have *meant* by 'effective'. Nevertheless, it is very useful inasmuch as it puts (CT) in a somewhat different light than usual: it can be seen as posing the problem whether the precise mathematical definition of recursiveness stemming from Gödel and Kleene really comprises 'recursions of the most general kind'. For, as Péter has pointed out,

> ... the reason for which general recursion was called 'general' consists in the circumstance that we are unable to imagine any more general kinds of recursions ([203], p. 223).

2. THE FORMALIZATION OF DIAGONALIZATION

As Hilbert pointed out in 1928, his problem of the 'finitization' of Dedekind's isomorphism result actually presupposed the completeness of the existing formulations of the rules for pure predicate logic. As we have seen, there had been serious doubt about this possibility, which Gödel's dissertation [99] ended on a happy note. Gödel's investigation also brought out an important connection between the completeness of elementary logic and the *Entscheidungsproblem*, for as he points out, his completeness theorem

> ... kann ja als eine Art der Entscheidbarkeit aufgefasst werden (jeder Ausdruck des engeren Funktionenkalkuls kann entweder durch endliche viele Schlüsse [mit ganz bestimmten konkret aufgezählten Schlussregeln deduzieren] als allgemein giltig erkannt oder seine Allgemeingiltigkeit durch eine Gegenbeispiel widerlegt werden) ([99], p. 4).

But he then adds that a strictly 'intuitionistic' completeness proof for logic could only be given by a positive solution to the *Entscheidungsproblem* itself, and he explicitly warns in this connection against using too general a '*Lösbarkeitsbegriff*' by way of rejecting the following kind of use of his completeness theorem to settle the *Entscheidungsproblem*. To tell whether a given formula *A* is valid, start enumerating all possible deductions of Gödel's complete axiomatization of logic, looking for one ending in *A*, and simultaneously, try to find, using "*alle überhaupt erdenklichen Hilfsmitteln*", counterexamples to *A*. Assuming the law of the excluded middle, we must sooner or later find either the proof or the counterexample. Gödel regarded it as questionable whether such a *Lösbarkeitsbegriff*' had any meaning at all. Now what is interesting about this hypothetical 'method' for solving the *Entscheidungsproblem* is that it is essentially the very same argument used many years later by Kalmar against (CT).[21] Gödel's point, of course, was that although his completeness theorem allowed one to be sure that this "British museum" search would eventually find a proof of *A* if it *was* valid, the arbitrary "conceivable" means employed to find counterexamples to *A* are too vague and general to permit any specific conclusion concerning the fate of this simultaneous search. Kalmar's formulation would instruct us here to try to *prove* the existence of such counterexamples by 'arbitrary correct means', but this is hardly any better, or even really any different.

Gödel, however, in his incompleteness paper a year later was able to add an interesting point of clarification for the problem raised by this argument; in fact, he was able to isolate its weakness very sharply. Thus, in his 1931 paper he proves (Theorem IX) that in any formal system including number theory and any further set theory desired, there exist undecidable problems of elementary predicate logic, "that is, formulas of the restricted function calculus for which neither validity nor the existence of a counter-example is provable" ([255], p. 612). He then explains why this does not conflict with his completeness theorem:

> ... I showed that every formula of the restricted function calculus either can be proved to be valid or has a counterexample. However, by Theorem IX the existence of this counterexample is *not* always provable in the formal systems we have been considering (*ibid.*, p. 612).

We see that this sharpens his 1930 critique of the hypothetical method above for solving the *Entscheidungsproblem:* all the '*erdenklichen Hilfsmitteln*' (or Kalmar's 'arbitary correct proofs') we would need for establishing counterexamples could never be formalized in any given formal system, however strong the assumptions it may postulate, and so they could hardly be as effectively enumerable as the possible proofs for *A* itself in that hypothetical method. This is the 'protection' referred to earlier (p. 9 above) that the incompleteness theorem provides for (CT): any attempt, such as Kalmar's or the one sketched by Gödel, to solve the decision problem effectively, runs into this roadblock of extreme generality. Whether it is one of the *greatest possible* generality – that is the question we must examine.

We shall enumerate Gödel's discoveries in a way that can be directly compared with the earlier work (D1)–(D6) and (H1)–(H6) of Dedekind and Hilbert respectively.[22] To this end, we formulate Gödel's fundamental breakthroughs as follows:

(G1) Formal system *A* with axioms (P1), (P2), and a provably *complete* set of axioms and rules for elementary logic with identity.

(G2) Proof that A is consistent if its intuitionistic counterpart is consistent.[23]

(G3) A is extended to P by addition of rule of induction and rules for *higher* logic.

(G4) Proof that P is deductively incomplete; in particular, Dedekind's isomorphism result (D4) cannot be 'finitized' in Hilbert's sense. (cf. p. 161, above).

(G5) Proof that the consistency of P cannot be proved in P.

(G6) (i) All *general recursive* functions are formally calculable in P.

(ii) All *general recursive* functions can be given explicit *arithmetical* definitions using just addition, multiplication, and *elementary* logic with identity.

It would be hard to exaggerate the importance of these contributions for the program of formalized number theory. And as in Dedekind, *the decisive breakthrough lies in the recursion theory*. Thus, in (G6) one finally had the basis for a proof that the system Z of Hilbert and Bernays was both expressively and experimentally complete: amidst the tendency to speculate whether the existence of outstanding unsolved problems such as Fermat's theorem either explain or are explained by the incompleteness, there is also the tendency to forget that before Gödel nobody realized that Fermat's theorem could even be fully *expressed* in Z. Gödel showed for the first time according to Hilbert and Bernays

> ... dass der Formalismus des Systems (Z) nicht nur ... schwierige Probleme der Zahlentheorie zu formulieren imstande ist, sondern dass er überhaupt *eine Formalisierung der gesamten Zahlentheorie liefert*. Es sind nämlich in diesem Formalismus bereits alle die Funktionen vertretbar, welche durch Rekursiongleichungen eingefuhrt werden, und zwar auch unter Zulassung der (in der Rekursiven Zahlentheorie betrachteten) Erweiterungen des Rekursionsschemas ([106]), pp. 381–2).

We recall (p. 162 and p. 184ff.) that these matters, and not deductive completeness, were felt to be of prime importance for their program. That Gödel finally made it possible to establish the experimental completeness of Hilbert's formal arithmetic has been noted by Kreisel with all the force and irony it deserves:

> ... before Gödel's work it was not even realized that all multiply recursive predicates could be defined in first order arithmetic, even when symbols for primitive recursive functions are added; although multiply recursive predicates were accepted finitist. In fact, the schemata for recursions of higher type which Hilbert considered ... could, at least at first sight, be expected to 'transcend' even analysis. On the other hand, as we

know now, the *proofs* in present-day number theory can indeed be rather easily formalized in Hilbert's own formal system Z of arithmetic by means of devices introduced by Gödel. So, while in Hilbert's days the empirical evidence (from 'ordinary' number theory) for the adequacy of Z was slight, at the present time it would be overwhelming! ([13], p. 174).

Hilbert and Bernays see the significance of (G6) (ii) for formalism in the fact that it allows one, in an elementary system with recursion axioms for 'plus' and 'times', to formalize Dedekind's proof for his recursion theorem, thereby *eliminating the function and set quantifiers from his and other standard proofs of it.*[24]

The decisive technique which enabled Gödel's proof of incompleteness [64] to succeed where earlier attempts, such as Finsler's had failed, was his arithmetization of his formal system F. Of course, this in turn required his basic advances in recursion theory; in particular, he defined for the first time in a precise way the class of *primitive recursive* functions by means of which he was able to arithmetize the metamathematics of his system. The two central constructions of his arithmetization are

(GA) (i) a primitive recursive *proof-predicate* $P(x, y)$ for F,
(ii) a primitive recursive *substitution function* $s(x, y)$ for F,

which explicitly and rigorously establish the theses (HT) and (KT) of Hilbert and König, respectively. (GA) enables Gödel to arithmetize, and by (G6)(i) also to completely formalize in F, the kind of diagonal argument which Finsler had only sketched conceptually. Thus Gödel composes his proof-predicate with the diagonalization $s(y, y)$ of his substitution function to form $P(x, s(y, y))$. Then, if the (formalization of the) formula $(x) - P(x, s(y, y))$ has the g.n. p, the formalized diagonal argument shows that if F is consistent, $(x) - P(x, s(\bar{p}, \bar{p}))$ must be undecidable in F. However, in his 1934 Princeton lectures he pointed out that this diagonal construction and result generalize to *any* (arithmetized syntactical) predicate which can be expressed in F. Thus, if the expression $P(s(x, x))$ has g.n. q, $r = s(q, q)$ and A_r is the sentence with g.n. r, we can show by the formalized diagonal argument that A_r is provably a 'fixed-point' for the predicate $P(x)$, saying of itself, 'I have property P', i.e.,

(4) $\quad \vdash_F A_r \leftrightarrow P(\bar{r})$

If P is the predicate 'not-provable', then A_r is the familiar Gödel-sentence which is unprovable in F unless F is inconsistent. Such a sentence has the interesting property that it may be read in two different ways: first, as the very long, complex, and obscure statement of F about *numbers* that A in fact is, or, secondly, and by virtue of the mapping between language and metalanguage, as if it were a very short, simple, and clear statement A_r' *about* F and its *proofs*. Moreover, since read as A_r' it says 'I am unprovable', we need only the simple (from the point of view of metalanguage) hypothesis that all provable sentences of F hold true, as well as truth preservation by the mapping, to establish A_r metamathematically. This shows that in order to prove A_r in F we have to add new axioms to F; in particular, Con(F), expressing the consistency of F, will suffice, showing furthermore that Con(F) was not in fact provable in F as long as F is consistent. This also nicely settles Hilbert's problem about 'finitizing' Dedekind's isomorphism of Peano models.

But for mechanism, the importance of Gödel's formalization is that *he has brought Finsler's undecidable sentences down from 'rein Gedankliche' and put them back into the formal system.* In short, he *formalized Finsler's diagonal argument*. This is important for two reasons: (i) now machines can compute undecidable sentences, and (ii) such sentences promise to make it hard, as we saw in Chapter I, to refute the claim that machines can do everything that we can do effectively. Later I shall argue that Gödel's unprovable sentences are nothing but 'guardian angels' protecting this very claim.

If we try nevertheless to turn an unprovable A_r against mechanism, we get stuck, as Putnam [212] has emphasized, on the fact that we do have

(4)′ $\quad \vdash_F \mathrm{Con}(F) \rightarrow A_r.$

So we must also prove Con(F), which is unlikely if F is very complicated, according to Putnam. Lucas [173] tries to get unstuck by asserting his own consistency, while Nagel and Newman [193] have raised the question of whether Putnam [212a] does not dissipate the force of his argument after all by his acceptance of Gentzen's consistency proof. What Lucas overlooks is the ambiguity of the term 'the formal expression of F's consistency': Con(F) is only one of many such expressions, some of which *are* provable in F, without, of course satisfying (4)′. In fact, a study of Lucas' explanations of his own consistency makes it clear that what he means is *not* formalized by Con(F): he describes it as halting, tentative, and self-correcting! Are there any formal expresions in F of this kind of 'belief in consistency'? I do not know, but will only remark now that if F is Peano arithmetic, then it will have a statement of consistency whose 'assertive force' lies somewhere between Lucas' and Con(F). Thus, if $P(x, y)$ is the normal 'r.e. expression of proof' of Feferman [78] and $k = \ulcorner 1 = 0 \urcorner$, then Con($F$) is $(x) \neg P(x, \bar{k})$, which like Gödel's A_r remains unprovable despite the provability of each of its numerical substitution instances. However, if we abbreviate $\neg P(x, \bar{k})$ as $A(x)$, we *do* get

$$(4)'' \qquad \vdash_F (n)(\exists p)P(p, s(\ulcorner A(x) \urcorner, \ulcorner n \urcorner)),$$

which is a nice provable expression of consistency after all. But while it seems to cover what Lucas described as his consistency, it does *not* yield A_r.[25] But this only works for number theory, and Lucas might reply that at this level he knows Con(F) because he can follow Gentzen's proof. But then we should have been discussing something more complex than number theory to begin with after all, whereupon Putnam's point will loom large again.

A seemingly more sophisticated incompleteness argument against mechanism has been formulated by Kreisel [155]. Following him, we let $\mathscr{P}$ stand for a 'restricted or preferred class' of proof principles, not necessarily formalized (e.g. Hilbert's '*finiten Schliessen*'), and F for a formal system which is '*intended* to be about (abstract) objects which have no place in $\mathscr{P}$' (e.g. some formal set-theory). Finally, let $A_{\mathscr{P}}$ stand for that proper subclass of formulae of F which are meaningful for $\mathscr{P}$, and for $A \in A_{\mathscr{P}}$ we simply let A denote its own interpretation in $\mathscr{P}$. $P(x, y)$ is the proof-predicate defined in $\mathscr{P}$ for F. Kreisel then formulates a 'generalized' Hilbert problem as one of showing, by means in $\mathscr{P}$, that the reflection principle

$$(*) \qquad (\exists x)P(x, \ulcorner A \urcorner) \rightarrow A$$

holds for all $A \in A_{\mathscr{P}}$. This problem contains the consistency problem as a special case. Such a $\mathscr{P}$-proof of $(*)$ could be regarded as a 'reduction', in the following sense. In general, an x for which $P(x, \ulcorner A \urcorner)$ holds will be a (Gödel-number of a) proof which contains formulae not in $A_{\mathscr{P}}$, i.e. a 'non-constructive' proof. Such formulae may refer to abstract (infinite) objects. But from a $\mathscr{P}$-proof of $(*)$ it follows that, if x is a proof of A in F, then there is a $\mathscr{P}$-proof of A, i.e. one containing only constructive or finitary assertions and "thus one may say that *all reference to abstract meaning* (involved in the intended interpretation of F) *has been eliminated*" ([155], p. 234). As for mechanism, Kreisel had meanwhile indicated that

Probably the major attraction of formalization was that it suggested the possibility of a *mechanistic theory* of human reasoning, in particular, that the propositions A above not only can be decided by means of the formal rules, but that something like the repeated application of such rules is all that goes on even if we consciously think of reasoning differently; more precisely, that the higher nervous system consists of a *mechanism* whose behavior is given by the formal rules, as an electronic computer is a mechanism whose physical behavior realizes certain mechanical laws (the 'instructions' which it is given) (*ibid.*, p. 226).

But now from Gödel's two theorems it follows that for suitably strong F, *if F codifies $\mathscr{P}$*, then $(*)$ is not $\mathscr{P}$-provable (for all $A \in A_{\mathscr{P}}$). Kreisel now goes on to claim that such a $\mathscr{P}$ would be required for 'analogues' to mechanism:

Also one would need such a $\mathscr{P}$ to support the following analogue to the mechanistic theory [quoted above]: the true reason why we accept F is the elementary proof of $(*)$, and the derivations in $\mathscr{P}$ are what really goes on in reasoning (*ibid.*, p. 236, note).

So Gödel's theorem appears to thwart even this reduction. In both cases the point would presumably be that it shows the impossibility of reducing "references to abstract meaning" to formal or 'finitary' manipulations. But does such an application of Gödel's results hold any more water than those previously considered? No.

In the first place, Kreisel's argument is already weak on the metamathematical level alone, for it depends on the assumption that F *completely codifies* $\mathscr{P}$. But this is *just* what Gödel's theorem renders highly implausible, as was pointed out long ago by Gödel himself. After all, many accept Gentzen's proof as finitary. In any event, the concept of finitary proof remains without any explication comparable to what (CT) provides for that of finitary definition.

Secondly, if F is indeed a strong set theory, then all we seem to really have is our *belief* in $(*)$, which is hardly based on any *proof*. Indeed, Gödel's independence proof for $(*)$ can hardly imply anything about the mental or physiological processes leading to such belief. And even if $(*)$ were established by the most restricted $\mathscr{P}$ you like (and, in particular, it was not independent for sufficiently strong F), it still would not follow that 'derivations in $\mathscr{P}$ are what really goes on in reasoning'. In fact, nothing at all would follow about the mechanism of the higher nervous system, any more than anything about its structure follows from the way we 'consciously think of reasoning'. It would be absurd to imagine that we could introspect effectively about our neurophysiology, period – no matter what results turn up in logic. But it is equally difficult to imagine what a mechanist interested in modelling neurophysiological structure could learn from some $\mathscr{P}$-proof of $(*)$, even one which established the decidability of F for good measure. Indeed, *it would not even follow from the* *decidability of all mathematics that the behavior of the nervous system was somehow 'recursive'.* Consider, for example, an instance of (∗) which is neither $\mathscr{P}$-provable nor derivable in F, but can be established by an extension $\mathscr{P}'$ of our methods. Now we have no reason to suppose that the physiological accompaniments of proofs of such instances of (∗) are different in any important and recognizable way, say, more 'abstract', than those of $\mathscr{P}$-provable instances. Indeed, if thinking does in general involve something 'going on in the nervous system', why should we imagine this to be any less so when we think 'abstractly'?

Actually one could argue that our knowledge of the truth of A_r is really based on more concrete reasoning than is usual in number theory, *for it does not refer to 'numbers' at all, but rather to a syntactical interpretation of* A_r, i.e. to its relations to certain formal manipulations. The usual locution that we establish A_r by 'intuitive means' although it is not formally decidable really gets things almost upside down. For to establish something 'intuitively' in any theory presumably means that we appeal in some rather unsystematic way to our intuitions about the *objects* whereof the theory treats, i.e. that we intuit somehow that it holds in the 'intended model'. But anyone who has ever expanded one of these A_r into the primitive notation of number theory soon realizes the futility of trying to 'see' it as a truth 'about numbers'. But Gödel shows us how to see it as a *syntactical truth about symbols*. 'Seeing the truth' of unprovable sentences is often motivated by the 'ω-incompleteness' of formal systems. Thus $(x)A(x)$ may be unprovable in F even though we have

$$\vdash_F A(\bar{1}),\ \vdash_F A(\bar{2}),\ \vdash_F A(\bar{3}),\ldots$$

for every numerical instance. Seeing the *provability* of each of these infinitely many instances is what is taken for 'seeing the truth of $(x)A(x)$'. But in these cases *F will also prove just what we see*: (4)″ will always hold for such $A(x)$. Returning to Kreisel's argument, we know from Löb [167] that any sentence A is provable in F just in case (∗) is provable in F for that A, hence that (∗) is unprovable in F for our $(x)A(x)$. So nothing more would be 'really going on' in our proof of (∗) for $(x)A(x)$ than in our proof for $(x)A(x)$ itself, which we have just seen to involve something rather more syntactical than abstract. To involve something which, in fact, is 'really going on' in F itself.

Gödel has pointed out that (4) leads straight to Tarski's theorem on the indefinability of truth. Given any predicate P of F, call the r of (4) a provable *truth-point* for P: if we knew P's logical behavior only at such points we could not distinguish it from a 'truth-predicate' for F, for it would so far provably satisfy Tarski's criterion

$$\text{(T)} \qquad P(\ulcorner A \urcorner) \leftrightarrow A$$

on truth definitions. But by (4) *every* P has provable truth-points, hence every $\neg P$ must also have them. By the formal liar paradox then, no P of F has *every* point as a truth-point. In particular, every P has provable *non*-truth-points, and thus provably fails to satisfy (T) for all A. Now Lucas sees this result as crucial, for "when we are comparing minds with calculating machines, the relevant assessment of their performances will be in respect of their truth and falsity. And in this crucial respect there is a difference between minds and machines. Machines do not have an adequate concept of truth, as Tarski's theorem shows" ([177], p. 310). Since we can 'see the truth' of Gödel's one quantifier unprovable A_r, it seems plausible to explain incompleteness by our ability to operate with a truth concept not available in F. But the unprovability of A_r is *not* due to the lack of a suitable truth concept in F, for Hilbert-Bernays [107] have shown that for every n there is a predicate T_n of F, defined with $n + 1$ quantifiers, which provably satisfies (T) for all A^n with $\leqslant n$ quantifiers, i.e.

$$(T_n) \qquad \vdash_F T_n(\bar{r}) \leftrightarrow A^n_r.$$

So F can define an infinite sequence of ever more comprehensive 'local truth-predicates' for itself, but even T_1 is already important, for it defines truth for all sentences with one quantifier, and, indeed, *these include the only kind of undecidable sentences which we have yet been able to see as 'intuitively true'*. It has been suggested that the human mind cannot handle more than four or five quantifier alternations, hence Rogers emphasizes that although we have proved the existence of arithmetical truths of arbitrary quantifier complexity, nevertheless, "we have done so by diagonalization arguments *rather than direct insight*" ([221], p. 323; italics mine). In any event, the Hilbert-Bernays result (T_n) shows that we can hardly be satisfied with the 'semantic' explanation of the difference between man and machine. For all sentences that will ever matter F has an adequate concept of truth. What F does not have, of course, is any effective means of *deciding* which sentences are true, even those with one quantifier. But this is simply because there *is none, for man or machine*. It is non-computability and not semantics which explains the 'limitations' of F. To see this clearly, we need only formulate explicitly the connection which has emerged between the concepts of formality and computability. Where f is any function, we write $\bar{f}$ for the wff of F expressing f. Then F is *formal* just in case only computable functions f are expressible by wffs $\bar{f}$ which *provably* represent f in F, i.e. for each f:

$$(n)(m)[f(n) = m \Rightarrow \vdash_F \bar{f}(\bar{n}) = \bar{m}] \Rightarrow f \text{ is computable.}$$

Suppose now that h is a non-computable function expressible in F by applying quantifiers to decidable relations. It then follows from this by *classical* logic that there will exist numbers n_0 and m_0 for which

$$h(n_0) = m_0,$$
$$\nvdash_F \bar{h}(\bar{n}_0) = \bar{m}_0.$$

These are just the sparks that fly off F as its deductive processes grind away trying to formally compute a non-computable function: to say that some evaluations of h are true but formally unprovable is simply to say that h is non-computable. For by (CT), there is *no* effective method for computing h, formal or otherwise. The point about incompleteness is not that formal systems are 'missing some of our intuitions', but rather that the processes they *are* capable of expressing may be effectively undecidable, according to (CT), *no matter how many of our intuitions they might formalize.*

A standard and eloquent statement of the view that incompleteness shows that we 'cannot formalize our intuitive number concept' is due to Weyl:

> ... if the logic of the infinite is taken into account, it seems impossible to axiomatize adequately even the most primitive process, the transition $n \rightarrow n'$ from an integer n to its follower n'. As K. Gödel has shown, there will always be constructively evident arithmetical propositions which cannot be deduced from the axioms however you formulate them, while at the same time the axioms, riding roughshod over the subtleties of the constructive infinite, go beyond what is justified by evidence. We are not surprized that a concrete chunk of nature ... challenges our analysis by its inexhaustibility and incompleteness; ... However, it is surprising that a construct created by the mind itself, the sequence of integers, the simplest and most diaphanous thing for the constructive mind, assumes a similar aspect of obscurity and deficiency when viewed from the axiomatic angle ([266], p. 445).

But this common view, so plausible at first sight, does not hold up on closer examination. To see why not, we consider some weaker formal systems than Hilbert's Z.

Let us first consider the strict axiomatization P_m of Dedekind's theory of the Peano model which uses *monadic* second-order logic, the so-called 'monadic second-order successor arithmetic', and has 0 and $'$ as its only primitives. Despite having only monadic predicates available in the language of P_m, we can introduce the binary relation $a = b$ of identity by definition. The non-logical axioms of P_m are then:

(P1) $a' \neq 0,$
(P2) $a' = b' \rightarrow a = b,$
(P3)′ $(A)\{[A(0) \wedge (t)(A(t) \rightarrow A(t'))] \rightarrow (t)A(t)\}.$

Despite its second-order comprehension axiom, the restriction to monadic predicates might seem to prevent P_m from formalizing anything of interest at all. However, a number of useful and familiar relations between numbers besides identity can be introduced by definition, such as order and congruence to any specific modulus n.[26] Addition itself, as a binary function, cannot be defined in P_m, but an interesting variety of other processes can be.

The principal interpretation of P_m is the model $\langle N, 0, '\rangle$ where N is the natural numbers, and it formalizes considerably more about them than Weyl's 'primitive processes'. Moreover, P_m is complete and decidable so that N and those functions and relations on N that it *can* express appear here without 'obscurity and deficiency'. However, it is also possible to interpret N as the points of a discrete time scale and the predicates as 'basic processes' varying in time between two 'states' T and F ('on' or 'off'). The course in time of a finite number of these can be simply called a 'process', and it turns out that the theories of both *nerve nets* and *finite automata* are models of these 'processes'. A basic problem in these interpretations is to determine whether processes (finite automata) exist satisfying given conditions expressible in P_m. The crucial cases, which cannot be solved by the comprehension axiom, are those where a predicate is defined by recursion in terms of itself. Thus, if $\mathscr{L}(a, E(t);\ t < a)$ means that the wff $\mathscr{L}$ contains the predicate variable E whose argument t is bounded by a, then we want to know if we can prove in P_m the existence of a predicate E satisfying

$$E(a) \leftrightarrow \mathscr{L}(a,\ E(t);\ t < a).$$

A recursion theorem can indeed be proved in P_m for every such wff $\mathscr{L}$, namely,

(5) $(\exists P)(z)[P(z) \leftrightarrow \mathscr{L}(z, P)].$

The proof is similar to Dedekind's, with differences due to the restriction to monadic predicates.[27] This recursion theorem is the key tool in proving the completeness and decidability of P_m. We could now explain this lack of 'obscurity and deficiency' in P_m by reference to the fact that *the finite automata whose behavior it describes have no such obscurity and deficiency in their behavior*. Their halting problems, e.g., are all effectively solvable. But we shall see that Gödel's (4) leads to a very powerful existence theorem for automata whose behavior may be so complex as to be unpredictable – just what the doctor ordered for mechanism.

Consider now the first-order theory A of addition with the axioms (P1), (P2), recursion axioms for addition and the induction schema. A formalizes the process of addition of numbers, and those definable in terms of it. But A is complete and decidable, so the number sequence shows no 'aspect of obscurity and deficiency' when viewed from this 'axiomatic angle'. True, there are still familiar functions and relations over N not expressible in A, but N itself is intact, in all its infinitude, and A completely describes the two most basic processes on N, namely, counting and adding. So it can't be just the infinite number sequence *itself* that defies axiomatic description. In fact, it is also present and available for both addition and multiplication in Tarski's complete and decidable algebra, but cannot be isolated from the real field by any formula of his language. Of course, if we add the recursion axioms for multiplication to A we do get some 'obscurity and deficiency'. Must we say then that as long as we are content to just add our numbers we can keep them in perfect axiomatic focus, but that as soon as we start to add and multiply them they somehow suddenly elude such focus and become accessible only to the constructive mind? No, the number sequence itself doesn't go out of focus, but rather the processes now definable from 'plus', 'times', and logic: by (G6) (ii) every general recursive function is now definable, and due to the presence of unbounded quantifiers, so are non-recursive functions. But such functions hardly represent processes which are 'simple' and 'diaphanous' to the constructive mind, for according to (CT) there *can be no effective method whatever for dealing with these processes*. So they better give rise to some 'obscurity and deficiency': otherwise (CT) would hardly be plausible (cf. p. 198).

The familiar claim that incompleteness revealed the fundamental limitations in the axiomatic method in mathematics is strikingly at odds with the historical fact that this method in mathematics has mushroomed spectacularly since 1931. Would a dose of mathematical logic in the training of all mathematicians have changed this? No. The axiomatic method can be questioned, but hardly this way: much more to the point are the following words of Dana Scott:

> In a way the axiomatic method has been *too* successful. Propositional Calculus, First-order Logic, Elementary Geometry are all completely axiomatized. Incompleteness creeps into Number Theory, Second-order Logic, Set Theory, but few people seem to worry. The axioms we have for these theories go very far, and the unprovable truths are all terribly 'unnatural'. Most of us understand better what *can* be shown rather than what *cannot*; hence there is a psychological gain in self-confidence as soon as some *axioms* are written down ([231], p. 262).

Again, the highly theoretical nature of this incompleteness that 'creeps' into number theory. In fact, one could argue that the incompleteness really has nothing to do with number theory per se, for both it and the incompleteness that creeps into second-order logic can both be derived already from the undecidability that has meanwhile crept into our complete axiomatization of first-order logic, or alternatively, from the unpredictable behavior of the machines it can describe. But it is no more my intention than it was Scott's to belittle the incompleteness theorems – though many mathematicians do just that – but rather to explain them in a 'positive' and natural way. For I believe it is only when presented as 'negative' results that they begin to strike one as 'unnatural', for, as we have noted, our axiomatizations of number theory are indeed experimentally complete. I would like to now use the historical perspectives developed above, as well as an interesting remark by Bernays, to explain them positively, for indeed,

> Dieses bedeutende Theorem ist übrigens keineswegs bloss als ein negatives Ergebnis zu beurteilen, vielmehr hat es im Bereich den Beweistheorie eine ähnliche Rolle wie etwa die Entdeckung der Irrationalzahlen im Bereich der Arithmetik ([22], p. 11).

In fact, the pythagorean discovery of incommeasurability was also an incompleteness result, showing that the Greek system of rational arithmetic could not completely represent all geometrical ratios. Their arithmetic was incomplete with respect to describing continuous geometric magnitude. The Greeks never introduced irrational *numbers*, but were aware that incommeasurable geometric ratios could be indefinitely approximated by an infinite series of rational ones, and hence had a dim notion of limit. Just as the logically satisfactory completion of the real number system – whose irrationals are the limits of sequences within the incomplete system of rationals – depended on, and was necessary for, a satisfactory explication of the central notion of a continuous function, so too we see now that

'ω-incomplete' formalisms for number theory having proofs for each sentence of some infinite series $A(\bar{1})$, $A(\bar{2})$, $A(\bar{3})$, . . . though not for its 'limit' $(x)A(x)$ – its system of proofs being rather like the rationals – were essential for the satisfactory explication of the central notion of a *computable* function. Indeed, only this incompleteness helps to insure that these computable functions are not, analogously to the uncountability of the reals, *effectively enumerable*. And just as incommeasurability was the discovery that the euclidean algorithm for finding the least common multiple of two segments did not always terminate, in particular not when applied to the diagonal and side of a euclidean square, so too incompleteness was the discovery that certain algorithms depending on provability in formal systems do not always terminate, in particular that any algorithm for

$$f(n) = \begin{cases} 1, \text{ if } \vdash_F A_n, \\ 0, \text{ if } \vdash_F \neg A_n, \end{cases}$$

is necessarily partial for suitable F – not terminating for Gödel's undecidable sentences. And just as there had been numerous unsuccessful attempts to rigorously introduce the irrational numbers before Weierstrass and Dedekind, so too there had been the equally unsuccessful and paradoxical attempts of Hessenberg [115], Richard [219], and Finsler [82] to introduce such sentences before Gödel's final success. Of course, it was not originally his purpose to clarify the computable function by means of a theory having such sentences, but, in retrospect, it may as well have been: I will argue later that undecidable sentences are nothing but the guardian angels of computability theory, specifically of Turing's thesis. They are also paradigms of theoretical entities: they are never actually seen themselves with the naked eye – being too long and far away from ordinary arithmetical experience – but are known only through a metamathematical code giving them a meaning which, although 'unnatural', is more manageable than their natural number-theoretical meaning would be. We hide the undecidables themselves behind codes, much like Nietzsche's peacock who hides his tail from every eye and calls it his pride.

Finally, in analogy with Tarski's question about elementary geometry, we ask: what is 'elementary arithmetic'? We have seen that, historically, there were two quite different conceptions. First, there was the older algorithmic conception which prevailed for centuries, and, much more recently, the more theoretical deductive conception of Dedekind, stemming from Gauss, which identifies the 'science of arithmetic' with the set of theorems derivable from his explication of the natural number system as a Peano model. The incompleteness that creeps into any effective formalism for Dedekind's conception is only a kind of 'invisible protective shield' for the apparently complete development of the algorithmic conception contained therein, a very remarkable development indeed. Few mathematicians thought there was any chance for it. Equally remarkable was the fact that all of the theorems of Gauss's 'higher arithmetic' are formalizable in a suitable such formalism. As Scott remarked, our axioms here go very far. This brings us to the decisive issue raised by Gödel's work for mechanism: just how 'general' are his recursions in (G6)? Mechanists can only hope that he was completely justified in so calling them.

3. THE GENERALIZATION OF RECURSION

Gödel begins his 1934 lectures [100] by defining a 'formal mathematical system' as one for which there is a 'finite procedure' for telling whether a formula is either meaningful or an axiom, and also whether one formula is an immediate consequence of another by the rules. Next he gives, preparatory to the arithmetization of the metamathematics of his formal system, a precise description of the primitive recursive functions (at the time he just called them 'recursive functions'), and he then relates this precise concept to that of his 'finite procedures':

> Recursive functions have the important property that, for each given set of values of the arguments, the value of the function can be computed by a finite procedure (*ibid.*, p. 43).

This is essentially the 'easy half' of the thesis (HT)′ of Hilbert and Skolem. Gödel then adds a footnote in which he expresses his belief in a *suitably generalized converse*:

> The converse seems to be true, if, besides recursions according to the scheme [of primitive recursion], recursions of other forms (e.g. with respect to two variables simultaneously) are admitted. This cannot be proved, since the notion of finite computation is not defined, but it serves as a heuristic principle (*ibid.*, p. 44).

Now, since at the end of his lectures he gives a precise definition of a class of functions which he calls *general recursive*, it seems that this 'heuristic principle' is really (CT) itself. Indeed, we may wonder why we now speak of 'Church's thesis' rather than 'Gödel's thesis'. Gödel's editor, Professor Davis, wondered about the same thing to Gödel, whose letter of explanation is paraphrased by Davis along with a revealing summary of the whole situation:

Since Gödel's characterization of a 'formal mathematical system' uses the notion of . . . a 'finite procedure' for carrying out the rules, an exact characterization of what constitutes a finite procedure becomes a prime requisite for a completely adequate development. Church's thesis identifies the functions which can be computed by a finite procedure with the class of general recursive functions. In the present article Gödel shows how an idea of Herbrand's can be modified so as to give *a* general notion of recursive function . . . Gödel indicates that he believed that the class of functions obtainable by *recursions of the most general kind* were the same as those computable by a finite procedure. However, Dr. Gödel has stated in a letter that he was, at the time of these lectures, not at all convinced that *his* concept of recursion comprised all possible recursions, and that in fact the equivalence between his definition and Kleene's . . . is not quite trivial ([61], pp. 39–40; italics mine).

So what Gödel was claiming in 1934 was essentially the thesis (R)′ on p. 188 equating effectiveness with the most general recursions, which we saw emerging in the 1920's from various earlier and more fragmentary proposals concerning the significance of recursion. The penultimate proposal leading to Gödel's own precise definition, however, was supplied by Herbrand, who also believed in (R)′, and whose explication of 'most general recursion' was formulated in a letter to Gödel, who in turn quoted it in his Princeton lectures as follows:

(H) "One may attempt to define this notion as follows: if ϕ denotes an unknown function, and $\psi_1, \ldots, \psi_k$ are known functions and if the ψ's and the ϕ are substituted in one another in the most general fashions and certain pairs of the resulting expressions are equated, then if the resulting set of functional equations has one and only one solution for ϕ, ϕ is a recursive function" (*ibid.*, p. 70).

Shortly before his death, however, Herbrand published another account of functions computable according to 'intuitionistic procedures', which for him was synonymous with 'effective'. In fact, this second account actually occurs as part of the very description of a "theory . . . which is a *formal* translation of classical arithmetic" ([255], p. 623) which Herbrand was trying to prove consistent. The primitives are 'identity', 'zero', and 'successor' and the crucial axioms are those of 'Group C' introducing functions. As examples of schemata permissible for introducing new functions from old he cites simple recursions and nested double recursion, but his general formulation of this group of axioms runs as follows:

(H)′ "We can also introduce any number of functions $f_i(x_1, \ldots, x_n)$ together with hypotheses such that . . . *Considered intuition-*

> *istically, they make the actual computation of the* $f_i\,(x_1\,,\,\ldots,\,x_n)$ *possible for every given set of values, and it is possible to prove intuitionistically that we obtain a well-determined result* (*ibid.*, p. 624).

Herbrand explains the locution 'considered intuitionistically' in a footnote: "This expression means: when they are translated into ordinary language, considered as a property of integers and not as a mere symbol" (*ibid.*). Presumably this just means: informally. In any case, we see that Herbrand's 'formal translation' of arithmetic is hardly a formal system in today's sense. Indeed, this very fact is exploited by Herbrand to show that Gödel's result (G5) on consistency proofs does not apply to his arithmetic, for one can only carry out Gödel's argument, says Herbrand, for a 'definite theory', whereas his own theory does not have this property because of the impossibility of 'describing outright' all the functions which can be introduced by means of his axiom group *C*, viz. (H)′:

> One could object that it is perhaps possible to describe at once all the schemata included in *C* for the construction of functions. This would mean that it is possible to describe outright all intuitionistic procedures for the construction of number theoretic functions. But this is impossible. For, if this were the case, we could, by an intuitionistic procedure, number all the functions of one variable, $f_1(x), f_2(x), \ldots, f_n(x), \ldots$; then $f_x(x) + 1$ would be an intuitionistically defined function that would not be among the functions previously listed, hence a contradiction . . . but to carry out Gödel's argument, we have to number all objects occurring in proofs; we are thus led to construct the function $f_y(x)$ of two variables; this justifies what we were saying above, namely, that it is impossible, in an arithmetic containing the hypotheses *C*, to formalize Gödel's argument about this arithmetic (*ibid.*, p. 627).

So Gödel's problem was how, for the sake of describing outright 'every recursive function' – and also of the generality of his (G5) – to avoid Herbrand's iteration and diagonalization. His point of departure was Herbrand's (H). What was still missing, however, in (H) – and in (R)′ itself for that matter – was an account of the *rules* by which the values of a function are to be determined from a 'general recursion'. We cannot remain content with (H) alone, for there might well be a set of recursion equations such as it describes, for which a unique solution *exists* – in the sense of classical mathematics – but whose values, each and every one, could not actually be derived in a finite number of steps by rules fixed in advance.[28] In short, the 1892 requirements (F)(i)-(ii) of Hilbert on effective algorithms must surely be satisfied also. Our recursion equations won't really be effectively usable unless we are told explicitly how to use them.

Gödel fixes two rules in advance, one of substitution and one of replacement, and then defines a (total) function to be *general recursive* if its values can all be derived in a finite number of steps from a set E of equations as described in (H) by these rules.[26] Now once Gödel had given a precise description of the permissible equations and chosen a symbolism in which to express them, it was easy to see that one could effectively enumerate all sets E of equations that could *possibly* define a general recursive function: but it was by no means clear – and this was the fallacy lurking in Herbrand's argument for the impossibility of describing all computable functions – that one could effectively identify those sets E which *actually did* define such a function. After all, the claim that such an E does define a general recursive f has roughly the general form

$$(6) \qquad (n)(\exists m)[E \vdash f(\bar{n}) = \bar{m}],$$

and such propositions, when suitably arithmetized in Gödel's sense might well turn up as undecidable sentences in very strong formal systems. Indeed, Gödel had already demonstrated that the application of quantifiers to the arithmetization of symbolic deductive relations and processes was a prime source of such sentences. So if, for example, every formal system satisfying Gödel's incompleteness theorem did so also for the arithmetizations of some propositions of form (6), then the effective iteration claimed in Herbrand's argument *would be every bit as incredible as the use of his completeness theorem to solve the Entscheidungsproblem that Gödel warned against in his dissertation* (cf. pp. 189ff.). At this point we turn to the work of Kleene [146] who actually verified this hunch amidst his searching examination of Gödel's general recursion concept.

We recall that Gödel failed to assert (CT) in 1934 only because of doubt that his recursions were the most general; in particular, the proof by Kleene that his own class of recursive functions based on the minimalization principle was identical with Gödel's class was, as Gödel later noted, 'not quite trivial'. Kleene's functions, the so-called μ-recursive functions, are defined as the smallest class (i) containing the successor, projection, and constant functions, (ii) closed under composition and primitive recursion, and (ii) containing

$$(\mu) \qquad f(n) = \mu x[g(n, x) = 0]$$

for each primitive recursive g for which $(n)(\exists x)[g(n, x) = 0]$ holds. Thus, unlike Gödel's general recursive functions, the μ-recursive functions are not defined by reference to any *formalism* for expressing equations whose calculations proceed by fixed rules governing the use of that formalism: like the primitive recursive functions, they are each defined by a finite number of applications of fixed definitional schemas, starting from a given set of simple initial functions. In particular, no explicit reference is made to any rules for calculating the values of these recursions. Now, as we have noted (cf. p. 176), the absence of any explicit rules raises no great problem for the primitive recursive functions, for the structure of their definitions is such as to make it intuitively obvious that their values can always be calculated in a foreseeable number of steps. In the case of (μ), however, which is analogous to the use of the archimedean axiom in geometric constructions, no particular method of calculation is as obviously implicit in this definition of f. Thus, it would be reassuring to know that this f could always be defined recursively by a set of Gödel's equations and its values derived by his rules. Kleene showed that this could be done as follows: given a set of equations for $g(n, x)$, add two more equations defining a new function h as follows:

$$h(0, n, x) = x,$$
$$h(z + 1, n, x) = h(g(n, x + 1), n, x + 1).$$

Kleene then shows that the f of (μ) can be recursively defined by the additional equation:

$$(\mu)' \qquad f(n) = h(g(n, 0), n, 0),$$

whereupon its values become derivable by Gödel's rules, rendering the method of computation for f officially and uniquely fixed.[30]

In order to show that all of Gödel's functions are μ-recursive, Kleene adapts Gödel's method of arithmetization to the formalism for defining equations and its rules. First he constructs, analogously to Gödel's proof-predicate, a primitive recursive predicate $T(e, n, x)$ which holds when e is the g.n. of a set E of equations with the principal function symbol $\bar{f}$, and x is the g.n. of a derivation from E of the equation $\bar{f}(\bar{n}) = \bar{p}$, where no $y < x$ has this property. Then he constructs a primitive recursive $U(x)$ which extracts the value p from the proof of this equation. Thus he is able to construct a normal form for Gödel's functions:

(7) (i) $f(n) = U(\mu x T(e, n, x))$,

(ii) $(n)(\exists x) T(e, n, x)$.

This explicit normal form (7)(i) for each Gödel function shows that they are all μ-recursive. It also leads to Kleene's important enumeration theorem for partial recursive functions.

Kleene then takes up the diagonalization problem and proves formally that the diagonal procedure, applied to a recursive sequence of recursive functions, does not lead outside the class of recursive functions. This leads to the theorem that the set of g.n.'s e for which (7)(ii) holds is not recursively enumerable. In short, the general recursive functions are not r.e. Now, in any formal system F satisfying Gödel's incompleteness theorem the predicate $T(e, n, x)$ is formalizable, say by $\mathscr{T}$, and hence there must exist a number e such that

(6)′ (i) $(n)(\exists x)T(e, n, x)$,
(ii) $\nvdash_F (x)(\exists y)\mathscr{T}(e, x, y)$.

For otherwise we could recursively enumerate all theorems of F having the form $(x)(\exists y)\mathscr{T}(\bar{n}, x, y)$ to obtain such an enumeration of all recursive functions.[31] This shows that there will always be true propositions of form (6) whose formalized arithmetizations cannot be proved in any given formal system, confirming the weakness we noted in Herbrand's argument for the impossibility of describing all effective functions. Indeed, these true but unprovable sentences (6)′(i) are *just the guardian angels which look after the formalization of effectiveness found in any suitable F*: they protect (CT) from refutation by the diagonal argument. If we try to construct Herbrand's diagonal function for such an F, we find that we can effectively generate only an infinite sequence of *provably* recursive functions: the diagonal function *will* then be formally computable in F but it will not be provably total, i.e. the g.n. of its formal algorithm will satisfy (6)′. This resolves the Richard paradox for effectiveness in favor of (RP)″(ii).[32] These results also explain Kleene's motivation for the definition of Gödel's recursive functions. Known examples of recursive definitions had always displayed two features: (i) they were defined by a finite set of equations, and (ii) they were 'definitions by mathematical induction, in one form or another'. Gödel, however, chose to incorporate only feature (i) into his definition of 'recursive function' and Kleene nicely explained why:

This choice may seem unexpected, since the word 'recursive' has its root in the verb 'recur', and mathematical induction is our method for handling recurrent processes. The meaning of the choice is not that feature (ii) will be absent from any particular recursion, but that it is transferred out of the definition itself to the application of the definition. To show by finitary means that a given scheme has feature (i), except in trivial situations, one will presumably have to make use of mathematical induction somehow. But in defining the totality of general recursive functions, we forego the attempt to characterize in advance in what form the intuitive principle of induction must manifest itself. By Gödel's theorem we know that the attempt at such a characterization by the formal number-theoretic system is incomplete ([149], p. 274).

This is comparable to the use of Hilbert's rule of 'infinite induction' as opposed to ordinary formal induction discussed before (see p. 153 ff.). One presumably has to use some kind of induction to prove that each of the infinitely many premises of Hilbert's rule hold, i.e. to apply the rule. Hence 'induction' is merely transferred out of the formulation of the rule to its application. In both cases, this transfer can be motivated by undecidability.

Finally, Kleene proves a stronger incompleteness theorem somewhat closer to Gödel's. The diagonal predicate $(x)\text{-}T(n, n, x)$ is easily shown to be non-r.e. From this it follows, by an argument similar to our informal one for (4) above, that for every suitable formal system F there is a number p such that

(6)″(i) $(x)\text{-}T(p, p, x)$,

(ii) $\nvdash_F (x)\mathscr{T}\neg(\bar{p}, \bar{p}, x)$.

We lose no generality, however, if we assume that this is the T-predicate for *Turing machines*.[33] Thus (6)′ and (6)″ show that the source of incompleteness may be assumed to lie in *the inability of any formal system to completely describe the behavior of machines.*

The predicate $(x)\text{-}T(n, n, x)$ figures in Kalmar's argument against (CT). Since $(\exists x)T(n, n, x)$ is an r.e. predicate, Kalmar proposes to 'compute' it as follows: given any n, search through $T(n, n, 1)$, $T(n, n, 2)$, . . . for such an x and simultaneously try to prove by 'arbitrary correct means' that no such x exists. If it exists, our search will turn it up. Hence, if this 'method' is not successful, as (CT) implies, there must exist, according to Kalmar, a number p such that $(x)\text{-}T(p, p, x)$ is true, yet unprovable by any correct means at all, which is an 'implausible consequence' of (CT). Hoering [131a] has analyzed Kalmar's argument and showed that he actually needs for this consequence of (CT) the additional assumption that the set of numbers n for which $(x)\text{-}T(n, n, x)$ is 'provable by correct means' is effectively enumerable. But this is very implausible, if not meaningless, in its own right, and doubly so in the light of (6)″, for F could include all the set theory you wish (if consistent). The assumption of the effective enumerability of all intuitively correct proofs, as well as certain variants of it, also plays a role in attempts to refute (CT) by means of intuitionistic principles. For discussion, consider

$$(7)' \qquad (f)(\exists e)(x)(\exists p)[T(e, x, p) \wedge f(x) = U(p)]$$

which, read classically, says falsely that all functions are recursive. But the intuitionist reads (7)′ as saying that all constructive or effective functions are recursive: for him there *are* no others. Thus arises the possibility, allegedly foreclosed to classical mathematics, of seeing (CT) as a mathematical statement, capable of proof or refutation like any other. In fact, Kreisel and others have proved the consistency or inconsistency of (7)′ with various proposals for explicating Brouwer's ideas about the subjective nature of mathematics. The main interest has been in principles implying the negation of (7)′, for "on the basis of the intuitionistic foundations of mathematics a strictly mechanistic characterization of effectivity does not seem probable".[34] For Kreisel, (CT) is essentially an 'axiom of reducibility' for constructive mathematics, and hence a 'promising' area of such research is any that "looks abstract enough not to be reducible to mechanical operations".[35] Such an area is the theory of Brouwer's 'creative subject' introduced to prove certain mathematical theorems by means of functions defined to depend on whether 'the creating subject has experienced the truth' of a given proposition, and Kripke has introduced a method for formalizing such proofs. The primitive is $\Vdash_n A$, read "the creative subject has evidence for A at stage n," and the following axioms are asserted:

$$\begin{aligned} \text{(CS)} \quad &\text{(i)} \ \Vdash_n A \vee \neg \Vdash_n A \\ &\text{(ii)} \ \Vdash_n A \to A \\ &\text{(iii)} \ (\Vdash_n A \wedge m < n) \to \Vdash_m A \\ &\text{(iv)} \ A \to \neg\neg(\exists n) \Vdash_n A \end{aligned}$$

(i) states the decidability of $\Vdash_n$, while (ii) states that it has the force of proof. (iii) says that the subject never forgets, while (iv) asserts that it is absurd that he will never have a proof of A if it holds.[36] Given any A we can now define a function which is effective by (i):

$$\alpha(n) = \begin{cases} 1, \text{ if } \Vdash_n A, \\ 0, \text{ if } \neg \Vdash_n A. \end{cases}$$

One can now derive, using (CS) and intuitionistic logic, the so-called *Kripke schema*:

$$\text{(KS)} \qquad (\exists\alpha)\{[(\exists n)(\alpha(n) \neq 0) \to A] \wedge [(n)(\alpha(n) = 0) \leftrightarrow \neg A]\},$$

which plays the role of a comprehension principle for functions in the intuitionistic systems. It has been shown by Kleene that his proof that

$(x) - T(n, n, x)$ is not r.e. can be formalized in such systems, and by Kripke that by taking this predicate as the A in (KS) one gets a proof of the negation of (7)′. Of course, intuitionistically, this only means that it is absurd that all functions can be proved to be recursive, *not* that we can actually exhibit an effective computation method for $(\exists x)T(n, n, x)$ or any non-recursive function! Indeed, the intuitionistic examination of Church's thesis has been comprised solely of showing that it is or isn't *formally consistent* with this or that intuitionistic formal system – which could hardly have impressed Brouwer. The irony of this has caused some writers to soften claims originally made that this argument 'refutes Church's thesis'.[37] As for (KS) itself, the only known basis for it is (CS), which, as we mentioned, is fraught with serious problems of interpretation. In particular, it implies that all possible proofs can be arranged in an ω-sequence, which acutely conflicts with the impredicative nature of intuitionistic implication.[38] At any rate, we are left, as Kreisel admits, with "the question whether *constructive rules f defined without reference to such an hypothetical ω-order of mathematical proofs are equivalent to recursive functions*" ([156], p. 128). But this also raises the question of whether there are provable existence principles which would enable us to prove the existence of recursive functions satisfying any conditions which our intuition could vouchsafe as effectively satisfiable. This brings us back to Kleene's work.

Outside of the fact that no counterexamples to (CT) are actually known, the most convincing evidence for it is commonly believed to be the provable equivalence of several explications of effectiveness which differ widely both in formulation and in their intuitive and conceptual background. These various explications can, however, be seen as falling under a small number of conceptually similar groups, each containing several variant formulations of an underlying intuitive idea, and the equivalence proofs then can be seen as showing the 'stability' of these ideas themselves, quite apart from their connection with effectiveness. At least three such ideas are (i) 'recursiveness', (ii) 'representability' in a 'formal system', and (iii) 'computability' by a 'machine' or 'algorithm'. A major step in the stabilization of recursiveness was taken by Kleene in his proof of the equivalence of μ-recursiveness with Gödel-recursiveness sketched above. However, if we are to really understand the evidence for (CT), and hence (CT) itself, we must examine equivalence proofs between concepts falling under different conceptual groups, at least to the point of isolating key ideas, for they have too often been taken for granted, with the result that discussions of (CT) tend to end just where they should really begin. As an example, I will discuss another equivalence proof of Kleene's, perhaps the most decisive of all such proofs, namely his proof of the equivalence between recursiveness and λ-definability, a concept which falls under group (ii) above. In particular, I believe that in this way we can better still appreciate how deeply (CT) depends on the Gödel incompleteness construction.

We have already mentioned the close connection between Church's system of λ-conversion and the combinatory logic of Schönfinkel and Curry. As it happened, the original aim of both of these developments was the formulation of a logic adequate to the foundations of all mathematics without having to introduce such *ad hoc* distinctions as Russell's 'theory of types' to avoid the paradoxes. It was Kleene [145] who undertook the detailed study of number theory in the λ-system, with particular attention to those functions of the integers which were 'formally definable' in it. Church had, in fact, already suspected that these 'λ-definable' functions might provide a good approximation of effectiveness.[39] Because the conversion rules in the λ-system had, as in combinatory logic, been broken down into such 'microscopic processes', the development was quite laborious, but Kleene soon found that a certain class of self-referential constructions, similar to that of the Gödel incompleteness constructions in (4), were of basic importance. In fact, they were similar in essential respects to the diagonal fixed-point combinators discovered by Curry in his study of the Russell paradox in combinatory logic (cf. p. 184ff). This was true to some extent already in building up to the primitive recursive functions, and much more so as he tried to λ-define functions 'recursive in various more general senses'. These diagonal constructions are much too elaborate, as they appeared in the λ-system itself, to present here. However, Kleene [147] was soon able to distill out a basic theorem that summarized all of them, and which forms the basic tool in the proof that all recursive functions are λ-definable: it is called his recursion theorem.

In addressing himself to the question of what functions are λ-definable, Kleene emphasizes that it was only a special case of the larger problem concerning the 'self-reflexiveness' of the λ-formalism: "what functional relationships among well-formed expressions can be expressed by well-formed formulas?" ([147], p. 343). After first selecting a suitable class of wffs to stand for the integers and the initial functions, and then establishing closure under composition, Kleene proves the above-mentioned theorem which takes care of both primitive recursion and minimalization as special cases. Briefly, he shows that, given any λ-formula F with two free variables, one can construct another λ-formula L such that

(5)′ $\quad L(x)$ conv $F(x, L)$, for all x.

Kleene explains that this result

> . . . constitutes a schema for circular definition. Given any set of conditions of dependence of any entity $L(x)$ on the variable natural number x and L itself, if the set can be expressed in the λ-notation by a formula F, a formula L satisfying the conditions in terms of the equivalence relation A conv B can be found. To do this it need not be known that the conditions actually determine a function $L(x)$ (*ibid.*, p. 347).

In order to explain Kleene's use of this result, as well as its connection with Gödel incompleteness contained in (4), we present it in the form in which Kleene [148] proved it for the formalism of recursive functions, and in connection with Kleene's further extension of that class to the *partial recursive* functions. Thus, if in the definition of Gödel's general recursive functions, we relax the requirement that a unique value be derivable for every argument from a set of equations to requiring only that at most one such value be derivable, we obtain the class of partial recursive functions which includes the general recursive functions as a subclass of special cases of partial functions whose defining equations happen to yield values for every natural number as argument. The partial recursive functions turn out to be precisely those computed by Turing machines. Hence everything that follows can be regarded as really about them. An immediate advantage of working with this wider class is that the normal form theorem (7) now yields Kleene's enumeration theorem: as z runs through the integers, the predicates $U(\mu yT(z, x, y))$ and the corresponding one for functions of two variables now yield a recursive enumeration

(E)′ $\quad \phi_1(x), \phi_2(x), \phi_3(x), \ldots, \phi_z(x), \ldots,$
$\quad\quad \psi_1(x, y), \psi_2(x, y), \psi_3(x, y), \ldots, \psi_z(x, y), \ldots,$

with repetitions of all partial recursive functions of one and two variables. The fact that (E) contains repetitions of *each* function, in fact infinitely many, is essential. We shall see that it is impossible to effectively eliminate this 'redundancy' from the system (E)′ of partial recursive functions.

We return now to our discussion on pp. 179–80 of the requirements which our effectiveness intuition imposed on the hypothetical system (E) of all effective partial functions. The enumeration requirement (e) on p. 179 is satisfied now, for $U(\mu yT(z, x, y))$ is indeed a partial recursive function of two variables, say $\psi_u(z, x)$, and we have the following analogue to Turing's universal machine:

(8) $\psi_u(z, x) \simeq \phi_z(x)$, for all z, x.

If we now diagonalize to form $\phi_z(z) + 1$ we do not get outside the enumeration (E)′: if this function is ϕ_d we get only that $\phi_d(d)$ must be undefined. Kleene now proves a fundamental theorem which formally verifies our intuitive requirement $(s)''$ on closure under substitution, sometimes called the 'parameterization' theorem, of which we shall need just one special case. Kleene formally constructs a primitive recursive function s satisfying

(9) $\psi_z(k, y) \simeq \phi_{s(z,k)}(y)$, for all z, k.

This result constitutes an adaptation and generalization of Gödel's substitution function to the formalism of recursive functions, and enables Kleene to formally satisfy the strong form $(d)'''$ of diagonalization on p. 180. Kleene now composes a given partial recursive $\psi(x, y)$ with this diagonalization $s(x, x)$ to form another partial recursive $\psi_d(x, y) = \psi(s(x, x), y)$. Taking $x = d$ gives $\psi_d(d, y) = \psi(s(d, d), y)$. But taking $z = k = d$ in (9) gives also $\psi_d(d, y) = \phi_{s(d,d)}(y)$. Hence $\phi_{s(d,d)}(y) = \psi(s(d, d), y)$. Setting $r = s(d, d)$, we have the theorem that for any partial recursive ψ there is a number r such that

(10) $\phi_r(x) \simeq \psi(r, x)$.

This is a special case of Kleene's *recursion theorem* (sometimes called the 'fixed-point' theorem: given a recursive f, define the partial recursive $\psi(x, y) = \phi_{f(x)}(y)$ and the r of the theorem will satisfy $\phi_r(y) = \phi_{f(r)}(y)$, a fixed point for f considered as a mapping of the partial recursive functions). These results, whose proofs are closely related to Gödel's (4), are the analogues to Kleene's own (5)′ for the recursive functions. We might call the r of (10) a 'recursion number' for ψ to emphasize its role in satisfying the 'arbitrary recursion' which ψ can be used to set up. The proof of (10) essentially transfers, to the formalism for defining recursive functions, Gödel's formal diagonal argument. There is also, as in Gödel's construction, a 'self-referential' aspect to these constructions which we will discuss in connection with Turing machines in the next section. Thus the "Gödel machine" M_r with g.n. r, which computes ϕ_r, will, given an input x, first print out its own g.n. r to the left of that input and then apply the machine which computed ψ to the two numbers r and x now on his tape. That the formalization of the self-reference associated with the traditional paradoxes has turned out to be so intimately related with the 'self-definition' involved in definitions by recursion – indeed, on the metamathematical level on which the Gödel-numbering plays an essential role they are virtually identical – seems to me to constitute one of the most peculiar and interesting conceptual connections revealed by modern logic. At first sight, it would seem that these two kinds of self-reference had nothing to do with each other. In particular, it seems to indicate that self-reference or 'reflexiveness' is an essential property of effective processes in general. I shall say a few words about this also in the next section. Meanwhile, Kleene has argued that (10) and related results are the decisive support needed for (CT) and I would now like to examine the basis for this. I will begin with two examples not explicitly mentioned by Kleene.

First, let us see that (10) does indeed supply that basic protection needed for (CT), namely, the undecidability results. Since we have hitherto appealed to Gödel's incompleteness theorem for this kind of 'rear guard action', this will also enable us to see that (10) does indeed include as special cases results having the 'negative' significance usually attributed to Gödel's result. As is well known, the basic undecidability result underlying the 'limitative' theorems is the unsolvability of the 'halting problem', i.e. determining effectively whether $\phi_z(x)$ is defined for arbitrary z, x. Were it solvable there would be a recursive $h(z, x)$ taking the value 1 or 0 as $\phi_z(x)$ was or was not defined. But trivially then, $h(z, x)$ would be the totalization of the partial recursive ψ defined by

$$\psi(z, x) = \begin{cases} \text{undefined if } \phi_z(x) \text{ defined,} \\ 0, \text{ if } \phi_z(x) \text{ undefined,} \end{cases}$$

which is immediately contradicted by the recursion number r supplied by (10) for this ψ. The diagonal argument commonly used to prove this result is already built into the proof of (10) itself. Many related undecidability results can be obtained with similar ease, all showing that some method (say, involving diagonalization) contemplated for escaping from (E)′ would actually involve recursively undecidable steps.

The second kind of support given by (10) to (CT) is the circumstance pointed out above that it actually played the decisive role in establishing the most difficult and striking equivalence proof between two proffered explications of effectiveness yet known. In fact, Kleene has pointed out that several other equivalences of this kind are best established in this way, namely, by first establishing that the recursion theorem holds for each characterization. Rogers [220] has used the recursion theorem to sharpen the known equivalence results to isomorphism results. After explicating the intuitive concept 'Gödel-numbering of the partial recursive functions', he applies the recursion theorem to prove that any two such Gödel-numberings are isomorphic in the sense that there exists a recursive one-one map between them in such a way that corresponding numbers index the same partial recursive function. As an application he shows that from any known equivalence proof, say between Turing machines and Post normal systems, there exist such isomorphisms for which corresponding formal objects define identical partial recursive functions. We might compare this situation with Dedekind's attempt to defend his thesis (DT) by applying his recursion theorem to prove the isomorphism between any two Peano models. What is as yet lacking of course, for the case of (CT), is an agreed-upon 'Peano model for effectiveness'. However, the task of axiomatizing informal recursive function theory has only been taken up very recently. Already it appears that the 'uniformly reflexive structures' of Wagner are going to play a central role, and that the two crucial axioms characterizing these structures are that (the generalized versions of) our (8) and (9) hold for them. The latter imposes the 'uniformity' and (10) then gives these structures their 'reflexiveness'.[40] Friedman [94] has obtained some relative categoricity theorems in this area, as well as minimality results to the effect that a structure satisfying certain axioms must contain at least the partial recursive functions. The possibility of proving stronger results in this direction, say that any such structure satisfying all the axioms we could think of – along the lines of our intuitive discussion on pp. 176–80, above, – for effectiveness contained exactly the partial recursive functions, seems to me to weaken the basis for common opinion that (CT) could never be proved from the standpoint of classical mathematics. In any case, we could sum up the situation as follows:

In the beginning was the finest flower of the constructive mind, the intuitive number sequence. Dedekind gave in (DT) an adequate informal mathematical explication, using the finest flower of the non-constructive mind, set theory. Hilbert tried to make Dedekind's analysis acceptable to the constructive mind through formalization, raising three questions: (i) Can Dedekind's adequacy proof for (DT) be 'finitized' by this formalisation? (ii) Can his recursion theory be finitized in this way? (iii) Can his recursion theory be generalized to encompass all effective functions?

Gödel and Skolem have answered (i) negatively. But this apparently is only the very cheap price we have to pay for the remarkable fact that Gödel was able to lead us to answer (ii) and, apparently (iii), affirmatively. We can explain (ii) further by examining another argument for (CT) from (10) by Kleene.

We recall that Dedekind was able to prove an existence theorem (2), p. 51, whose only intuitive justification had been the closure principle (2)″ for effectiveness. In view of the difficulties encountered in providing a consistent foundation for the set and function quantifiers used by Dedekind, many have looked for more elementary constructive replacements for (2) which could nevertheless be shown to formally encompass a wider effective domain. Kleene observes that (8) and (10) provide this. Thus, assume that g is an effective total function of the system (E)′. Then we can compose it with the universal function ψ_u of (8) to define a partial recursive ψ as follows:

$$\psi(z, x) = g(\psi_u(z, x - 1)).$$

Taking now a recursion number r supplied by (10) for this ψ yields

$$\phi_r(x) = g(\psi_u(r, x - 1)),$$

and taking $z = r$ in (8) gives

$$\phi_r(x) = g(\phi_r(x - 1),$$

which recovers Dedekind's recursion clause. We could call M_r here the 'Dedekind machine' for g. To include the boundary clause we need only define

$$\psi(z, x) = \mu y\{(x = 0 \wedge y = a) \vee [x > 0 \wedge y = g(\psi_u(z, x - 1))]\},$$

and the recursion number r for this ψ provides the solution to (2). We can say that Dedekind's theorem has been 'finitized' in the sense that, because of our Gödel-numbering and the presence of ψ_u, we can reduce assertions of function existence to those asserting the existence of *numbers*. These recursion numbers, of course, provide in and of themselves only for the existence of *partial* recursive functions satisfying a given recursion. A separate inductive argument is needed to prove, if desired, the totality of ϕ_r. For our treatment of Dedekind's recursion it is indeed obvious; in fact, the simple 'fallacious' argument (2)′ on p. 53, above, will do now, for we already know by (10) that a (partial) recursive function *exists*: only the extent of its domain of definition is left open. In the general case, however, the separate argument required to show the desired totality may require a transfinite induction of a non-trivial kind.[41] For a more complicated example, we can apply the analogues of (8) and (10) for tertiary functions to solve an Ackermann-type double recursion whose crucial 'nesting clause' is $h(x+1, y+1) = h(x, h(x+1, y))$, by defining the partial recursive

$$\sigma(z, x, y) = \sigma_u(z, x-1, \sigma_u(z, x, y-1));$$

then by recursion we have

$$\phi_r(x, y) = \sigma_u(r, x-1, \sigma_u(r, x, y-1)),$$

and by universality of σ_u

$$\phi_r(x, y) = \phi_r(x-1, \phi_r(x, y-1)).$$

The full recursion with boundary clauses would be handled analogously as above. One very soon gains a sense of the striking power of (10) to provide solutions to the most arbitrary recursion equations one can think of, provided only they are expressible by partial recursive functions. In some cases, one may well initially doubt whether there could actually be Turing machines which computed their solutions, so that (10) supports (CT) in the interesting sense of providing confirming instances in cases where counterexamples might have seemed to momentarily lurk. Kleene, however, goes further and suggests that (10) can be used to prove the (partial) recursiveness of *any* (partial) effective process. In short, that it establishes the thesis:

(KT)′ Every effective partial function is partial recursive.

Kleene later formulated his argument for this in terms of Turing machines:

> A great stock of intuitively computable functions are known... Likewise we have a great stock of methods or operations (for obtaining new intuitively computable functions from others) which are paralleled by operations for building new Turing machines from given Turing machines. If there were a function which is intuitively computable but not Turing computable, it would have to be 'inaccessible' by any process of building up toward it from this stock of functions and operations already mastered ... the *recursion theorem* [is] applicable to the building process in general. It is hard to imagine how one could give a description, or a set of instructions, for a computation procedure that a human computer could follow, except by putting the description together out of simpler elements, already known, and then it would come under this theorem ([150], p. 240).

Now this argument is not decisive as it stands, for one can imagine that there might exist computation procedures that were 'put together' from known simpler elements, but only by some powerful novel method not covered by the theorem. What makes it convincing, however, is that the theorem covers several forms of the only method that history has yet suggested as a candidate for such an escape, namely, diagonalization. Indeed, Kleene's recursion theorem is the *strongest principle known for formally proving the partial recursiveness of intuitively effective partial functions* – and hence also for proving the existence of Turing automata. That we have discovered a class closed under known effective operations including diagonalization without even having to distinguish types or orders is seen by Gödel himself as a 'miracle': according to him the importance of recursiveness

> ... is largely due to the fact that with this concept one has for the first time succeeded in giving an absolute definition of an interesting epistemological notion, i.e. one not depending on the formalism chosen. In all other cases ... such a demonstrability or definability, one has been able to define them only relative to a given language, and for each ... it is clear that the one thus obtained is not the one looked for. For the concept of computability, however, ... the situation is different. By a kind of miracle it is not necessary to distinguish orders, and the diagonal procedure does not lead outside the defined notion ([101], p. 84).

As we have seen, this miracle, which is essential to mechanism, is possible precisely because of his so-called incompleteness theorems.

4. CHURCH'S THESIS AND ARTIFICIAL INTELLIGENCE

Turing himself, in contrast to Kleene's purely mathematical appeal to closure properties, tried to justify the identification of effectiveness with Turing-computability by a direct analysis of the effective human computation of Kronecker's 'computable real numbers' – a kind of Kantian deduction of the *a priori* conditions of effective computation and symbol manipulation. To motivate his approach we consider the relevance of Church's thesis to psychology.

Dennett has argued that a non-circular psychology must make no appeal to unexplained 'intelligence', which is indeed the basic *explicandum* of psychology: if it posits atomic tasks requiring intelligence, it must posit 'undischarged homunculi' to perform them, hence beg the question. But then, Dennett argues,

> If Church's thesis is correct then the constraints of mechanism are no more severe than

the constraints against begging the question in psychology, for any psychology that stipulated atomic tasks that were 'too diffIcult' to fall under Church's thesis would be a theory with 'undischarged homunculi ([68], p. 181).

Hence, he concludes, if non-circular psychology is possible at all, then (CT) would provide "almost total support for the claim that *artificial* intelligence deals in fact with all *possible* intelligence", and not just *mechanically* realizable intelligence. In fact, (CT) faces the same problem as psychology itself, since "the intuitively computable functions mentioned in Church's thesis are those that 'any fool can do', while the admissible atomic functions of a psychological theory are those that presuppose *no* intelligence" (*ibid.*). Indeed, an effective procedure is precisely one whose execution begs no question, but to develop a non-circular theory of effective human computation Turing had to face the fact that a 'human computer' does need intelligence – to follow rules formulated in a language he must *understand*.[42] His solution was to describe a hypothetical machine – his model of the human computer – which need only scan, print, and erase symbols on a tape along which it moves. Since these atomic tasks presuppose no intelligence, it follows that a non-circular psychology of computation is possible. But does the psychology of computation comprehend *all* behavior? More specifically, is the main cybernetical claim

(C) Every 'precisely described' human behavior can be simulated by a suitably programmed computer.

implied by (CT)? Dennett speaks for many when he says that (C) "is, or boils down to, some version of" (CT), say Turing's

(T)′ Every humanly effective computation procedure can be simulated by a Turing machine.

which did *inspire* (C), as is brought out in the following description by Minsky of the doubt dispelled by Turing's analysis:

Turing's paper must be viewed against the intellectural background of a variety of ideas concerning descriptions and processes ... *What processes can be described*? Surely the notion of description entails some language. Could any one fixed language admit descriptions of all describable processes? Can there be procedures which are, somehow, well defined, yet cannot be described at all? It could be argued that there might exist processes whose communication requires the transmission of a mental attitude ... which cannot be captured in any finite number of words – which must remain intuitive ([184], p. 104).

Babbage's formulations also seemed to reduce (C) to (T)′, but this seems to presuppose that all behavior is somehow a form of computation.[43] As we saw, this was assumed by Hobbes for what we might call all cognitive behavior. But (C) does not imply that every human behavior *can* be 'precisely described', any more than (T)′ says that every function is effective.

Turing's work was nicely complimented by the McCulloch-Pitts theory of 'neural networks' which answered another argument against mechanism, namely, that the functions of the human *nervous system* were too complex to be performed by any mechanisms. Their models are constructed from simple 'formal neurons' (logical gates) by simple connections including feedback loops, and can simulate many important psycho-neural functions. They show that a neural network becomes, when suitably supplemented with scanners for reading a tape and motor control to move and print on it, an 'organism' equivalent to a Turing machine, also conversely that

> Every Turing machine is realized by a suitably supplemented neural network.

which McCulloch and Pitts saw as a 'psychological justification' of (T)′, for "if a number can be computed by an organism, it is computable" by a Turing machine.[44] This, of course, is not decisive, but it is important for showing that 'organisms' synthesized 'from the inside' out of neural elements can realize the machines Turing arrived at only by analyzing the human computer 'from the outside' without any neurological hypotheses. Turing did, however, make one very general assumption about the internal constitution of the human computer, namely:

(F) The human computer is capable of only *finitely* many internal (mental) states.

which Turing justified by arguing that otherwise the mental states of the human computer "will be 'arbitrarily close' and will be confused". A more explicit formulation of Wang argues that (F) is a consequence of what he calls the 'principle of finiteness':

> . . . because these states must be somehow stored in the mind, in order that they can all be ready to be entered upon. An alternative way of defending this application of the principle of finiteness is to remark that since the brain as a physical object is finite, to store infinitely many different states, some of the physical phenomena which represent them must be 'arbitrarily' close to each other and similar to each other in structure. These items would require an infinite discerning power, contrary to the fundamental physical principles of today ([259], p. 93).

Post (1924) found the first undecidability theorems but hesitated to publish them until he could establish their generality by a "psychological analysis of the mental processes involved in combinatory mathematics" ([209], p. 338). Even after reaching his general concept of a 'normal system', he warned that "for full generality a complete analysis would have to be found of all possible ways in which the human mind could set up finite processes for generating sequences." A key assumption in Post's analysis is that his 'problem solver' is restricted to finitely many rules, which he says is evident from the fact that "the system of symbolizations in question is essentially to be a human product and each symbolization a human way of describing the original mathematical state." Meanwhile, Turing found his closely related machines, where the finitude of rules is replaced by (F), which connects (CT) with classical doctrines of the finiteness of the mind itself. Post valued (F) for its direct emphasis on the "obvious limitations" of our admittedly "dynamic mental world" surrounded by an "outer symbol space", and felt that his own metaphysical deduction of (CT) could be avoided if Turing's (F) could "bear up under adverse criticism", some of which has recently come from Gödel.[45]

For Gödel (T) means that "mental procedures cannot carry any further than mechanical procedures", but he objects that it depends on (F), in which Turing "completely disregards" that

(G) "Mind, in its use, is not static, but constantly developing".

as seen from its 'development' of set theory by ever stronger axioms of infinity. Of course, his own second incompleteness theorem makes it doubtful that the mind will ever be able to prove even the consistency of these axioms, let alone their truth. In any event, Gödel granted that (F) holds "*at each stage* of the mind's development", but says that

(G)′ " . . . there is no reason why this number [of mental states] should not converge to infinity in the course of its development" ([259], p. 325).

though he admits that to "accelerate and uniquely determine" this development would require a much deeper grasp of the mind than we now have. Thus Gödel admits that (F) is true *now*, but thinks it may fail in the 'limit' of our development. He also admits that Turing's argument becomes valid under two more assumptions: (i) there is no mind separate from matter, and (ii) the brain functions like a digital computer.[46] Though (ii) is very plausible to Gödel, he sees (i) as a prejudice that will be scientifically refuted, perhaps by showing that "there aren't enough nerve cells to perform the observable operations of the mind" ([259], p. 326).

Now we can surely agree with (G), but it hardly implies (G)′, nor even that the *number* of states – as opposed to their increasingly efficient use – is increasing at all. Recall here the point emphasized by M. Lockwood in his critique of Abelson's argument for the infinity of mental states, namely, that a given computation is humanly achieved only be passing through a *sequence of states*. Indeed, a universal machine can apparently compute any function we can with a finite number of states that remains forever fixed. So whether Turing 'disregarded' (G) is irrelevant. In fact, Post often emphasized (G), but still regarded (F) as obvious. The question is: what reason is there to suppose that (F) will ever fail? Gödel gave none. We might have seen Royce's idealistic interpretation of Dedekind's infinity theorem as supporting Gödel's vision in (G)′, but not even the theorem itself has survived the paradoxes. Even postulating that mind – now admitted as existing separately from matter – becomes capable of infinitely many states in its final development would not yet overthrow (T)′, for it would still have to be shown that we could make *effective use* of all these states. Indeed, this is just the challenge of Turing's argument for (F): to show that these states would not be 'confused'.[47] An infinite mind, by itself, gets us no closer to the construction of an effective non-Turing computable function.

Gödel, we recall, wanted to refute mechanism with his incompleteness theorem, but its generality depended on (T)′ to explicate what formal systems *are*. However, he came to realize that (T)′ was itself a strong, even if restricted, form of mechanism as we have several times emphasized. Moreover, Turing's argument for (T)′ clearly does establish *something*. What? According to Gödel and Kreisel, it only established

(T)″ Every *mechanically* effective computation procedure can be simulated by a Turing machine.

What then is the implied difference between 'humanly' and 'mechanically' effective? According to Wang's formulation of Gödel's views, our "intuitive notion does not require that a mechanical procedure should always terminate or succeed. A sometimes unsuccessful procedure, if sharply defined, still is a procedure, i.e. a well-determined namner of proceeding" ([259], p. 84). In other words, Gödel believed that Turing's analysis only establishes Kleene's generalization (KT)′ of Church's thesis for *partial* recursive functions. For the notion of mechanical procedure, Turing's analysis is "both correct and unique", showing that this notion "has the same meaning for the intuitionists as for the classicists" (*ibid.*). However, for the "more complex concept of always terminating procedures", the intuitionist and classical meanings may diverge since in this case the intuitionist asks for a *constructive* proof of (6)′(i). But *do* they diverge: are there general recursive functions whose totality cannot be constructively proved? Such an example would seem to refute the intuitionistically interpreted converse of Church's thesis:

> If a total function is Turing computable, then it is humanly effectively computable.

But Kleene has argued cogently that this rests on a misunderstanding:

> In defending this implication to an intuitionist, or to any other kind of constructivist who considers an algorithm to exist only when it is proved by his standards that it always works, we only ask him to accept the following: if the hypothesis that a function is Turing computable holds by *his* standards, so does the conclusion. Put thus, it is hard to see how it can be questioned. Only if one allows a non-constructive interpretation of the hypothesis, and yet insists on a constructive interpretation of the conclusion, is the converse of Church's thesis in doubt ([150], p. 241).

We know of no machine *known* to always halt, but only non-constructively so. Nevertheless, Wang sees a difference between the classical and constructive notions: whereas Turing only adequately explicated such notions as 'mechanical procedures' or 'algorithms':

> The related concept of effective procedures or constructive procedures, meaning procedures which can in the most general sense be carried out, suggests somewhat different elements which are related to the difference between mental and mechanical procedures, and the question as to the method by which a Turing machine ... is seen to be one which defines a Turing computable ... function ([259], p. 85).

But whether a procedure literally 'can in the most general sense be carried out' does not depend on whether or not its termination can be constructively guaranteed in advance, *but only on the execution of its atomic tasks*. If I can actually follow some procedure, however general, step by step, this would surely not be changed by forgetting that termination had been constructively proved. I may feel less like I am wasting my time if I believe an algorithm will terminate, but I would be no less able to *effectively execute these steps* than if I could not prove it, nor any more so if I could prove it in the most constructive way imaginable. Nor does Wang explain his puzzling suggestion that such questions are related to differences between mental and mechanical procedures. Indeed, if Hobbes was right, there *is* no such difference. But even if there is, why should mental procedures require any more assurance of their termination than mechanical ones? An intuitionistic proof of 6′(i) is surely no contribution to psychology.

Acording to Kreisel, Turing's (F) is "not unrelated" to the mechanist and formalist *petitio principii* of assuming that "our thoughts are less complicated than the objects, the infinite sets we think about" ([157], p. 320). He suggests rather that "human computations are more 'complicated', or, better, more abstract than the objects on which they operate – our thoughts may be more complicated than the objects thought about" (*ibid.*). But Turing argued that (F) poses no undue restriction on computation since "the use of more complicated states of mind can be avoided by writing more symbols on the tape" ([61], p. 136), so some clarification of the "complexity" of a thought or state of mind is sorely needed. In case of thoughts about *other thoughts* Kreisel's suggestion may well prove to be as paradoxical as Dedekind's *Gedankenwelt*, having already the odd consequence that one can only think about simpler thoughts. It might indeed be better to call our thoughts 'more abstract' than their objects – as long as these are not themselves thoughts – but this hardly affects Turing's (F), for it is independent of the way states are realized. States can be defined quite abstractly as classes of behaviorally equivalent histories, without compromising the motivation for (F) in the least.Indeed, Turing emphasized that (F) is essentially equivalent to "the fact that human memory is necessarily limited": it cannot distinguish all possible histories.[48] In any case, Kreisel produces no counterexample to (T)′, despite some interesting discussion of the possible relevance of recursive progression results to the problem.

A more plausible source of possible counterexamples to (T)′ would seem to be the translations of natural languages. Machine translation of them has remained more of a dream than a reality. Could this be due to a machine's inability to deal with *meaning* as such? As opposed to syntax. Indeed, as we have mentioned, Turing's analysis succeeded just because it *circumvented* the problem of how a computer can understand ordinary language. Translation, however, is a human procedure which deals with meaning if any procedure does: given a sentence in, say, German, we look for one in English with the same meaning. But is this a humanly effective procedure? Quine's principle of the indeterminacy of translaton would present a problem. Let xTy stand for 'y is a translation of x', c for a choice function, and $\mathscr{E}_g = \{e \mid gTe\}$. If Quine is right, then $f(g) = c(\mathscr{E}_g)$ is not only not an effective 'translation function', it is not really a function at all. Indeed, no objective disagreement could arise about what values f takes on, for it is essentially just a 'translation manual', and Quine [213] has argued that other such manuals f' can be found *inconsistent* with f, but still agreeing with f in stimulus meaning. According to Quine's principle, there are no objective facts which could decide between f and f'.

Gödel also expected the refutation of mechanism in biology by a theorem showing that human bodies would be too *improbable* on mechanistic grounds.[49] This is related to the classical teleological argument: an intelligent designer is needed to explain purposive systems, for their chance mechanical emergence is too improbable. Machines were teleological systems designed by humans for various purposes, by following designs in their minds. Such conscious rule-following was necessary to the existence of machines, for matter by itself, Broad argued, has no tendency to "fall into these extraordinary special arrangements". A causal explanation of purpose would make teleology unnecessary for both man and machine, but the human rule following acts needed to explain machines are themselves unpredictable and hence admit only *post hoc* explanation, as opposed to machine behavior which is predictable just because it is only *governed* by rules. So in addition to the challenges over predictability and reproduction (cf. Chapter I), mechanism must show that machines can follow as well as be governed by rules, and that they could arise in the most improbable ways.

In order to examine Turing's 'paper machines' with respect to these problems, let us imagine, as a *Gedankenexperiment*, that we have just designed, without any background knowledge of the metamathematics of arithmetic, the first seven such machines M_1, M_2, M_3, M_4, M_5, M_6, M_7 for the purpose of computing the functions of 'identity', 'successor', 'addition', 'odd-even', 'subtraction', 'doubling', and 'multiplication', respectively.[50] Unlike Gödel's numbering introduced for the theoretical purpose of proving incompleteness, our numbers are merely convenient names for these machines. Later, when our improving techniques for constructing new machines from old ones encourages the belief that we can find machines to compute most any 'ordinary' arithmetic function, we extend our 'bookkeeping' system to cover all machines in an arbitrary lexicographical ordering. It would be obvious, but apparently insignificant, that if there is one machine to compute a function, there are infinitely many. The need to distinguish between ϕ_z and ψ_z would be seen, but one of them would usually be thought of as the 'appointed purpose' of M_z, the others as 'extraneous' to it.[51] One would note that any M_z can in addition to its ψ_z, appointed or not, be used to compute infinitely many $\phi_{z'}$

appointed to other machines by simply fixing a parameter on its tape.[52] Indeed, it would even be noticed that, in some cases, by a kind of lucky accident, this parameter was actually identical to z' itself, e.g. as $\phi_7(1, x) = \phi_1(x)$ already holds for the first seven machines. Such cases would not initially be thought of as anything but 'random pairs' of machines depending on the chance agreement of *independent* proceses, viz. M_7's behavior and our numbering. But they would remind anyone contemplating the horizons of all 'computable functions' that the 'universal function' $u(z, x) = \phi_z(x)$ was obviously (partially) effective, for *us*.[53] For we know our own bookkeeping system and also how to 'hand simulate' any machine, i.e. how to follow its rules (quadruples). But, as we have argued in Chapter I, the idea that one of these very *machines* could compute $u(z, x)$ would strike us initially as utterly incredible. After all, *we* had adopted, for our own purpose and unbeknownst to our machines, an artificial numbering of them: would it not seem as if only by the most highly improbable kind of 'pre-established harmony' could one of them just happen to be so constituted that it could infallibly 'interpret our code'? That every time it was applied to the tape

$$(Tu) \qquad \underset{q_0}{\overbrace{\boxed{1}\boxed{1}\boxed{\dots}\boxed{1}\boxed{1}}^{z+1}}\boxed{B}\overbrace{\boxed{1}\boxed{1}\boxed{\dots}\boxed{1}\boxed{1}}^{x+1},$$

it would wind up with exactly the same result as M_z applied to just the $x + 1$ block? To be sure, something like this occurred with our occasional random pairs of machines, but there was surely nothing to this: one could easily see, e.g., that M_7 was not 'simulating' M_1 purposively, or following its rules. Indeed, M_7 could not possibly know what they were: that it got the same result, applied to M_1's number and x, as M_1 got when applied to just x, was clearly a lucky accident of our bookkeeping. Accidents will happen. Moreover, M_7 does not bear this random relation to any of our other original machines. So how on earth could any machine be lucky enough to have such a relation to *every* machine?

A further difficulty is related to the problem of machine self-reproduction: if one machine can simulate another, then it must be more complex, for it must somehow contain the complete program of the machine to be simulated, plus additional instructions for executing this program.[54] The accidental simulations of random pairs seem to show this 'degeneracy' principle very clearly: thus M_7, our multiplier, is clearly more complex than the identity machine M_1 it was lucky enough to simulate. Or consider our doubler M_6: it computes $\psi_6(y) = 2y$, and so would be randomly simulated by, say, a machine M_z computing $\psi_z(x, y) = xy - 4y$, which would clearly be more complex than M_6. On the other hand, if M_8, say, happened to compute $\phi_8(y) = y + 17$, then M_6 would in turn randomly simulate the simpler machine M_8. Such experiments would soon convince us that we would, as Minsky puts it, "need at least a sequence of more and more complex machines for the execution of more and more complex procedures".[55] Meanwhile, we are faced with the *human* computability of the universal function $u(z, x)$: a *single* human mind can simulate the computations of any M_z, however complex it may be.

This is why Turing felt that his discovery of the universal machine M_u was so important for both (T)′ and cybernetics generally. It was a striking confirmation of (T)′, showing that its most improbable-looking consequences could yet be verified. M_u also showed that the distinction between rule-following and rule-governed behavior can be made for machines. In fact, every machine is governed by its quadruples; nevertheless, once M_u's 'compiler' subroutine recovers a description of the quadruples (rules) $Q_1, \ldots, Q_k$ of M_z from the tape (Tu) above, M_u then operates on the tape

$$(Tu)' \qquad \boxed{Q_1'}\boxed{B}\boxed{Q_2'}\boxed{B}\boxed{Q_3'}\boxed{\cdots}\boxed{Q_k'}\boxed{B}\overbrace{\boxed{1}\boxed{1}\boxed{\cdots}\boxed{1}}^{x+1},$$

where Q_i' is M_u's coded description of the quadruple Q_i of M_z, and its further behavior is literally a matter of *following* these rules of M_z. M_u is 'conscious' of these rules in the sense that they are stimuli on its tape. Indeed, for Turing it was the universal machine M_u – which according to (T)′ can simulate any effective process by itself – that opened up the possibility of a machine model of man. Philosophers tend to overlook this when discussing the relevance of undecidability to mechanism, often talking as if mechanism required a single *special purpose* machine for its model of man. But Turing's universal machine already contains, via programming, every conceivable machine, lending great flexibility to the mechanist model. Just as a man can only perform this or that computation or task when given suitable instructions or education, so with Turing's universal machine M_u. Thus, Benaceraff [12] has tried to apply the incompleteness theorem to show that, if a man (who has learned Gödel's calculation) is a machine, we can never know *which* machine. If he is a universal machine, however, the conclusion should rather be that we could not always know which *program* he was using. This brings us once again to the claim that Gödel's (4) refutes mechanism.

Post was not only the first to prove undecidability, he was the first to use it against mechanism.[56] In fact, it was in the midst of his search for a metaphysical deduction of (T)′ that it occurred to Post that it might tell against mechanism:

> We see that a *machine* would never give a complete logic; for once the machine is made *we* could prove a theorem it does not prove ([209], p. 417).

Reflection on this (now familiar!) argument, and the vision it conjures up of an open transfinite sequence of extensions of a formal system, convinced Post, however, that this "creative process" could be "completely symbolized". Indeed, he began to sense at this point the existence of a general principle for reducing creative processes to mechanical ones:

> The creative germ seems not to be capable of being purely presented but can be stated as consisting in constructing ever higher types. These are as transfinite ordinals and the creative process consists in continually transcending them by seeing previously unseen laws which give a sequence of such numbers. Now it seems that this complete seeing is a complicated process mostly subconscious. But it is not given till it is made completely conscious. But then it ought be constructible purely mechanically ([209], p. 423).

Post isolates the final claim here – that any law we become 'completely conscious' of can be mechanically constructed – as his 'fundamental assumption', which he calls the 'Axiom of Reducibility for Finite Operations'.[57] Reflection on this principle – his version of (T)′ – caused Post to have second thoughts about his refutation of mechanism:

> The conclusion that man is not a machine is invalid. All we can say is that man cannot construct a machine which can do all the thinking he can. To illustrate this point we may note that a kind of machine-man could be constructed who would prove a similar theorem for his mental acts ([209], p. 423).

Here Post had already come close to the heart of the problem facing the anti-mechanist in *this* context: if he applies the Gödel argument straight away, he will end up proving the existence of a machine which by his argument could not be modeled by any machine.[58] It was no accident that Post became "completely conscious" of his reducibility version of (T) while critically examining the undecidability argument against mechanism, for this is just what defuses the argument. Let us examine this more closely now in connection with the more recent formulations of Turing and Lucas.

Turing observed that it is sufficient to consider his own result on the halting problem for his machines, namely, that the 'halting function' $h(z, x)$ is not computable by any machine, i.e. $h \neq \psi_z$, for any z. Moreover, given any machine M_z, we can actually calculate a number, the 'Achilles heel' of M_z, which will show the inability of M_z to solve the halting problem. But now, the argument goes, M_z can't be a model of *me*, for I have just stumped M_z with a question it can not answer. And, as Lucas puts it, "I can take on all comers". Hence no machine M_z could possibly model me. And all thanks to Gödel, for inspection of our calculation of the Achilles' heel of M_z will show it to be a formal diagonal argument going back to him. Turing granted that the feeling of superiority we derive from being able to pose a question to any M_z which we know beforehand it will answer wrongly or not at all "is no doubt genuine", but claimed that it is only a "petty triumph" scored over one machine at a time, there being "no question of triumphing simultaneously over *all* machines" ([251], p. 16). But Lucas had no trouble with this: "Nor does it signify that it is only *an* individual machine we have triumphed over: for the triumph is . . . over *any* individual machine that anyone cares to specify" ([173], p. 49). For "from the specification, I can calculate the Gödelian formula" ([174], p. 152). But these formulations make it clear that the real source of Lucas' feeling of superiority here is just the *effectiveness* with which he can find the Achilles' heel of any machine. This alone is what makes him confindent. But clearly then, Turing should have had the courage of his conviction in (T)′: if Lucas can effectively stump any machine, then by (T)′ there must be a *machine* which does this too! Let us examine such a machine in detail.

Because of closure under diagonalization there exists a partial recursive ψ_d defined by

$$\text{(d)} \qquad \psi_d(z, x) = \begin{cases} 1, \text{ if } \psi_z(x, x) = 0, \\ \text{undefined, if } \psi_z(x, x) \neq 0 \text{ or undefined.} \end{cases}$$

Hence, if we set $g(z) = s(d, z)$ for Kleene's function s of (8), we have

$$\text{(8)}' \qquad \psi_d(z, x) = \phi_{g(z)}(x), \text{ with } g \text{ recursive.}$$

Assume now that some machine M_p could compute the halting function

$$\text{(h)} \qquad \psi_p(x, y) = \begin{cases} 1, \text{ if } \phi_x(y) \text{ defined,} \\ 0, \text{ if } \phi_x(y) \text{ undefined.} \end{cases}$$

Taking $z = p$ in (8)′ gives $\psi_d(p, x) = \phi_{g(p)}(x)$, which together with (d) implies $\phi_{g(p)}(x)$ defined $\Leftrightarrow \psi_p(x, x) = 0$, which for $x = g(p)$ gives

(g) $\qquad \phi_{g(p)}(g(p))$ defined $\Leftrightarrow \psi_p(g(p), g(p)) = 0$.

But taking now $x = y = g(p)$ in (h) gives also $\psi_p(g(p), g(p)) = 0 \Leftrightarrow \phi_{g(p)}(g(p))$ undefined, contradicting (g). Letting M_t compute $g(z)$, we now have a machine simulating Lucas' Gödel-calculations: for any p, M_t calculates a number $g(p)$ showing that ψ_p cannot satisfy (h) and hence that M_p cannot solve the halting problem. For M_t always calculates $g(p)$ to satisfy (g). Like Lucas, M_t can find the 'Achilles' heel' in any machine which underlies the Gödel incompleteness. But Lucas has a trump card here: he can do this for M_t *also*. Can M_t do this? Of course: if we put t on M_t's tape it will proceed as calmly as ever to calculate $g(t)$, which is now a number showing that M_t *itself* cannot solve the halting problem, i.e.,

(g)′ $\qquad \phi_{g(t)}(g(t))$ defined $\Leftrightarrow \psi_t(g(t), g(t)) = 0$.

So with *just one* t-block on its tape, M_t calculates a number $g(t)$ which, when placed *twice* on its tape, would cause itself to print 0 just in case $M_{g(t)}$ would halt when applied to just a $g(t)$-block on its tape. This leaves nothing to be desired by way of Gödel calculations. Of course, the computation of this 'Gödel-function' $g(z)$ which assigns an Achilles' heel to each machine M_z is only one of the jobs a human computer occupies himself with, as is indeed the case with Turing's model M_u of the human computer: M_u simulates every other machine in addition to M_t.

Kleene [150] has pointed out that Gödel's essential discovery was, in effect, just the existence of such machines as M_t computing the function $g(z)$. In short, he *mechanized the old diagonal argument for incompleteness*, which, before Gödel, had seemed to be the exclusive possession of a mathematical mind like Finsler's. This makes it all the more strange that his work should now be imagined to refute mechanism, for he established for the first time precisely that, from the proposition 'I can find a limitation in any given machine', it by no means follows that I am not a machine. He put mechanism back in the running. Note also the importance of the undecidability of M_u's halting problem for (T)′: otherwise the function

$$d(x) = \begin{cases} \phi_z(z) + 1, \text{ if } M_z[z] \text{ halts}, \\ 0, \text{ otherwise}, \end{cases}$$

would be a clearly effectively computable function which M_u could *not* compute. The existence of a universal machine is essential to Turing's thesis, and the undecidability of its halting problem is essential to its universality.

Such is the basic dilemma confronting anti-mechanism: just when the constructions used in its arguments become effective enough to be sure of, (T)′ then implies that a machine can simulate them. In particular, it implies that our very behavior of applying Gödel's argument to arbitrary machines – in order to conclude that we cannot be modelled by a machine – *can indeed be modelled by a machine*. Hence any such conclusion must fail, or else we will have to conclude that certain *machines* cannot be modelled by any machine! In short, anti-mechanist arguments must either be ineffective, or else unable to show that their executor is not a machine.

We now reconsider, in light of the machines M_t and M_u, the argument from predictability against mechanism considered in Chapter I. Though he presumably regards the Gödel argument as much stronger, Lucas apparently buys this argument also, for in commenting on Turing's suggestion that machines might reach a level of complexity beyond which they ceased to be predictable, he admitted that

> Although it sounds implausible, it might turn out that above a certain level of complexity, a machine ceased to be predictable, even in principle... It would begin to have a mind of its own when it was no longer entirely predictable and entirely docile, but was capable of doing things which we recognized as intelligent... But then it would cease to be a machine, within the meaning of the act ([173], p. 58).

On the other hand, in answering Benaceraff he objected that

> Only if, at least in principle, my programme could be known and all my actions infallibly predicted and mechanistically explained, can mechanism worry us ([174], p. 150).

But Turing's mechanism, which worries Lucas considerably, would *not* imply that all his actions could be 'infallibly predicted', for the halting theorem shows precisely that Turing's model M_u *is* unpredictable – *even when you know its program.* For undecidability just means that no effectively computable h infallibly satisfies

$$h(z, x) = \begin{cases} 1, \text{ if } M_u[z + 1\ B\ x + 1] \text{ halts,} \\ 0, \text{ if } M_u[z + 1\ B\ x + 1] \text{ doesn't halt.} \end{cases}$$

In short, the function $g(z)$ of M_t is simply to find errors in the attempts of any machine to predict in general what M_u will do, and by (T)′, no other effective means of prediction are available to us. But what could it mean for all my actions to be 'infallibly predicted' by a *non*-effectively computable method, within the meaning of the act of prediction? In fact, Turing's theorem shows that, in general, our predictions of M_u's behavior, as in the case of human rule following acts, must be *post hoc*: the only 'method' for telling what M_u will do is to wait and see. But M_u clearly is still a machine, and if man is a machine anything like M_u, this would tend to confirm La Mettrie's thesis that man is a machine which is too complex to define *a priori*. This also realizes a shadowy vision of Babbage, and even one of von Neumann – inspired by Gödel's theorem – of an automaton so complex that its behavior was "asymptotically infinitely longer to describe" than the automaton itself ([196], p. 54), which Gödel himself explained as following in a letter to von Neumann's editor Burks:

> ... what von Neumann perhaps had in mind appears more clearly from the universal Turing machine. There it might be said that a complete description of its behavior is infinite because, in view of the non-existence of a decision procedure predicting its behavior, the complete description could only be given by an enumeration of all instances ... The universal Turing machine, where the ratio of the two complexities is infinity, might then be considered to be the limiting case of other finite mechanisms. This immediately leads to von Neumann's conjecture ([196], p. 56).

Gödel admits this imposes "the finitistic way of thinking" on descriptions, so this is perhaps problematic. But prediction is another matter, and *eo ipso* effective, as I have argued in Chapter I.

By generalizing M_u to a 'universal constructor' and adapting the diagonal argument, von Neumann was also able to realize, as already mentioned in Chapter I, his vision of self-reproducing machines, thereby removing in principle the classical argument against mechanism. Finally, Myhill [191] combined von Neumann's result with Gödel's (4) and Kleene's (10) to construct an infinite 'evolutionary' sequence of machines in which each machine constructs the next one, its 'offspring', and each offspring proves more theorems of arithmetic than its 'parent'. In fact it has been suggested by Rogers that Kleene's recursion theorem (10)

> ... displays the intellectually central features of a self-reproduction proof and is directly relevant to fallacious arguments (by 'infinite regress') that self-reproducing machines cannot exist[59] ([221], p. 189).

Such examples point to another basic fallacy in the mathematical arguments against mechanism: they forget that these undecidability results for machines have no direct bearing on their relation to us, but *deal solely with relations between the machines themselves*. They deal, as it were, with private, in-house relations, difficulties, and misunderstandings that arise between machines. Thus the halting theorem says just that no machine can react to universal machines in a certain way, and Myhill's evolutionary sequence concerns machines which outsmart other machines (viz. their parents). This point is especially clear when we reflect on Kleene's theorem (10), which may be seen as a warning that it will be hard to distinguish universal from non-universal machines, and hence 'real' from 'accidental' simulation. For it says that no matter *what* machine you examine there will be infinitely many other machines forming random pairs with it that it will *seem* to 'simulate'. In short, every machine acts infinitely often *as if* it is simulating others, and hence 'emulates' the universal machine. This makes it harder to determine experimentally whether a given machine is universal since initial favorable tests need never be due to actual universality: it may only be emulating M_u. This is similar to Gödel's (4) which can be read as saying that *every* formal predicate has provable 'moments of truth' when it accidentally emulates the truth-predicate for the formal system, i.e. it provably satisfies Tarski's criterion on truth definitions for infinitely many sentences. Instead of these rather 'static' simulation relations between machines, it is possible to apply (10) to find machines standing in more 'dynamic' relations, i.e. relations where one machine's behavior depends on what another machine *does*. Thus we can define a partial recursive function ψ by

$$\psi(z, x) = \begin{cases} 1, \text{ if } (\exists y)[T(x, z, y) \wedge U(y) = 1], \\ \text{undefined, otherwise,} \end{cases}$$

and the recursion number r supplied by (10) now satisfies

$$\phi_r(x) = \begin{cases} 1, \text{ if } (\exists y)[T(x, r, y) \wedge U(y) = 1], \\ \text{undefined, otherwise.} \end{cases}$$

This Gödel machine M_r, given x on its tape, first prints out its *own* number r to the right of x and then applies its *universal sub-routine* to the tape $[x + 1\ B\ r + 1]$, waiting to see how M_x will respond to his name (i.e. r). If M_x responds to r by printing a 1, then the Gödel machine will print 1; otherwise it never halts. In short, M_r responds, by printing 1, to *any* machine M_x if M_x will so respond to him, like a person who says hello to just those who say hello to him. If we reverse the z and x in the definition of ψ above, and apply (10) again, we get a b satisfying

$$\phi_b(x) = \begin{cases} 1, \text{ if } (\exists y)[T(b, x, y) \wedge U(y) = 1], \\ \text{undefined otherwise.} \end{cases}$$

The Gödel machine computing ϕ_b is much more 'introspective': given x, it first prints out its own number b to the left of x; then its universal sub-routine recovers a coded description $C(M_b)$ of its *own* quadruples on its tape and then *applies* them to z; if and when this process terminates with a 1, then M_b itself prints 1. Thus this Gödel machine has a sub-routine in which it *simulates itself*. We could call it a 'Babbage machine' since it realizes Babbage's vision of a machine which "eats its own tail".[60]

Such examples illustrate the theoretical significance that Gödel's constructions have for mechanism. They have led us to our strongest known existence theorem for machines, which can be used to establish the existence of pathological and unpredictable machine behaviors that earlier mechanists could only dream about, and seemingly only idly so. Whereas Gödel had only revitalized the Hilbert program, he and Turing have vitalized the mechanist one, period, by showing *a priori* that its abstract machine concept has many of the appropriate kinds of properties for modelling living, reproducing, rule-following, self-reflecting, accident-prone, and lucky creatures. The versimilitude of this model, however, can only be established empirically by programming.

We may conclude our critique of the anti-mechanist use of Gödel with a remark on the relation between the two forms of mechanism we have discussed, namely, the thesis (M) that man *is* a universal Turing machine, and the thesis (C) of artificial intelligence that all of man's precisely describable behavior can be *simulated* by a suitably programmed computer. The main point is that *neither* of them entails the other. For if human beings are machines, they are machines which have had billions of years of chemical, biological and cultural evolution to adapt to their environments and to each other, and there is no *a priori* reason for supposing that we could, as Turing put it, 'speed up' evolution that much with mere programming. The cognitive operations we humans can program on the spot may never quite match those that evolution has had billions of years to program. In particular, evolution has produced machines, if machines we are, whose behavior is maintained by highly developed motivational systems which are conspicuously lacking in the machines that we manufacture and whose needs we administer to entirely. On the other hand, we might be able to machine simulate through programming every human behavior that we *can* describe precisely without it being the case that we are machines. For a non-mechanical system may well exhibit many mechanically describable behaviors, while exhibiting others that we can hardly describe at all, presumably because they are the product of a non-mechanical system. Von Neumann once suggested vision as an example of human behavior which we may not be able to completely and precisely describe, but we have seen that significant progress has been made in computer vision nevertheless. The question is: how can we show that a system is *non*-mechanical?

To show that man is not an abstract (universal) Turing machine it would be sufficient to show that at least one of the following conditions is false:

(i) Man is capable of only a finite number of internal (mental or physical) states $q_i \in Q$.

(ii) Man is capable of discriminating only a finite number of external environmental states $s_i \in S$.

(iii) Man's memory is described by a function f from pairs $\langle q_i, s_i \rangle$ to Q.

(iv) There is a finite set B of atomic human behaviors, including some which may be identified with states s_i of S, which they effect (as in printing a symbol), and each $\langle q_i, s_i \rangle$ determines a unique element of B. A human (molar) behavior is comprised of a finite sequence of atomic behaviors of B, some of which have neural dimensions, while others may be molar behaviors in their own right.

We have discussed Turing's finiteness condition (i) at some length and found its plausibility and resources considerable. Part of what we mean in calling man an 'abstract' machine is just that we abstract away from the particulars of his internal states, e.g. whether they are mental or physical, requiring only their finiteness. Turing's second finiteness condition (ii) seems as plausible as (i), but differs from it insofar as it would follow from a finiteness condition in physics itself to the effect that there *were* only finitely many states of the environment there to be discriminated. Together the two finiteness conditions (i) and (ii), which are just more explicit versions of the theses (H_1) and (H_2) of Hilbert's finite *Einstellung*, imply that the infinite can be eliminated from psychology even if it remains necessary in physics.

The recursive memory function required by condition (iii) might not exist, but it is not clear how one could show this effectively. It has been suggested that quantum effects might play an important role in the dynamics of neural events. If so this could conceivably have the consequence that some neural states do not determine any particular new neural state. But such a failure of deterministic memory could hardly be ascribed to any conscious 'freedom' of the mind to remember what it wants to. We argued in the first chapter that this quantum kind of indeterminism only goes to strengthen the mechanist point of view generally. And there are formulations of indeterministic machines ready at hand to accomodate this possibility. The deterministic behavior function of condition (iv), however, presents a more inviting target for free-will arguments. Indeed, there are many arguments in the literature purporting to show that people can always choose to act differently from however they may be predicted to act, provided they are privy to such prediction. But condition (iv) only implies the predictability of the next *atomic* behavior, and we have seen that this by no means implies the effective predictability of molar behavior in general, but rather leads to an unpredictability of behavior which is nevertheless compatible with the strictest determinism. Whereas conditions (i) and (ii) would free psychology of the infinite, conditions (iii) and (iv) would relieve it of any need to postulate freedom.

Condition (iv) does, however, pose the problem of analyzing human behavior into some kind of natural units or atoms (cf. Dennett's 'atomic tasks'). Classical behaviorist psychology looked hopelessly for some kind of atomic reflexes, but Skinner [234a] has formulated the problem of determining 'units of behavior' within the framework of his more general operant conditioning in a way that still respects the 'continuity of behavior' evidenced by our success in shaping it operantly. If some such analysis as this cannot be carried through, it is hard to see how condition (iv) can be fulfilled, and hence it is quite possible that man is not a machine.[61] And yet it is equally hard to imagine anything but Descartes' mechanical organisms existing at the dawn of evolution. The failure of mechanism would thus leave us with the problem of fitting the emergence of some mind either actually infinite or absolutely free into evolutionary theory. We have argued, however, that even the truth of mechanism would not necessarily imply that by intelligent programming we could educate Turing's 'child-machine' as well as evolution itself would. Turing did, however, formulate a context in which the imitation of a highly evolved human skill by a machine might possibly succeed through programming, his so-called 'imitation game'. The point here is that some of the most highly evolved and characteristic human traits, such as love, are associated with the relations and differences between men and women. The question has been asked whether there are any real emotional or intellectual differences between them beyond the obvious biological ones that are not mere cultural artifacts. Turing considered a game in which a person tries to distinguish a man from a woman solely by the quality of their answers to questions. Such a game would be trivial, however, unless at least one of them was allowed the free use of another characteristically human trait, namely the attempt to deceive through false answers. Turing considers the case where the man is allowed this all too human latitude, and asked whether *his* place could be taken by a suitably educated machine. In short, Turing's question is whether programming can imitate a man's highly evolved ability, such as it may be, to make himself by deceit emotionally and intellectually indistinguishable from a woman. To my knowledge very little has been done on this very difficult problem, which is connected with issues having much more currency now than when Turing wrote. It is also clear to me that none of the many discussions of Turing's game even begin to address the questions he actually raised with it. Most of them are in fact as naive as the machine he hoped to educate for the man's deceitful place in the game.[62]

NOTES

[1] Of course, certain important consequences of Dedekind's recursion theorem, notably his isomorphism of any two Peano models, have then to be relativized to the universe of the formal set theory. In particular, the result fails when countable models of the theory are considered.

[2] This is discussed in more detail below.

[3] See Borel [29], p. 213.

[4] Borel and du Bois-Reymond often used the diagonal method for the construction of fast-growing functions in growth-rate arguments. See du Bois-Reymond [72] and Borel's remarks in [166], pp. 12ff. This use of diagonalization is prominent in recursive function theory, especially the branch dealing with complexity and speed of computation.

[5] That is, $g(x) = f_x(x)$ must also be a (partial) unary function of (E). The Richard's Paradox is avoided by the partiality of g.

[6] Perhaps no principle of logical inquiry has been more used yet less discussed in the history of logic from Aristotle to Frege. Thus, as late as Boole, substitution principles are used apparently without any conscious awareness. Even Frege, though he is conscious of using such principles in his axiomatic formulations of propositional and predicate logic, never explicitly lists them as actual rules of inference. Indeed, he claims, for instance, that *modus ponens* is the *only* rule of inference in formulation of propositional logic. See [255], p. 3.

[7] See Schönfinkel [255], p. 357; Curry was a student of Hilbert (see below in text).

[8] The application of a combinator x to an argument y is written (xy). When x represents a function, (xy) can be interpreted as the result of substituting y for the first variable in x. Thus, if f represents a unary function, (fx) means what is ordinarily written $f(x)$. (fx) may then be either applied to other arguments y yielding $((fx)y)$, or taken as an argument of another function g yielding $(g(fx))$. Application is not associative, so that $(g(fx))$ is

not the same as $((gf)x)$. The expression $fx_1 \dots x_n$ is often used to abbreviate $((\dots(fx_1)\dots)x_n)$. A binary function $F(x, y)$ is construed as a unary function f which, when applied to an argument x, has as its value the function (fx) which, when applied to the argument y has $F(x, y)$ as its value, i.e. $F(x, y) = ((fx)y)$, or just fxy for short. Schönfinkel introduced the combinators S and K governed by the rules:

$$Sxyz = xz(yz),$$
$$Kxy = x.$$

Thus, if x represents a binary function f and y a unary function g, then $Sxyz$ means what is ordinarily written as $f(z, g(z))$. K is the 'constant function': if x represents a constant, K is the constant function whose value is x for every argument y. See Stenlund [240], whose formulation I have largely followed.

[9] See Stenlund [240], pp. 14ff.

[10] See note 8 for an explanation of S and K.

[11] Stenlund remarks that "the difference between the λ-calculus and the theory of combinators is parallel to the situation in set theory where the comprehension axiom can be replaced by class operations" ([240], p. 16).

[12] His original definition was $W = SS(SK)$.

[13] See Curry [58], pp. 184–5 and pp. 154–6 for the interesting history of these developments. To briefly explain this, we need to define the 'composition' combinator $B = S(KS)K$, which will satisfy the rule $Bfgx = f(gx)$. Curry's original 'paradoxical combinatory' Y is then defined by $Y = WS(BWB)$, where W is the diagonalizing combinator defined in the previous note. Y is now called a 'fixed-point' combinator since it can be proved to satisfy the theorem $YX = X(YX)$, for every X. Thus, given any combinator X, Y produces a fixed-point for X. This is the combinatory analogue for the fixed-point theorems of Gödel and Kleene discussed below in the text. In fact, it is the essential construction used by Kleene [147] in his very important proof discussed below that all recursive functions are λ-definable. Bernays [20] has emphasized the importance of these results for the formalist program.

[14] See Ackermann [2], pp. 10ff., and von Neumann [195], pp. 298–9. The essential problem was in establishing sufficient '*Kontrollierbarkeit*' over the possible recursions admitted as axioms, by means of assigning increasing ordinals to increasingly complex recursive definitions.

[15] This approach led to a certain arbitrariness in the specification of their formal systems for arithmetic. Thus, after introducing all the ordinary Peano-Dedekind axioms, Ackermann continued:

Zu diesen ein für allemal festehenden Axiomen können noch Definitions-axiome für Funktionen hinzukommen. Sämtliche Funktionen werden durch Rekursion definiert. Bei der Aufstellung dieser Axiome ist eine gewisse Willkür gestattet. Wir schreiben nur ein Schema vor, in das sich alle diese Axiome einordnen lasse müssen ... Wie viele derartige Axiome man gebraucht ... und welche Gestalt diese in einzelnen haben, ist der Willkür überlassen ([2], p. 5).

[16] This has been claimed, though without explicit textual support, by Steen [237a] and K. Mainzer.

[17] See Péter [203] for an extensive discussion of this theme.

[18] This would have been a fair description of Skolem's work of 1923 but Hilbert apparently was not acquainted with Skolem's paper at that time.
[19] If one is really looking for the dream of Hilbert's which was undermined by Gödel's work, then this is it. There is nothing in this world he wanted more than to use his proof-theory to solve what he considered the most beautiful of all mathematical problems, Cantor's continuum problem, but his approach required that there be *uncountably* many effective functions defined by various recursions. As I shall argue below, when the significance of Gödel's incompleteness construction began to emerge, especially in the work of Kleene, it pointed inevitably to (CT), viz. to the *countability* of all effective functions.
[20] In particular, Hilbert's 'second lemma' (see [255], p. 391). This lemma itself, incidentally, is very much in the spirit of (CT), for the latter can be thought of as a 'reducibility' principle asserting that however complex and high-type function and set quantifiers you use in defining an effective function, this definition can always be reduced to one expressible in a very simple first-order formalism. This theme is discussed below in a more cybernetical context.
[21] See Kalmar [140].
[22] I shall follow the formulations of Gödel's 1934 lectures rather than his 1931 paper.
[23] This was proved simultaneously and independently by Gödel and Gentzen in 1932. See Davis [61], pp. 75–82.
[24] See Hilbert-Bernays [106], pp. 420ff., for a full discussion.
[25] Unless, of course, F is inconsistent.
[26] See Siefkes [233], pp. 8ff.
[27] See Siefkes [233], pp. 45ff.
[28] The following example of such a recursion is given by Rogers [221], who credits it to Kreisel. Where $T(x, y, z)$ is the primitive recursive Kleene-predicate for recursive functions we define:

$$g(x, y, z) = \begin{cases} 1, & (\exists p)[T(x, x, p) \wedge p \leq y), \\ 2z, & \text{otherwise.} \end{cases}$$

If we now add, to a set of recursion equations for g, the additional equation

$$f(x, y) = g(x, y, f(x, y + 1)),$$

it follows, by Kleene's result, that the predicate $(\exists y)T(x, x, y)$ is recursively undecidable, that the function f just defined cannot be recursive. On the other hand, it can be shown that f is the unique solution for the set of equations just described. It follows that the values of f cannot be computed from these equations by any fixed set of effective rules – assuming (CT).
[29] See Gödel [100], p. 71.
[30] See Kleene in [61], p. 259 for details and discussion.
[31] See Kleene [146], Theorem XIII; also Kleene [150], pp. 257, 259.
[32] This shows the fallacy of Kreisel's claim in [157] that a complete formalization would *have established* Church's thesis: a complete formal system for arithmetic in which all recursive functions were definable would lead to an *easy* refutation of Church's thesis by the diagonal method, for in such a case the diagonal function would surely be effectively computable, but not even *definable* in F, if consistent – hence a counterexample to Church's thesis.

The resolution of the Richard paradox effected by Church's thesis may be seen as 'dialectical' in Hegel's sense. We begin historically with the thesis (RP) (ii) of the countability of the effective functions, but later the diagonal method led to the antithesis of their *un*countability. Finally, Church's thesis has led to a 'synthesis' of the two: for while it decides on (RP) (ii) as the literal truth, it finds a 'kernel of truth' in (RP)(i) also, namely that the effective functions are not *effectively* countable. Curiously, Zermelo [278] regarded Gödel's non-r.e. set of unprovable sentences as uncountable (cf. [107]). See Webb [261], p. 164 for Church's temporary identification of these two notions.

33 $T(m, x, y)$ will then mean that y is the Gödel-number of a computation of the machine with Gödel-number m which began with the input x.

34 Van Dalen [86], p. 261. This opinion is widespread among intuitionists. Brouwer occasionally spoke of 'finite algorithms' for calculation but completely ignored (CT). Of course, van Dalen *is* right: Church's thesis *does* seem improbable, as I have tried to bring out in several ways. In fact, Bowie [31] has tried to refute it by arguing that, since any physical device for computing functions by means of a random element *might* compute any (!?) function at all, the probability that it computes a recursive function is zero. But this argument no more leads to an explicit counterexample than does the intuitionistic metamathematical approach. Indeed, one may regard Bowie's remark not so much as an argument against Church's thesis as simply another way of emphasizing what a *strong* thesis it is. As an argument against the thesis it would be every bit as ironic as the intuitionistic ones, depending as it does on the vague notion of 'randomness' whose best explication in some contexts depends precisely on this thesis itself. Lucas [176a], pp. 122–3 claims that only humans and not machines can introduce any element of randomness, but this seems rather implausible in the face of quantum mechanics and undecidability. (cf. Chapter I.) See also Ross [223] for a good critical discussion of Bowie's argument, and Berg and Chihara [15] for some misunderstandings which both Bowie and Ross fall prey to.

35 Kreisel [156], p. 128.

36 I am following van Dalen [86]. Troelstra [250] contains an excellent discussion of some serious problems besetting the interpretation of $\Vdash_n A$. The axioms (CS) are generally not themselves added to the intuitionistic number-theory formalisms; rather their principal role is that of justifying the Kripke schema given below, which follows from (CS) and the following function definition by intuitionistic logic. The axiom (CS) (i) is also supposed to insure that this definition is effective – assuming, of course, the intuitionistic meaning of disjunction. But this seems to beg the question. In any event, this definition does not lead to any effective non-recursive functions after all.

37 See Troelstra [250], pp. 100ff.

38 See Troelstra [250], pp. 100ff.

39 See Church's remarks in Davis [61], p. 100.

40 See Strong [241] for an extensive survey of efforts in this direction. See also Bird [25] for a good introduction to the related development, stemming largely from D. Scott, of basing the whole theory of computation on Kleene's fixed-point and recursion theorems. Thomas [248] defuses Kleene's argument for (CT) based on them.

41 See Takeuti [243], where the recursion theorem is employed to sharpen the consistency proofs for number theory and sub-systems of analysis. The key step in such proofs is to replace the ordinary formal induction axioms with the effective ω-rule and then show that cuts can always be effectively removed from the resulting infinite proof-trees. (Con-

sistency then follows trivially from this cut-elimination property.) Takeuti makes a rather complicated use of the recursion theorems to prove the existence of a recursive function which, applied to any (description of an) infinite proof-tree, always eventually yields a (description of a) cut-free infinite proof. Transfinite induction then has to be used to prove that the Turing machine computing this function always halts, i.e., that this function is total.

[42] See Minsky [184], pp. 106ff. for an illuminating discussion of the problem this posed for early investigators of effectiveness.

[43] See Turner [253], p. 86 for a good critical discussion of this assumption.

[44] [181], p. 35.

[45] Post did not say where this criticism came from. Kreisel [157] claimed that Gödel would soon publish a "forthcoming note" in *Dialectica* on his objections to Turing's arguments for (T)′, but it has not appeared. Meanwhile Wang [259] quotes and paraphrases Gödel extensively on the matter. In particular, it contains a long note, from which we quote in the following, that Wang says will soon appear appended to Gödel's 1934 lectures in a new edition of [61].

[46] See Wang [259], p. 326.

[47] It is not clear to me why Gödel's mental states would not also become confused in the eventuality of (G)′, for even *if* mind exists separately from matter, our thought is seemingly 'channeled' through our brains. This is the classical cartesian problem of the interaction between mind and body.

[48] See Turing [61], p. 117; also Minsky [184], pp. 15ff.

[49] See Wang [259], p. 326.

[50] As examples, we take $M_1 = \{q_0 1 B q_0\}$, $M_2 = \{q_0 1 R q_0\}$, $M_3 = \{q_0 1 B q_0, q_0 B R q_1, q_1 1 R q_1, q_1 B R q_2, q_2 1 B q_2\}$ $M_6 = \{q_0 1 B q_1, q_1 B R q_2, q_2 1 B q_3, q_3 B R q_4, q_4 1 R q_4, q_4 B R q_5, q_5 1 R q_5, q_5 B 1 q_6, q_6 1 R q_6, q_6 B 1 q_7, q_7 1 L q_7, q_7 B L q_8, q_8 1 L q_1, q_1 1 L q_1\}$. For the quadruples of our multiplier M_7, see Davis [60], pp. 12ff. We assume that each machine begins in the starting state q_0, scanning the leftmost stroke 1 on its tape. An input of n is represented by $n + 1$ strokes in successive cells of the tape, which we call an 'n-block". An input of n and m to, say, our adder M_3, is represented by an n-block and an m-block on the tape separated by a single blank 'B'. An *output* of n is just n (possibly scattered) 1's on the tape. Naturally, each of these machines performs very different tasks if they begin in arbitrary states on arbitrary parts of arbitrary tapes. Also they may contain finite auxilliary alphabets.

[51] Thus, e.g., we can imagine, as would be plausible, that M_6 was *designed* to double numbers, i.e. to compute $\phi_6(x) = 2x$. However, it will compute something no matter *how many* separated input-blocks we put on its tape: with 2 such inputs, M_6 computes $\psi_6(x, y) = 2x + y + 1$. But, in general, we would not initially be interested in these 'by-products' of M_6.

[52] Thus, e.g., if we fix a 5-block on M_3's tape, it can be *used* to compute $\psi_3(5, x) = 5 + x$. Note, however, that we do *not* have $\phi_3(x) = 5 + x$ but rather $\phi_3(x) = x$, as the reader may easily verify by inspecting the quadruples given for M_3. And if we fix a 2-block on M_7's tape we can use it to compute $\psi_7(2, x) = 2x = \phi_6(x)$, for by definition we have $\psi_7(x, y) = x \cdot y$.

[53] That is, we know we can compute $u(z, x)$ effectively *when it is defined.* Note that we always have $\phi_z(x)$ defined $\Leftrightarrow$ $M_z[x + 1]$ halts, where the right side is short for the

machine M_z, started in state q_0 scanning the leftmost stroke on a tape containing *only* an $x + 1$ block, eventually halts. Also $\psi_z(x, y)$ defined $\Leftrightarrow M_z[x + 1\ B\ y + 1]$ halts.

54 See Moore [187] and Singh [234], pp. 185–6 for a discussion of this problem.

55 See Minsky [184], pp. 106ff.

56 He did both already in 1924!

57 *Ibid.*, p. 424. This anticipates Kreisel [156].

58 Chihara [51], p. 562, makes a similar point very neatly.

59 Whether or not (10) brings out any *physically* significant feature is quite another matter. In fact, Wigner [273] has used quantum mechanical calculations to argue against the physical plausibility of von Neumann's [196] results – which did not use (10). Specifically, Wigner tries to show that a self-reproducing system depends, from the quantum mechanical point of view, on the satisfiability of a very large number of equations in relatively few unknowns, so that "it would be a miracle" (*ibid.*, p. 205) if they were satisfied. Arbib [4] points out, however, that one of the assumptions made by Wigner in his calculations – roughly speaking, that self-reproducing systems have no special structure – is very implausible for such systems. But even if Wigner *could* get his result from biologically reasonable assumptions, it would not be clear that the 'miracle' he would see in the satisfiability of his equations is any greater than that of the satisfiability of Kleene's equations (10). Indeed, the solvability of the latter is closely related to the miracle Gödel saw in the failure of diagonalization to lead outside the recursive functions. See Hofstadter [132] for a revealing discussion of the relation between self-reproducing machines and the molecular biology of self-reproducing systems.

60 These consequences of Kleene's (10) may make the following explanation of its significance easier to understand:

> Now suppose that we are asked to write a program B whose outputs are to depend in a prescribed way not only on... but also on B itself... We may object that we cannot reasonably be asked to write a program without knowing in advance what this program is supposed to do; and in the present case what the program is supposed to do depends on the – yet unwritten – program itself. A programmer confronted with such a task might well feel like the wise men of Babylon when the were commanded by Nebuchadnezzar not only to interpret his dream but to guess what the dream was. Like them he might object that this was not the kind of problem he could fairly be expected to solve ([11], p. 274).

But this is just the sort of problem that (10) enables us to solve.

61 In a penetrating note on Church's thesis, Parikh [196a] points out both the importance and difficulty of isolating a suitable set of atomic acts for generalizing or extrapolating this thesis to cover such seemingly effective everyday tasks as cooking from a recipe.

62 An exception is Hofstadter [132], which also contains a full and useful discussion of Turing's thesis. I regret that this remarkable book did not appear in time to take fuller advantage of its discussions of certain topics treated herein, particularly that of machine vision.

CONCLUDING SUMMARY

In the beginning mechanism seemed as untenable for our reproducing bodies as for our calculating minds. It took centuries to find mechanisms for the two simplest computations of our minds, but it seemed unlikely that they could be found for the higher mental calculations, and inconceivable that a single mechanism could ever encompass all of its calculations. At the center of these calculations lay the finest flower of the constructive mind, the intuitive number sequence (N). Dedekind formulated the mathematical foundation (DT) for this flower – as well as a theory of recursions on it undreamt of in earlier theories of calculation – by means of the finest flower of the non-constructive mind, set theory. But for Dedekind this latter flower, especially its infinite petals, grew only in the mind itself, leading to a temporary engagement of mathematics and idealism. But paradoxes were also discovered to grow here, causing Dedekind to remark that "human reason totters". Constructivists like Kronecker reclaimed their flower, claiming now that it – together with its effective functions – should provide the foundation for the rest of mathematics. Cantor's diagonal argument seemed to refute this, but Richard showed that this also led to paradox. Yet it also led to recursions far transcending any formal principle like Dedekind's. Finally, after having opened the door to new vistas and a lofty goal for mechanism in geometry, Hilbert attempted – by means of compromise reminiscent of Kant's between empiricism and rationalism – to make Dedekind's foundation acceptable to the constructive mind through *formalization*. But his initial formalisms, embedded in the newly crystalized elementary logic, seemed like fingers in the dyke in the face of the higher recursions unleashed by Ackermann and the formally undecidable propositions decided in Finsler's mind – both on the wings of diagonalization. Still he raised two important questions: (i) could Dedekind's completeness theorem be finitized in the sense of showing that this new logic could mechanically extract from elementary axioms all the theorems promised by it, i.e. those true in all Peano models; (ii) could Dedekind's recursion principle be generalized to encompass all *possible* recursions? For they seemed to encompass effectiveness. Gödel intervened to show that the old isomorphism could not be so finitized by formalizing the diagonal argument, bringing Finsler's undecidable propositions down from '*rein Gedankliche*' and relocating them in formal systems – much as Aristotle is said to have brought Plato's forms down from heaven and put them back into matter. This enabled him to define a class of formal recursions that deserved to be called 'general', and Kleene showed, by means of new kinds of undecidable sentences, that effective diagonalization could no longer escape from Gödel's class, thereby inviting (CT). Kleene then showed how to replace self-referential Gödel sentences by analogous Gödel-machines, forging a link between self-reference and recursion. By transplanting the formal diagonal argument for sentences to formalisms for machines themselves, its pathology can now service cybernetical models directly: after all, sentences are all talk and no action. Turing discovered a remarkable class of machines, having one for computing each of Gödel's most general recursions, in fact. And no matter how you effectively code them, there is always a machine in their midst that breaks that code and simulates all of them, even itself, thereby resurrecting some dreams of Babbage, and showing that machines can follow rules as well as be governed by them. And when Gödel's transplanted and mechanized diagonal argument is applied to this machine, it only shows it to be what most people thought no machine could be: unpredictable. This provides a striking gloss on La Mettrie's thesis that man is a machine that is too complicated to be defined or identified *a priori*. And when von Neumann generalized its operations from the simulation to the construction of other machines, that argument showed further its ability to reproduce itself. So varied has the evidence become for Turing's principle that many regard it as implying the cybernetical hypothesis itself, which is the modern operational form of mechanism. Gödel himself had hopes for the eventual failure of Turing's critical underlying finiteness hypothesis for the calculating mind, resurrecting shades of Dedekind's fallen *Gedankenwelt* proof, but meanwhile mechanists can thank their lucky stars for his *proofs*. They have been almost as lucky as some of their machines. And though the mechanist program, as La Mettrie emphasized, can only be established empirically, if at all, we see that the *a priori* logical results about machines, stemming above all from Gödel, could not possibly have turned out any better for mechanism.

Finally, in connection with Hilbert's view on the importance of independence problems around which his formalism, both early and late, revolves, we note that our entire discussion, from beginning to end, has turned on trying to understand the significance of the discovery of this or that case of independence. We began (Chapter I) by noting that the independence of the two thermodynamic laws from classical mechanics led to a generalization of the machine concept. In particular, the impossibility of proving these laws from the axioms of mechanics – which physicists attempted to do just as geometers were trying to prove Euclid's postulate from his other axioms – implied the impossibility of proving machines to be in principle predictable, as well as the possibility of theoretical limitations on machine behavior, and the enrichment of mechanism by the element of chance. This latter enrichment of mechanism, as well as its finitistic element, was further strengthened by quantum theory. Then (Chapter II) we examined the independence of the laws of infinity and continuity from the laws of thought. In particular, the impossibility of proving the axiom of infinity and continuity from purely logical considerations implied the impossibility of proving that finite state machines must have fewer mental or physical states than humans. In Chapter III we continued this theme by arguing that the independence of the continuum hypothesis is like that of the parallel postulate after all in showing the impossibility of fine tuning of infinite structures once they are postulated. Further we noted that it is the independence of Desargues' theorem from plane descriptive or projective geometry – making machine proofs of it more difficult than they would otherwise be – which is really more like Gödel's incompleteness than that of Euclid's postulate. We then argued, in Chapter IV, that the independence of Gödel's sentences from arithmetic implied the possibility of machine self-reflection as well as the impossiblity of effectively predicting machine behavior in general. Finally we noted the dependence of this interpretation on Church's thesis, and examined the problem of proving this thesis. It is apparently independent of both classical and intuitionistic mathematics, which implies that even this limited form of mechanism comprises a fundamentally new hypothesis which cannot be established *a priori*, but only experimentally – just as in the case of geometrical independence.

Our discussion of abstract Turing machines has, of course, taken the problem of 'pattern recognition' largely for granted, but it is by no means obvious that a physical realization of a Turing machine can always be programmed to learn to recognize the global properties of arbitrarily degraded symbols. The problem here is to explicitly formulate something like the Gestaltist rules for good figures. Nelson [194] has meanwhile attempted to extend Church's thesis to the case of machine visual perception, arguing that a machine can learn to recognize a pattern-type just in case it is a recursive subset of all possible patterns that can impinge on a machine's tape. His treatment includes the case of degraded symbols, but it is not entirely clear what such abstract arguments imply for the visual capacities of physical realizations of Turing machines. We have seen, however, that machines can indeed be constructed which can see in a rudimentary fashion, and can even exploit this faculty to prove geometrical theorems and analyze scenes. Weizenbaum [232] points out that the general problem of computer vision "is in many respects fundamentally the same as that of machine understanding of natural language" (p. 182). The historical roots of this problem go back not to the issue of mechanism in arithmetic, but to the issue of blind mechanisms doing geometry that arose in Poincaré's critique of Hilbert's 'purely logical' approach to geometry. It is doubtful whether any sense can be made of saying that a man understands anything more of the 'semantics' of arithmetical symbols such as '||||||||' or '23817156' than a machine. But the symbolism of geometry is quite another matter: the effective use of Hilbert's "mnemonic symbols of space intuition", such as the perspective use of a Desargues figure to prove geometric theorems, demands some understanding of their meaning, i.e. of the spatial configurations they picture. Even the plane use of a Desargues figure to prove, say, the distributivity of field multiplication over addition, or of a Pascal figure to prove its commutativity, pose considerable challenges to the empirical intuition of current geometry machines. But Descartes feared that even our own space intuition was too indistinct to support his analytic geometry in three dimensions. Now that we have cultivated such intuition the question arises whether we can analyze it with sufficient clarity and distinctness to program our full use of it for machines. I believe that the ultimate test of artificial intelligence – at least the in case of mathematics – will come in this geometric arena which requires the coordinated use of both the perceptual and cognitive abilities of machines, as opposed to the relatively barren arithmetical arena in which machines need only think.

BIBLIOGRAPHY

[1] Abelson, Raziel, 'A Refutation of Mind-Body Identity', *Philosophical Studies* **XXI** (1970), 85–89.

[2] Ackerman, Wilhelm, 'Begründung des "tertium non datur" mittels der Hilbertschen Theorie der Widerspruchsfreiheit', *Mathematische Annalen* **XCIII** (1924), 1–34.

[3] Anderson, Alan Ross, ed., *Minds and Machines*. Contemporary Perspectives in Philosophy Series. Englewood Cliffs, New Jersey: Prentice-Hall, 1964.

[4] Arbib, Michael A., 'Some Comments on Self-Reproducing Automata', in *Systems and Computer Science*. Edited by John F. Hart and Satoru Takasu. Toronto: University of Toronto Press, 1967.

[5] Baird, A. W. S., 'Pascal's Idea of Nature', *Isis* **LXI** (1970), 297–320.

[6] Becker, Oskar, *Beiträge zur phänomenologischen Begründung der Geometrie*. 2, unveränderte Auflage, Frieburg: Max Niemeyer, 1973.

[7] Becker, Oskar, *Mathematische Existenz: Untersuchungen zur Logik und Ontologie mathematischer Phänomene*. Freiburg: Max Niemeyer, Halle a. d. S., 1927.

[8] Becker, Oskar, 'Das Symbolische in der Mathematik', in *Blätter für Deutsche Philosophie*. Bd. 1, Heft. 4 (1928), 329–348.

[9] Becker, Oskar, *Die Grundlagen der Mathematik in Geschichtlicher Entwicklung*. Freiburg: Karl Alber, 1954.

[10] Bell, E. T., *The Development of Mathematics*. New York: McGraw-Hill, 1945.

[11] Bell, John, and Machover, Moshe, *A Course in Mathematical Logic*. Amsterdam: North-Holland, 1974.

[12] Benacerraf, Paul, 'God, the Devil, and Gödel', *Monist* **LI** (1967), 9–33.

[13] Benacerraf, Paul and Putnam, Hilary, eds., *Philosophy of Mathematics: Selected Readings*. Prentice-Hall Philosophy Series. Englewood Cliffs, New Jersey: Prentice-Hall, 1964.

[14] Benda, Miroslav, 'On Strong Axioms of Induction in Set Theory and Arithmetic', in *Mathematical Logic in Latin America*. Edited by A. Arruda, R. Chuaqui, and N. C. A. da Costa. Amsterdam: North-Holland, 1980.

[15] Berg, Jonathan, and Chihara, Charles, 'Church's Thesis Misconstrued', *Philosophical Studies* **XXVIII** (1975), 357–362.

[15a] Berkeley, George, *Complete Works*, Vol. 1. London: Oxford University Press, 1901.

[16] Bernays, Paul, 'Über Hilberts Gedanken zur Grundlegung der Arithmetik', *Jahresbericht der Deutschen Mathematiker-Vereiningung* **XXXI** (1922), first section, 10–19.

[17] Bernays, Paul, 'Die Bedeutung Hilberts für die Philosophie der Mathematik', *Die Naturwissenschaften* **X** (1922), 93–99.

[17a] Bernays, Paul, 'Erwiderung auf die Note von Herrn Aloys Müller: "Über Zahlen als Zeichen" ', *Mathematische Annalen* **XC** (1923), 159–163.

[17b] Bernays, Paul, 'Grundbegriffe der reinen Geometrie in ihrem Verhaltnis zur Anschauung', *Die Naturwissenschaften* **XVI** (1928), 197–203.

[18] Bernays, Paul, 'Die Philosophie der Mathematik und die Hilbertsche Beweis-Theorie', *Blätter für deutsche Philosophie* **IV** (1930), 326–367.

[19] Bernays, Paul, 'Quelques points essentiels de la métamathématique', *L'enseignement mathématique* **XXXIV** (1935), 70–95.

[20] Bernays, Paul, 'Thesen und Bemerkungen zu den philosophischen Fragen und zur Situation der logisch-mathematischen Grundlagenforschung', in *Traveau du IXe Congrès international de Philosophie,* fasc. I-XII, A. S. I. nos. 530–541, Paris, 1937, pp. 104–110.

[21] Bernays, Paul, 'Sur les questions méthodologiques actuelles de la théorie hilbertienne de la démonstration', in *Les entretiens de Zurich sur les fondements et la méthode des sciences mathématiques, 6–9 decembre 1938*. Edited by F. Gonseth. Zurich: Leeman, 1941, discussion, pp. 156–161.

[22] Bernays, Paul, 'Zur Beurteilung der Situation in der beweistheoretischen Forschung', *Revue internationale de philosophe* **VIII** (1954), 9–13, discussion, 15–21.

[22a] Bernays, Paul, 'Betrachtungen über die Vollständigkeitsaxiom und verwandte Axiome', *Mathematische Zeitschrift* **63** (1955), 219–232.

[23] Bernays, Paul, "Die Mannigfaltigkeit der Direktiven für die Gestaltung geometrischer Axiomensysteme", in Bernays [24], pp. 142–154.

[24] Bernays, Paul, *Abhandlungen zur philosophie der Mathematik*. Darmstadt: Wissenschaftliche Buchgesellschaft, 1976.

[25] Bird, Richard, *Programs and Machines: An Introduction to the Theory of Computation*. London: Wiley, 1976.

[25a] Bishop, E., *Foundations of Constructive Analysis*. New York: McGraw-Hill, 1967.

[26] Blumenthal, Otto, 'David Hilbert', *Die Naturwissenschaften* **X** (1922), 66–72.

[26a] Bochenski, I. M., *A History of Formal Logic*. South Bend, Indiana: University of Notre Dame Press, 1961.

[27] Bohm, David, *Causality and Chance in Modern Physics*. New York: Harper Torchbook, 1961.

[27a] Bolyai, Wolfgang and Bolyai, Johann, *Geometrische Untersuchungen*. Edited by Paul Stäckel. Leipzig, 1913.

[28] Bolzano, Bernard, *Paradoxien des Unendlichen*. Philosophische Bibliothek, Bd. 99, Hamburg: Felix Meiner, 1975.

[28a] Boolos, George, 'On Second-Order Logic', *Journal of Philosophy* **LXXII** (1975), 509–527.

[29] Borel, Emile, *Space and Time*. New York: Dover Publications, 1960.

[30] Bowden, B. V., ed., *Faster than Thought: A Symposium on Digital Computing Machines*, London: Pitman, 1953.

[31] Bowie, Lee. G., 'An Argument Against Church's Thesis', *Journal of Philosophy* **LXX** (1973), 66–76.

[32] Boyd, Richard, 'Determinism, Laws, and Predictability in Principle', *Philosophy of Science* **XXXIX** (1972), 431–450.

[33] Bradley, F. H., *The Principles of Logic*, Vol. I. London: Oxford University Press, 1928.

[34] Broad, C. D., *The Mind and Its Place in Nature*. New York: Humanities Press, 1960.

[35] Brodén, T., 'Einige Worte über aktuelle mathematische Prinzipfragen', *Den Sjette Skandineviske Mat.-Kongr. i København 1925* (1926), 229–239.

[36] Brouwer, Luitzen E. J., Review of *Allgemeine Theorie der unendlichen Mengen und Theorie der Punktmengen*, by Arthur Schoenflies. *Jahresbericht der Deutschen Mathematiker-Vereinigung 23*, 2nd section, pp. 78–83.

[37] Brouwer, Luitzen E. J., 'Mathematik, Wissenschaft, und Sprache', *Monatshefte für Mathematik und Physik* **XXXVI** (1929), 153–164.

[38] Brouwer, Luitzen E. J., *On the Foundations of Mathematics*, in *Collected Works*, Vol. 1. Edited by A. Heyting. Amsterdam: North-Holland, 1974.

[39] Buck, R. C., 'Mathematical Induction and Recursive Definitions', *American Mathematical Monthly* **LXX** (1963), 128–135.

[40] Burks, Arthur W., 'Programming and the Theory of Automata', in *Computer Programming and Formal Systems*. Edited by P. Braffort and D. Hirschberg. Amsterdam: North-Holland, 1963.

[41] Burks, Arthur W., *Chance, Cause, and Reason: An Inquiry into the Nature of Scientific Evidence*. Chicago: Chicago University Press, 1977.

[42] Burks, Arthur W., ed., *Essays on Cellular Automata*. Urbana, Illinois: University of Illinois Press, 1970.

[43] Busemann, Herbert, 'Problem IV: Desarguesian Spaces', in *Development of Hilbert's Problems. Proceedings of Symposia in Pure Mathematics XXVIII* (1976), 131–141.

[44] Cantor, Georg, *Gesammelte Abhandlungen mathematischen und philosophischen Inhalts*. Edited by Ernst Zermelo. Hildesheim, Germany: Georg Olms, 1966.

[45] Carnap, Rudolf, *The Logical Syntax of Language*. International Library of Scientific Method. Translated by Amethe Smeaton. London: Routlege and Kegan Paul Ltd., 1937.

[46] Carnap, Rudolf, *The Logical Structure of the World: Pseudoproblems in Philosophy*. Translated by Rolf A. George. Berkeley and Los Angeles, California: University of California Press, 1967.

[46a] Carroll, Lewis, *Euclid and His Modern Rivals*. New York: Dover Publications, 1973.

[47] Cassirer, Ernst, *Determinism and Indeterminism in Modern Physics*. New Haven, Conn.: Yale University Press, 1956.

[48] Chaitin, Gregory, 'Information-Theoretic Limitations of Formal Systems', *Journal of the Association for Computing Machinery* **XXI** (1974), 403–424.

[49] Chaitin, Gregory, 'A Theory of Program Size Formally Identical to Information Theory', *Journal of the Association for Computing Machinery* **XXII** (1975), 329–340.

[50] Chaitin, Gregory, 'Randomness and Mathematical Proof', *Scientific American* **CCXXXII** (1975), 47–52.

[51] Chihara, Charles S., 'On Alleged Refutations of Mechanism Using Gödel's Incompleteness Results', *Journal of Philosophy* **LXIX** (1972), 509–526.

[52] Church, Alonzo, 'The Richard Paradox', *American Mathematical Monthly* **XLI** (1934), 356–361.

[52a] Church, Alonzo, Review, *Journal of Symbolic Logic* **XXX** (1965), 358–359.

[53] Chwistek, Leon, *The Limits of Science*. New York: Humanities Press, 1948.

[54] Cohen, Paul, *Set Theory and the Continuum Hypothesis*. New York: Benjamin, 1966.

[55] Coleman, William, *Biology in the Nineteenth Century: Problems of Form, Function, and Transformation*. New York and London: Wiley, 1971.

[56] Contro, Walter S., 'Von Pasch zu Hilbert', *Archive for History of Exact Sciences* **XV** (1976), 283–295.

[56a] Couturat, Louis, 'For Logistics', *Monist* **XXII** (1912), 481–523.

[57] Curry, Haskell B., 'An Analysis of Substitution', *American Journal of Mathematics* **LI** (1929), 363–384.

[58] Curry, Haskell B. and Feys, Robert, *Combinatory Logic*, Vol. I. Studies in Logic and the Foundations of Mathematics. Amsterdam: North-Holland, 1958.

[59] Daub, Edward E., 'Maxwell's Demon', *Studies in History and Philosophy of Science* **I** (1970) 213–227.

[60] Davis, Martin, *Computability and Unsolvability*. New York: McGraw-Hill, 1958.

[61] Davis, Martin, ed., *The Undecidable: Basic Papers on Undecidable Propositions, Unsolvable Problems and Computable Functions*. Hewlett, New York: Raven Press, 1967.

[62] Dedekind, Richard, 'Über die Einführung neuer Funktionen in der Mathematik', in *Richard Dedekind: Gesammelte mathematische Werke*. Edited by Robert Fricke, Emmy Noether, and Öystein Ore. Braunschweig: Friedr. Vieweg und Sohn Akt.-Ges., 1932.

[63] Dedekind, Richard, *Essays on the Theory of Numbers*. Translated by Wooster W. Beman. LaSalle, Illinois: Open Court Publishing Company, 1948.

[64] Dedekind, Richard, 'Über den Begriff des Unendlichen', Unpublished manuscript in the Niedersächsische Staats- und Universitätsbibliothek, Göttingen (Cod. Ms. Dedekind 13).

[65] Dedekind, Richard, *Was sind und was sollen die Zahlen*, 3rd edition. Brunswick: F. Vieweg und Sohn, 1911.

[66] Dehn, Max, 'Die Grundlagen der Geometrie in historischer Entwicklung', in *Vorlesungen über neuer Geometrie*. M. Pasch and M. Dehn. Berlin: Springer, 1926, pp. 185–271.

[66a] Dehn, Max, 'Die Legendre'schen Sätze über die Winkelsumme im Drieck', *Mathematische Annalen* **XLVIII** (1900), 404–439.

[66b] Dehn, Max, 'Hilberts geometrisches Werk', *Die Naturwissenschaften* **X** (1922), 77–82.

[67] Delong, Howard, *A Profile of Mathematical Logic*. Reading, Massachusetts: Addison-Wesley, 1970.

[68] Dennett, Daniel, 'Why the Law of Effect Will Not Go Away', *Journal of the Theory of Social Behavior* **V** (1975), 169–187.

[69] Descartes, Rene, *The Philosophical Works of Descartes*. Edited by E. S. Haldane and G. R. T. Ross, Vol. I and II. Cambridge, England: Cambridge University

Press, 1955.

[70] Detlefsen, Michael, 'On Interpreting Gödel's Second Theorem', *Journal of Philosophical Logic* **VIII** (1979), 297–313.

[71] Dreyfus, Hubert L., *What Computers Can't Do: The Limits of Artificial Intelligence*, revised edition. New York: Harper and Row, 1979.

[72] Du Bois-Reymond, Emil, *Über die Grenzen des Naturerkennens*. Berlin and Leipzig: Veit and Comp., 1882.

[73] Du Bois-Reymond, Paul, 'Über asymptotische Werte, infinitäre Approximationen und infinitäre Auflösungen von Gleichungen', *Math. Annalen* **VIII** (1875), 363–414.

[73a] Du Bois-Reymond, Paul, *Die allgemeine Funktionentheorie*, Vol. I. Tubingen, 1882.

[74] Eddington, Sir Arthur, *The Nature of the Physical World*. Cambridge, England: Cambridge University Press, 1929. (Reissued as an Ann Arbor Paperback, 1961.)

[75] Ehrenberg, W., 'Maxwell's Demon', *Scientific American* **CCXVII** (1967) 103–110.

[76] Enriques, F., *Historic Development of Logic*. New York: Russell and Russell, 1968.

[76a] Essler, Willhelm K., 'Über die Interpretation von Wissenschaftssprachen', *Philosophisches Jahrbuch* **LXXVII** (1970), 117–130.

[76b] Euclid, *Elements*, Vol I. Translated with introduction and commentary by Sir Thomas L. Heath. New York: Dover Publications, 1956.

[77] Fang, J., *Hilbert: Towards a Philosophy of Modern Mathematics II*. New York: Paideia Press, 1970.

[78] Feferman, Solomon, 'Some Completeness Results for Recursive Progressions of Theories (Ordinal Logics)', *Journal of Symbolic Logic* **XXIV** (1959), 312.

[78a] Feferman, Solomon, 'Arithmetization of Matamathematics in a General Setting', *Fundamenta Mathematicae* **XLIX** (1960), 35–92.

[79] Feferman, Solomon, 'Transfinite Recursive Progressions of Axiomatic Theories', *Journal of Symbolic Logic* **XXVII** (1962), 259–316.

[80] Fine, Arthur I., 'Note on Goodstein's "The Significance of Incompleteness Theorems"', *British Journal for the Philosophy of Science* **XV** (1964), 140–141.

[81] Finsler, Paul, 'Über die Grundlegung der Mengenlehre. Erste Teil', *Mathematische Zeitschrift* **XXV** (1926), 683–713.

[82] Finsler, Paul, 'Über die Grundlegung der Mathematik', *Jahresbericht Deutscher Mathematiker-Vereinigung* **XXXVI** (1927), 18 (kursiv).

[83] Finsler, Paul, 'Gibt es unentscheidbare Sätze?', *Commentarii Math. Helvetici* **XVI** (1944), 310–20.

[84] Finsler, Paul, 'Über die Berechtigung infinitesimalgeometrischer Betrachtungen', in *Aufsätze zur Mengenlehre*. Edited by G. Unger. Darmstadt, 1975.

[85] Finsler, Paul, 'Über die Grundlegung der Mengenlehre. Zweite Teil', *Commentarii Math. Helvetici* **XXXVIII** (1963–4), 172–218.

[86] Fraenkel, Abraham A., Bar-Hillel, Yehosua and Levy, Azriel, *Foundations of Set Theory* (with collaboration of Dirk van Dalen). Studies in Logic and the Foundations of Mathematics. Amsterdam: North-Holland, 1973.

[87] Frege, Gottlob, *The Foundations of Arithmetic: A Logico-Mathematical Enquiry into the Concept of Number*. Translated by J. L. Austin. New York: Philosophical Library, 1950.

[88] Frege, Gottlob, *Translations from the Philosophical Writings of Gottlob Frege*. Edited by Peter Geach and Max Black. Oxford: Basil Blackwell, 1960.

[89] Frege, Gottlob, *Grundgesetze der Arithmetik*. Hildesheim, Germany: Georg Olms, 1962.

[90] Frege, Gottlob, *Posthumous Writings*. Translated by Peter Long and Roger White, with the assistance of Raymond Hargreaves. Chicago, Ill.: University of Chicago Press, 1979.

[90a] Frege, Gottlob, *Philosophical and Mathematical Correspondence*. Abridged from the German edition by Brian McGuinness and translated by Hans Kaal. Chicago, Ill.: University of Chicago Press, 1980.

[91] Frege, Gottlob, *On the Foundations of Geometry and Formal Theories of Arithmetic*. Translated by Eike-Henner W. Kluge. New Haven, Connecticut: Yale University Press, 1971.

[92] Freudenthal, Hans, *The Language of Logic*. Amsterdam: Elsevier, 1966.

[92a] Freudenthal, Hans, 'Zur Geschichte der Grundlagen der Geometrie, zugleich eine Besprechung der 8. Aufl. von Hilberts "Grundlagen der Geometrie"', *Nieuw Archief voor Wiskunde* (4) **V** (1957), 105–142.

[93] Frey, Gerhard, 'Sind bewusstseinsanaloge Maschine möglich?', *Studium Generale* **XIX** (1966), 191–200.

[94] Friedman, Harvey M., 'Axiomatic Recursive Function Theory', in *Logic Colloquium '69 (Proc. Summer School and Colloq., Manchester, 1969)*. Amsterdam: North-Holland, 1971.

[94a] Fritz, Kurt von, 'The Discovery of Incommeasurability by Hippasus of Metapontum', *Annals of Mathematics* **XLVI** (1945), 242–264.

[95] Gauss, Carl Friedrich, *Disquisitiones Arithmeticae*. New Haven, Conn.: Yale University Press, 1968.

[96] Gelernter, H., 'Realization of a Geometry-Theorem Proving Machine', in *Computers and Thought*. Edited by E. A. Feigenbaum and J. Feldman. New York: McGraw-Hill, 1963.

[96a] Gelernter, H., 'A Note on Syntactic Symmetry and the Manipulation of Formal Systems by Machine', *Information and Control* **II** (1959), 80–89.

[97] Gentzen, Gerhard, *The Collected Papers of Gerhard Gentzen*. Edited by M. E. Szabe. Studies in Logic and the Foundations of Mathematics. Amsterdam: North-Holland, 1969.

[98] Gibson, A. B., *The Philosophy of Descartes*. New York: Russell and Russell, 1932.

[99] Gödel, Kurt, 'Über die Vollstandigkeit des Logikkalküls'. Unpublished thesis, University of Vienna, 1930.

[100] Gödel, Kurt, 'On Undecidable Propositions of Formal Mathematical Systems', in Davis [61], pp. 39–74.

[101] Gödel, Kurt, 'Remarks Before the Princeton Bicentennial Conference on Problems in Mathematics', in Davis [61], pp. 84–88.

[102] Gödel, Kurt, 'Über eine bisher noch nicht benützte Erweiterung des finiten Standpunktes', *Dialectica* **XII** (1958), 280–287.

[102a] Goldfarb, Warren D., 'Logic in the Twenties: The Nature of the Quantifier', *Journal of Symbolic Logic* **XLIV** (1979), 351–368.

[103] Good, I. J., 'Human and Machine Logic', *British Journal for the Philosophy of*

Science **XVIII** (1967), 145–146.

[104] Goodman, Nelson, *Problems and Projects*. Indianapolis, Ind.: Bobbs-Merrill, 1972.

[105] Goodstein, R. L., 'The Significance of Incompleteness Theorems', *British Journal for the Philosophy of Science* **XIV** (1963), 208–220.

[106] Grassman, Herman, *Lehrbuch der Arithmetik für höhere Lehranstalten*. Berlin, 1861.

[107] Grattan-Guinness, I., 'In Memoriam Kurt Gödel: His 1931 Correspondence with Zermelo on his Incompletability Theorem', *Historia Mathematica* **VI** (1979), 294–304.

[107a] Gunderson, Keith, 'Descartes, LaMettrie, Language, and Machines', *The Journal of the Royal Institute of Philosophy* **XXXIX** (1964), 193–222.

[108] Hadamard, Jacques, 'Cinq lettres sur la théorie des ensembles', *Bulletin de la Société mathématique de France* **XXX** (1905), 261–273.

[109] Hameister, E., 'Eine mathematikgeschichtliche Bemerkung zur Begründung der Proportionenlehre durch D. Hilbert', *Wissenschaftliche Zeitschrift der Technischen Hochschule Otto von Guericke Magdeburg* **XX** (1976), 323–325.

[110] Hawkins, David, *The Language of Nature: An Essay in the Philosophy of Science*. San Francisco: W. H. Freeman, 1964.

[111] Heidegger, Martin, *Kant and the Problem of Metaphysics*. Translated by James S. Churchill. Bloomington, Indiana: Indiana University Press, 1962.

[112] Helm, Bertrand P., 'The Critical Philosophy and the Royce-Bradley Dialogue', *Journal of the History of Ideas* **XI** (1973), 229–236.

[113] Henkin, Leon, 'On Mathematical Induction', *American Mathematical Monthly* **LXVI** (1960), 323–338.

[114] Herbrand, Jacques, *Logical Writings*. Edited by Warren S. Goldfarb. Cambridge, Mass.: Harvard University Press, 1971.

[114a] Hertz, Heinrich, *The Principles of Mechanics*. New York: Dover Publications, 1956.

[115] Hessenberg, Gerhard, *Grundbegriffe der Mengenlehre*. Göttingen: Vandenhoeck and Ruprecht, 1906.

[116] Hilbert, David, 'Über den Zahlbegriff', *Jahresbericht der Deutschen Mathematiker-Vereinigung* **VIII** (1900), 180–194.

[117] Hilbert, David, *Foundations of Geometry*. Translated by Leo Unger. La Salle, Illinois: Open Court, 1971.

[117a] Hilbert, David, *Vorlesung über die Grundlagen der Geometrie*. Gehalten von Prof. Dr. D. Hilbert, Göttingen, W-S 1898/99, ausgearbeitet von Dr. H. V. Schaper.

[117b] Hilbert, David, *Foundations of Geometry*. Authorized translation of E. J. Townsend. Chicago Ill.: Open Court, 1902.

[118] Hilbert, David, 'Mathematical Problems', *Bulletin of the American Mathematical Society* **VIII** (1901–2), 437–79.

[119] Hilbert, David, 'On the Foundations of Logic and Arithmetic', in Van Heijenoort [255], pp. 129–138.

[119a] Hilbert, David, 'Zur Begründung der elementaren Strahlungstheorie. Dritte Mitteilung', Göttinger Nachrichten 1914, pp. 275–298, *Physikalische Zeitschrift* **XV** (1914), 878–889. (Reprinted in Hilbert [128], pp. 238–257.)

[120] Hilbert, David, 'Axiomatisches Denken', *Mathematische Annalen* **LXXVIII**

(1918), 405–415. (Reprinted in Hilbert [128], pp. 146–156.)

[121] Hilbert, David, 'Neubegrundung der Mathematik (Erste Mitteilung)', *Abhandlungen aus dem mathematischen Seminar der Hamburgschen Universität* **I** (1922), 155–177. (Reprinted in Hilbert [128], pp. 155–177.)

[122] Hilbert, David, 'Die logischen Grundlagen der Mathematik', *Mathematische Annalen* **LXXXVIII** (1923), 151–165. (Reprinted in Hilbert [128], pp. 178–191.)

[123] Hilbert, David, 'Probleme der Grundlegung der Mathematik', *Atti del Congresso internazionale dei mathematici, Bologna 3–10 septembre 1928, Bologna* **I** (1929), 135–141.

[124] Hilbert, David, 'Die Grundlegen der elementaren Zahlenlehre', *Mathematische Annalen* **CIV** (1931), 485–494.

[124a] Hilbert, David, 'Knowledge of Nature and Logic'. Unpublished translation by Don Howard of 'Naturerkennen und Logic', reprinted in [128].

[125] Hilbert, David, 'Beweis des Tertium non datur', in *Nachrichten von der Gesselschaft der Wissenschaften zu Göttingen*, Mathematisch-physicalische Klasse, 1931, pp. 120–125.

[126] Hilbert, David, *Gesammelte Abhandlungen*, Vol. I. Bronx, New York: Chelsea Publishing Company, 1932.

[127] Hilbert, David, *Gesammelte Abhandlungen*, Vol. II. Bronx, New York: Chelsea Publishing Company, 1933.

[128] Hilbert, David, *Gesammelte Abhandlungen*, Vol. III. Bronx, New York: Chelsea Publishing Company, 1935.

[129] Hilbert, David and Bernays, Paul, *Grundlagen der Mathematik*, Vol. I. Berlin: Springer, 1968.

[130] Hilbert, David and Bernays, Paul, *Grundlagen der Mathematik*, Vol. II. Berlin: Springer, 1970.

[130a] Hilbert, David and Cohn-Vossen, Stephan, *Geometry and the Imagination*. Bronx, New York: Chelsea Publishing Company, 1956.

[131] Hobbes, Thomas, *Leviathan*, Parts I and II, New York: Liberal Arts Press, 1958.

[131a] Hoering, W., 'Absolut unentscheidbare Sätze der Mathematik', in *Logik und Logikkalkül*. Edited by Max Käsbauer and Franz von Kutschera. Munich: Karl Alber Freiburg, 1962, pp. 189–194.

[132] Hofstadter, Douglas R., *Gödel, Escher, Bach: An Eternal Golden Braid*. New York: Basic Books, 1979.

[132a] Hölder, Otto, Review of Grassman [106], *Göttingen gelehrte Anzeigen*, 1891, p. 591.

[133] Hume, David, *A Treatise of Human Nature*, 2 Vols. London: Everyman's Library, 1911.

[134] Huntington, Edward V., 'A Set of Postulates for Ordinary Complex Algebra', *Transactions of the American Mathematical Society* **VI** (1905), 209–218.

[135] Husserl, Edmund, *Ideas: General Introduction to Pure Phenomenology*. Translated by W. R. Boyce Gibson. London: George Allen and Unwin Ltd., 1931.

[136] Husserl, Edmund, *Formal and Transcendental Logic*. Translated by Dorian Cairns. The Hague: Martinus Nijhoff, 1969.

[137] Husserl, Edmund, *Philosophie der Arithmetik*. Edited by Lothar Eley. Husserliana. The Hague: Martinus Nijhoff, 1970.

[137a] Jaki, Stanley L., *Brain, Mind and Computers*. New York: Herder and Herder, 1969.

[138] Järnefelt, G. and Kustaanheimo, P., "An Observation on Finite Geometries", *Den 11 Skandinaviska Mathematikerkongress Lund, 1949*, pp. 166–173.

[139] Kac, Mark and Ulam, Stanislaw M., 'Mathematics and Logic: Retrospect and Prospects', in *Britannica Perspectives*, Vol. I. Encyclopaedia Britannica, Inc. Chicago: William Benton, 1968.

[140] Kalmar, Laszlo, 'An Argument Against the Plausibility of Church's Thesis', in *Constructivity in Mathematics: Proceedings of the Colloquium held at Amsterdam, 1957*. Edited by Arend Heyting. Amsterdam: North-Holland, 1959, pp. 72–80.

[140a] Kambartel, Friedrich, *Erfahrung und Struktur*. Bausteine zu einer Kritik des Empirismus und Formalismus, Frankfurt, 1968.

[141] Kanitscheider, Bernulf, *Geometrie und Wirklichkeit*. Berlin: Duncker und Humboldt, 1971.

[142] Kant, Immanuel, *Selected Pre-critical Writings and Correspondence with Beck*. Translated by G. B. Kerferd and D. E. Walford. New York: Barnes and Noble, 1968.

[143] Kaufmann, F., *Das Unendliche in der Mathematik und seine Ausschaltung. Eine Untersuchung uber die Grundlagen der Mathematik*. Leipzig and Berlin: F. Deuticke 1930.

[143a] Keferstein, Hans, 'Erwiderung auf vorstehende Abhandlung', Unpublished reply dated Nov. 17, 1890 to Dedekind [64], copy in Dedekind's hand in the Niedersächsische Staats- und Universitatsbibliothek, Göttingen.

[144] Keyser, J., 'The Axiom of Infinity: A New Presupposition of Thought', *Hibbert Journal* **II** (1903–4), 532–552.

[144a] Kitcher, Philip, 'Hilbert's Epistemology', *Philosophy of Science* **XLIII** (1976), 99–115.

[145] Kleene, Stephen C., 'A Theory of Positive Integers in Formal Logic', *American Journal of Mathematics* **LVII** (1935), 153–173, 219–244.

[146] Kleene, Stephen C., 'General Recursive Functions of Natural Numbers', Davis [61], pp. 236–253.

[147] Kleene, Stephen C., 'λ-Definability and Recursiveness', *Duke Mathematical Journal* **II** (1936), 340–353.

[148] Kleene, Stephen C., 'On Notation for Ordinal Numbers', *Journal of Symbolic Logic* **III** (1938), 150–155.

[149] Kleene, Stephen C., *Introduction to Metamathematics*. Princeton, New Jersey: D. van Nostrand, 1950.

[150] Kleene, Stephen C., *Mathematical Logic*. New York: John Wiley and Sons, 1967.

[151] Klein, Felix, *Geometry: Elementary Mathematics from an Advanced Standpoint*. Translated by E. R. Hederick and C. A. Noble. New York: Dover Publications, 1939.

[152] König, Julius, *Neue Grundlagen der Logik, Arithmetik und Mengenlehre*. Leipzig: Veit, 1914.

[152a] Koyré, Alexandre, *Newtonian Studies*. Chicago, Illinois: University of Chicago Press, 1965.

[153] Kreisel, Georg, 'Note on Arithmetic Models for Consistent Formulae of the Predicate Calculus II', *Proceedings of the XIth International Congress of Philosophy*, Bruxelles, 1953, Vol. 14, pp. 39–49.

[154] Kreisel, Georg, 'Mathematical Logic', in *Lectures on Modern Mathematics*, Vol. III. Edited by T. L. Saaty. New York: John Wiley and Sons, 1965.

[155] Kreisel, Georg, 'Mathematical Logic: What Has it Done for the Philosophy of Mathematics', in *Bertrand Russell: Philosopher of the Century*. Edited by R. Schoenman. London: Allen and Unwin Ltd., 1966, pp. 201–272.

[155a] Kreisel, Georg, 'Hilbert's Programme and the Search for Automatic Proof Procedures', in *Symposium on Automatic Demonstration*. Berlin: Springer, 1970.

[156] Kreisel, Georg, 'Church's Thesis: A Kind of Reducibility Axiom for Constructive Mathematics', in *Intuitionism and Proof Theory: Proceedings of the Summer Conference at Buffalo N. Y. 1968*. Edited by A. Kino, J. Myhill, and R. E. Vesley. Amsterdam: North-Holland, 1970, pp. 121–150.

[157] Kreisel, Georg, 'Which Number Theoretic Problems Can Be Solved in Recursive Progressions on Π_1-Paths through 0', *Journal of Symbolic Logic* **XXXVII** (1972), 311–334.

[158] Kronecker, Leopold, 'Über den Zahlbegriff', *Journal für die reine und angewandte Mathematik (Crelle)* **CI** (1887), 337–355.

[159] Kustaanheimo, Paul, 'A Note on a Finite Approximation of the Euclidean Plane Geometry', *Societas Scientiarum Fennica, Commentationes Physico-Mathematicae* **XV** (1953), 1–11.

[160] Kustaanheimo, Paul, 'Über die Versuche, ein logisch finites Weltbild aufzubauen', in *Proceedings of 2nd International Congress of the International Union for Philosophy of Science*, Zurich, 1954, pp. 59–65.

[161] Laing, Richard, 'Maxwell's Demon and Computation', *Philosophy of Science* **XLI** (1974), 171–178.

[162] La Mettrie, Julian Offray de, *Man a Machine*. La Salle, Ill.: Open Court, 1912.

[163] Larmour, Joseph, 'Energy', in *Encyclopeadia Brittanica*, Vol. VII. New York: Encyclopaedia Britannica Company, 1910.

[163a] Lear, Jonathan, 'Aristotle's Compactness Proof', *Journal of Philosophy* **LXXVI** (1979), 198–215.

[164] Leibniz, Gottfried Wilhelm, *Philosophical Papers and Letters*, 2 Vols. Translated and edited by L. E. Loemker. Chicago, Illinois: Chicago University Press, 1956.

[165] Leibniz, Gottfried Wilhelm, *New Essays Concerning Human Understanding*. Translated by A. G. Langley, 2nd edition. New York: Macmillan, 1916.

[166] Le Lionnais, F., ed., *Great Currents of Mathematical Thought*, Vol. I. *Mathematics: Concepts and Development*. Translated by R. A. Hall and Howard G. Bergmann. New York: Dover Publications, 1971.

[167] Löb, M. H., 'Solution of a Problem of Leon Henkin', *Journal of Symbolic Logic* **XX** (1955), 115–118.

[168] Locher, L., 'Die Finslerschen Arbeiten zur Grundlegung der Mathematik', *Comentarii mathematici helvitici* **X** (1938), 206–207.

[169] Locke, John, *Essay Concerning Human Understanding*, Vol. I. Edited by A. C. Fraser. Oxford: Clarendon Press, 1894.

[170] Lockwood, M., 'A Reply to Professor Abelson', *Philosophical Studies* **XXIV** (1973), 133–135.

[171] Lodge, Sir Oliver J., 'Continuity', *Nature* (1913), 33–48.

[172] Lorenzen, Paul, *Normative Logic and Ethics*. Manheim: Bibliographisches In-

stitut, 1969.

[173] Lucas, John R., 'Minds, Machines, and Gödel', in Anderson [3], pp. 43–59.

[174] Lucas, John R., 'Human Logic and Machine Logic: A Rejoinder', *British Journal for the Philosophy of Science* **XIX** (1967), 155–156.

[175] Lucas, John R., 'Satan Stultified', *Monist* **LII** (1968), 145–158.

[176] Lucas, John R., *The Freedom of the Will*. Oxford: Clarendon Press, 1970.

[176a] Lucas, John R., *The Concept of Probability*. Oxford: Clarendon Press, 1970.

[177] Lucas, John R., 'Metamathematics and the Philosophy of Mind: A Rejoinder', *Philosophy of Science* **XXXVIII** (1971), 310–313.

[178] Lyndon, Roger C., *Notes on Logic*. Van Nostrand Mathematical Studies. New York: American Book, Van Nostrand, Reinhold, 1966.

[179] Mahnke, Dietrich, 'From Hilbert to Husserl: First Introduction to Phenomenology, Especially that of Formal Mathematics', *Studies in History and Philosophy of Science* **VIII** (1977), 71–84, translated by D. L. Boyer.

[180] Martin-Löf, P., 'The Literature on von Mises' Kollektivs Revisited', *Theoria* **XXXV** (1969), 12–37.

[181] McCulloch, Warren S., *Embodiments of Mind*. Cambridge, Massachusetts: M. I. T. Press, 1965.

[182] Mehlberg, Henryk, 'The Present Situation in the Philosophy of Mathematics', in *Logic and Language: Studies Dedicated to Professor Rudolf Carnap on the Occasion of His Seventieth Birthday*. Dordrecht, Holland: D. Reidel, 1962.

[183] Meschkowski, Herbert, *Richtigkeit und Wahrheit in der Mathematik*. Mannheim, Germany: Bibliographisches Institut, 1971.

[184] Minsky, Marvin L., *Computation: Finite and Infinite Machines*. Englewood Cliffs, New Jersey: Prentice-Hall, Inc., 1967.

[185] Mises, Richard von, 'Die Wissenschaftliche Weltbild der Gegenwart', *Die Naturwissenschaften* **XVIII** (1930), 885–893.

[186] Mises, Richard von, *Probability, Statistics, and Truth*. Second revised English edition prepared by Hilda Geiringer. London: George Allen and Unwin, 1957.

[187] Moore, E. F., 'Mathematics in the Biological Sciences', *Scientific American* **CCXI** (1964), 148–164.

[188] Morrison, Philip and Morrison, Emily, eds., *Charles Babbage and His Calculating Engines: Selected Writings by Charles Babbage and Others*. New York: Dover Publications, 1961.

[189] Moss, J. M. B., 'Some B. Russell's Sprouts (1903–1908)', in *Conference in Mathematical Logic – London '70*. Edited by Wilfred Hodges. Berlin: Springer, 1972, pp. 211–250.

[189a] Mueller, Ian, 'Euclid's *Elements* and the Axiomatic Method', *British Journal for Philosophy of Science* **XX** (1969), 289–309.

[190] Müller, G. H., 'Über die unendlichen Induktion', in *Infinitistic Methods (Proceedings of the Symposium on Foundations of Mathematics, Warsaw 2–9 September 1959)*. Oxford, England: Pergamon Press, 1961, pp. 75–95.

[191] Myhill, John R., 'Abstract Theory of Self Reproduction', in *Essays on Cellular Automata*. Edited by Arthur W. Burks. Urbana, Illinois: University of Illinois Press, 1970.

[192] Nagel, Ernest and Newman, James R., *Gödel's Proof*. London: Routledge and

Kegan Paul Ltd., 1958.

[192a] Nagel, Ernest and Newman, James R., 'Putnam's Review of *Gödel's Proof*', *Philosophy of Science* **XXVIII** (1961), 209–211.

[193] Nathanson, Stephen L., 'Abelson's "Refutation" of Mind-Body Identity', *Philosophical Studies* **XXIII** (1972), 116–118.

[194] Nelson, R. J., 'On Mechanical Recognition', *Philosophy of Science* **XLIII** (1976), 25–52.

[195] Neumann, John von, 'Zur Hilbertschen Beweistheorie', *Mathematische Zeitschrift* **XXVI** (1927), 1–46.

[196] Neumann, John von, *Theory of Self-Reproducing Automata*. Edited and completed by Arthur W. Burks. Urbana, Illinois: University of Illinois Press, 1966.

[196a] Parikh, Rohit. 'Effectiveness', *Philosophical Forum*, forthcoming.

[197] Parsons, Charles D., 'Infinity and Kant's Conception of the "Possibility of Experience" ', in *Kant: A Collection of Critical Essays*. Edited by Robert P. Wolff. Anchor Books, Garden City, New York: Doubleday, 1967, pp. 37–53.

[198] Pattee, H. H., 'Can Life Explain Quantum Mechanics?', in *Quantum Theory and Beyond*. Edited by T. Bastin. Cambridge, England: Cambridge University Press, 1971.

[199] Peirce, C. S., 'The Logic of Number', *American Journal of Mathematics* **IV** (1881), 85–95.

[200] Peirce, C. S., 'Logic Machines', *American Journal of Psychology* **I** (1887), 164–171.

[201] Peirce, C. S., *Collected Papers*, Vol. II. *Elements of Logic*. Edited by Charles Hartshorne and Paul Weiss. Cambridge, Massachusetts: Harvard University Press, 1933.

[202] Peirce, C. S., *Collected Papers*, Vol. III. *Exact Logic*. Edited by Charles Hartshorne and Paul Weiss. Cambridge, Massachusetts: Harvard University Press, 1933.

[202a] Peirce, C. S., *New Elements of Mathematics*, Vol. II. Edited by Carolyn Eisele. Atlantic Highlands, N. J.: Humanities Press, 1976.

[202b] Peirce, C. S., New Elements of Mathematics, Vol. III. Edited by Carolyn Eisele. Atlantic Highlands, N. J.: Humanities Press, 1976.

[203] Péter, Rózsa, *Recursive Functions*. New York: Academic Press, 1967.

[204] Pogorelov, A. V., *Hilbert's Fourth Problem*, Translated by Richard A. Silverman. Washington, D.C.: V.H. Winston and Sons, 1979.

[205] Poincaré, Henri, *Science and Hypothesis*. New York: Dover Publications, 1952.

[206] Poincaré, Henri, *Science and Method*. Translated by Francis Maitland. New York: Dover Publications, n.d.

[207] Poincaré, Henri, *Mathematics and Science: Last Essays*. Translated by John W. Bolduc. New York: Dover Publications, 1963.

[208] Poincaré, Henri, 'The Foundations of Geometry' (Review of Hilbert [117b]), *Bulletin of the American Mathematical Society* **XII** (1903), 1–10.

[209] Post, Emil. L., 'Absolutely Unsolvable Problems and Relatively Undecidable Propositions: Account of an Anticipation', in Davis [61], pp. 338–433.

[210] Prenowitz, W., 'A Contemporary Approach to Classical Geometry', *American Mathematical Monthly* **LXVIII** (1961), 1–11.

[211] Proclus, *A Commentary on the First Book of Euclid's Elements*. Translated by

G. R. Morrow. Princeton, N.J.: Princeton University Press, 1970.

[212] Putnam, Hilary, 'Minds and Machines', in Anderson [3], pp. 72–85.

[212a] Putnam, Hilary, 'Review: Nagel and Newman, *Gödel's Proof*', *Philosophy of Science* **XXVII** (1960), 205–207.

[213] Quine, Willard van Orman, *Word and Object*. Cambridge, Mass.: M. I. T. Press, 1960.

[214] Rabin, M. O., 'On Recursively Enumerable and Arithmetic Models of Set Theory', *Journal of Symbolic Logic* **XXIII** (1958), 408–416.

[215] Raggio, A. R., 'Einige Betrachtungen zum Begriff des Spiels', *Kant Studien* **LVI** (1970), 227–237.

[215a] Raphael, Bertram, *The Thinking Computer: Mind Inside Matter*. San Francisco: Freeman, 1976.

[216] Rautenberg, Wolfgang, 'Hilberts Schnittpunktsätze und einige modell-theoretische Aspekte der formalisierten Geometrie', *Wissenschaftliche Zeitschrift der Humbolt-Universität zu Berlin, Mathematische-Naturwissenschaftliche Reihe* **XIV** (1965).

[217] Reid, Constance, *Hilbert*. New York: Springer, 1970.

[218] Resnik, Michael D., 'The Frege-Hilbert Controversy', *Philosophy and Phenomenological Research* **XXXIV** (1974), 386–403.

[218a] Resnik, Michael D., 'On the Philosophical Significance of Consistency Proofs', *Journal of Philosophical Logic* **III** (1974), 133–147.

[219] Richard, Jules, *Sur la Philosophie des Mathematiques*. Paris: Gauthier-Villars, Imprimeur-Libraire, 1903.

[220] Rogers, Hartley, Jr., 'Gödel Numberings of Parital Recursive Functions', *Journal of Symbolic Logic* **XXIII** (1958), 305.

[221] Rogers, Hartley, Jr., *Theory of Recursive Functions and Effective Computability*. McGraw-Hill Series in Higher Mathematics. New York: McGraw-Hill, 1967.

[222] Rosenbloom, Paul, *The Elements of Mathematical Logic*. New York: Dover Publications, 1950.

[223] Ross, David, 'Church's Thesis: What its Difficulties Are and Are Not', *Journal of Philosophy* **LXXI** (1974), 515–525.

[224] Royce, Josiah, *The World and the Individual*. New York: Dover Publications, 1959.

[225] Royce, Josiah, 'The Concept of the Infinite', *Hibbert Journal* **I** (1902), 21–45.

[226] Russell, Bertrand, 'The Axiom of Infinity', in *Essays in Analysis*. Edited by Douglas Lackey. New York: George Braziller, pp. 256–259.

[227] Russell, Bertrand, *Introduction to Mathematical Philosophy*. London: George Allen and Unwin, 1919.

[228] Schor, Dimitry, 'Neuer Beweis eines Sätzes aus den "Grundlagen der Geometrie" von Hilbert', *Mathematische Annalen* **LVIII** (1904), 427–433.

[228a] Schrödinger, Erwin, 'Über die Unanwendbarkeit der Geometrie im Kleinen', *Die Naturwissenschaften* **XXII** (1934), 518–520.

[229] Schütte, Kurt, 'Beweistheoretische Erfassung der unendlichen Induktion in der Zahlentheorie', *Mathematische Annalen* **CXXII** (1951), 369–389.

[230] Schwabhäuser, Wolfram, 'Über die Vollstandigkeit der elementaren euklidischen Geometrie', *Zeitschrift für mathematische Logik und Grundlagen der Mathematik*

II (1965), 137–165.

[231] Scott, Dana S., 'Background to Formalization', in *Truth, Syntax and Modality: Proceedings of the Temple University Conference on Alternative Semantics*. Edited by Hughes Leblanc. Amsterdam: North-Holland, 1973, pp. 244–273.

[232] Schoenfield, Joseph R., 'On a Restricted ω-Rule', *Buletin de l'académie polonaise des sciences* **VII** (1959), 405–407.

[233] Siefkes, Dirk, *Buchi's Monadic Second Order Successor Arithmetic*. Lecture Notes in Mathematics. 120. Decidable Theories I. Edited by G. H. Müller. Berlin: Springer, 1970.

[234] Singh, J., *Great Ideas of Information Theory, Language and Cybernetics*. New York: Dover Publications, 1966.

[234a] Skinner, B. F., *Science and Human Behavior*. New York: MacMillan Company, 1953.

[235] Smart, J. J. C., 'Gödel's Theorem, Church's Theorem and Mechanism', *Synthese* **XIII** (1961), 105–110.

[236] Smith, David E., *A Source Book in Mathematics*, Vol. I and II. New York: Dover Publications, 1959.

[237] Smith, Norman K., *New Studies in the Philosophy of Descartes: Descartes as Pioneer*. New York: Russell and Russell, 1963.

[238] Steen, S. W. P., *Mathematical Logic with Special Reference to the Natural Numbers*. Cambridge, England: Cambridge University Press, 1972.

[238a] Stein, Howard, 'On the Notion of Field in Newton, Maxwell, and Beyond', in *Minnesota Studies in the Philosophy of Science*, Vol. V. Minneapolis, Minn.: University of Minnesota Press, 1968.

[239] Steiner, Hans Georg, 'Frege und die Grundlagen der Geometrie', *Mathematisch-Physikalische Semesterberichte* **X** (1963), 175–186; **XI** (1964) 35–47,.

[240] Stenlund, Sören, *Combinators, λ-Terms and Proof Theory*. Dordrecht, Holland: D. Reidel, 1972.

[241] Strong, H. R., 'Algebraically Generalized Recursive Function Theory', *IBM Journal of Research and Development*, November (1968), 465–475.

[242] Szilard, Leo, 'On the Decrease of Entropy in a Thermodynamic System by the Intervention of Intelligent Beings', *Behavioral Science* **IX** (1964), 301–310.

[243] Takeuti, Gaisi, 'The Π_1^1-Comprehension Schema and ω-Rules', in *Proceedings of the Summer School in Logic, Leeds, 1967*. Edited by M. H. Lob. Lecture Notes in Mathematics. Berlin: Springer, 1968, pp. 303–331.

[244] Takeuti, Gaisi, *Two Applications of Logic to Mathematics* Iwanami Shoten and Princeton University Press, 1978.

[245] Tarski, Alfred, *A Decision Method for Elementary Algebra and Geometry*. Berkeley, California: University of California Press, 1951.

[246] Tarski, Alfred, *Logic, Semantics, Metamathematics: Papers from 1923 to 1938*. Translated by J. H. Woodger. Oxford: Clarendon Press, 1956.

[247] Tharp, Leslie H., 'Which Logic is the Right Logic', *Synthese* **XXXI** (1975), 1–21.

[248] Thomas, William J., 'Doubts About Some Standard Arguments for Church's Thesis', in *Logic, Language, and Probability*. Edited by Radu J. Bogden and Ilkka Niiniluoto. Dordrecht and Boston: D. Reidel, 1973, pp. 55–65.

[249] Trachtenbrot, B. A., 'The Definition of Finite Set and the Deductive Incom-

pleteness of the Theory of Sets', *American Mathematical Society Translations* **XXXIX** (1963), 172–192.

[250] Troelstra, A. S., *Principles of Intuitionism: Lectures Presented at the Summer Conference on Intuitionism and Proof Theory (1968) at SUNY at Buffalo, N. Y.* Lecture Notes in Mathematics. Berlin: Springer, 1969.

[251] Trudeau, Richard J., *Dots and Lines*. Kent State University Press, 1976.

[252] Turing, Alan M., 'Computing Machinery and Intelligence'. in *Minds and Machines*. Edited by K. Gunderson. Englewood Cliffs, N. J.: Prentice-Hall, 1964, pp. 4–30.

[252a] Turing, Alan M., 'Intelligent Machinery', in *Machine Intelligence*, Vol. V. Edited by Bernard Meltzer and Donald Michie. New York: American Elsevier, 1970.

[253] Turner, Merle B., *Realism and the Explanation of Behavior*. The Century Psychology Series. New York: Appleton-Century-Crofts, 1971.

[254] van Heijenoort, Jean, 'Logic as Calculus and Logic as Language', *Synthese* **XVII** (1967), 324–330.

[254a] van Heijenoort, Jean, 'Set-Theoretic Semantics', in *Logic Colloquium 76*. Edited by R. O. Gandy and J. M. E. Hyland. Amsterdam: North-Holland, 1977, pp. 183–190.

[255] van Heijenoort, Jean, ed., *From Frege to Gödel: A Source Book in Mathematical Logic, 1879–1931*. Cambridge, Massachussetts: Harvard University Press, 1967.

[256] Vartanian, Aram, *La Mettrie's L'Homme Machine: A Study in the Origins of an Idea*. Princeton, New Jersey: Princeton University Press, 1960.

[256a] Veblen, Oswald, 'A System of Axioms for Geometry', *Transactions of the American Mathematical Society* **V** (1904), 343–384.

[257] Veronese, Giuseppe, *Grundzüge der Geometrie von mehren Dimensionen und mehren Arten gradlichen Einheiten*. Leipzig: B. G. Teubner, 1894.

[258] Wang, Hao, *A Survey of Mathematical Logic*. Peking: Science Press, and Amsterdam: North-Holland, 1964.

[259] Wang, Hao, *From Mathematics to Philosophy*. New York: Humanities Press, 1974.

[260] Watson, A., 'Mathematics and Its Foundations', *Mind* **VII** (1938), 440–451.

[261] Webb, Judson C., 'Metamathematics and the Philosophy of Mind', *Philosophy of Science* **XXXV** (1968), 156–178.

[261a] Webb, Judson C., 'Gödel's Theorems and Church's Thesis: A Prologue to Mechanism'. In press, to appear in Vol. **XXXI** (1976) of *Boston Studies in Philosophy of Science*.

[262] Weizenbaum, Joseph, *Computer Power and Human Reason: From Judgement to Calculation*. San Francisco: Freeman and Company, 1976.

[263] Weston, Thomas, 'Kreisel, the Continuum Hypothesis, and Second Order Set Theory', *Journal of Philosophical Logic* **V** (1976), 281–298.

[264] Weyl, Hermann, 'Über die Definition der mathematischen Grundbegriffe', *Mathematisch-Naturwissenschaftliche Blätter* **VII** (1910), 93–95, 109–13.

[265] Weyl, Hermann, *Das Kontinuum. Kritische Untersuchungen über die Grundlagen des Analysis. Das Kontinuum und andere Monographien*. H. Weyl, E. Landau, and B. Riemann. New York: Chelsea Publishing Company, n.d.

[266] Weyl, Hermann, 'The Mathematical Way of Thinking', *Science* **XCII** (1940),

437–446.

[267] Weyl, Hermann, 'David Hilbert, 1862–1943'. *Obituary Notices of Fellows of the Royal Society* **IV** (1942–44), 547–553.

[268] Weyl, Hermann, 'David Hilbert and His Mathematical Work', *Bulletin of the American Mathematical Society* L (1944), 612–654.

[269] Weyl, Hermann, *Philosophy of Mathematics and Natural Science*. Princeton, New Jersey: Princeton University Press, 1949.

[270] Weyl, Hermann, 'Über den Symbolismus der Mathematik und mathematischen Physik', *Studium Generale* **VI** (1956), 219–228.

[271] Wiener, Hans, 'Grundlagen und Aufbau der Geometrie', *Jahresbericht der Deutschen Mathematiker-Vereinigung* **I** (1893), 90–93.

[272] Wiener, Norbert, *Cybernetics, or Control and Communication in the Animal and the Machine*. Cambridge, Mass.: M. I. T. Press, 1961.

[273] Wigner, Eugene P., 'The Probability of the Existence of a Self-Reproducing Unit', in *Symmetries and Reflections*. Cambridge, Mass.: M.I.T. Press, 1970. pp. 200–208.

[274] Wilder, Raymond L., 'Axiomatics and the Development of Creative Talent', in *The Axiomatic Method with Special Reference to Geometry and Physics*. Edited by Leon Henkin, Patrick Suppes, and Alfred Tarski. Amsterdam: North-Holland, 1959.

[275] Wilkes, M.V., 'Babbage as a Computer Pioneer', *Historia Mathematica* **IV** (1977), 415–440.

[276] Wilson, Edwin B., 'Logic and the Continuum', *Bulletin of the American Mathematical Society* **XII** (1908), 432–442.

[277] Wittgenstein, Ludwig, *Philosophische Bemerkungen*. Edited by Rush Rhees. Oxford: Basil Blackwell, 1964.

[278] Zermelo, Ernst. 'Über Stufen der Quantifikation und die Logik des Unendlichen', *Jarhesbericht der Deutschen Mathematiker-Vereinigung* **XLI** (1932), 85–88.

INDEX OF NAMES

INDEX OF SUBJECTS

SYNTHESE LIBRARY

Studies in Epistemology, Logic, Methodology, and Philosophy of Science

1. J. M. Bochénski, *A Precis of Mathematical Logic.* 1959.
2. P. L. Guiraud, *Problèmes et méthodes de la statistique linguistique.* 1960.
3. Hans Freudenthal (ed.), *The Concept and the Role of the Model in Mathematics and Natural and Social Sciences.* 1961.
4. Evert W. Beth, *Formal Methods. An Introduction to Symbolic Logic and the Study of Effective Operations in Arithmetic and Logic.* 1962.
5. B. H. Kazemier and D. Vuysje (eds.), *Logic and Language. Studies Dedicated to Professor Rudolf Carnap on the Occasion of His Seventieth Birthday.* 1962.
6. Marx W. Wartofsky (ed.), *Proceedings of the Boston Colloquium for the Philosophy of Science 1961-1962.* Boston Studies in the Philosophy of Science, Volume I. 1963.
7. A. A. Zinov'ev, *Philosophical Problems of Many-Valued Logic.* 1963.
8. Georges Gurvitch, *The Spectrum of Social Time.* 1964.
9. Paul Lorenzen, *Formal Logic.* 1965.
10. Robert S. Cohen and Marx W. Wartofsky (eds.), *In Honor of Philipp Frank.* Boston Studies in the Philosophy of Science, Volume II. 1965.
11. Evert W. Beth, *Mathematical Thought. An Introduction to the Philosophy of Mathematics.* 1965.
12. Evert W. Beth and Jean Piaget, *Mathematical Epistemology and Psychology.* 1966.
13. Guido Küng, *Ontology and the Logistic Analysis of Language. An Enquiry into the Contemporary Views on Universals.* 1967.
14. Robert S. Cohen and Marx W. Wartofsky (eds.), *Proceedings of the Boston Colloquium for the Philosophy of Science 1964-1966. In Memory of Norwood Russell Hanson.* Boston Studies in the Philosophy of Science, Volume III. 1967.
15. C. D. Broad, *Induction, Probability, and Causation. Selected Papers.* 1968.
16. Günther Patzig, *Aristotle's Theory of the Syllogism. A Logical-Philosophical Study of Book A of the Prior Analytics.* 1968.
17. Nicholas Rescher, *Topics in Philosophical Logic.* 1968.
18. Robert S. Cohen and Marx W. Wartofsky (eds.), *Proceedings of the Boston Colloquium for the Philosophy of Science 1966-1968.* Boston Studies in the Philosophy of Science, Volume IV. 1969.

19. Robert S. Cohen and Marx W. Wartofsky (eds.), *Proceedings of the Boston Colloquium for the Philosophy of Science 1966-1968.* Boston Studies in the Philosophy of Science, Volume V. 1969.
20. J. W. Davis, D. J. Hockney, and W. K. Wilson (eds.), *Philosophical Logic.* 1969.
21. D. Davidson and J. Hintikka (eds.), *Words and Objections. Essays on the Work of W. V. Quine.* 1969.
22. Patrick Suppes, *Studies in the Methodology and Foundations of Science. Selected Papers from 1911 to 1969.* 1969.
23. Jaakko Hintikka, *Models for Modalities. Selected Essays.* 1969.
24. Nicholas Rescher *et al.* (eds.), *Essays in Honor of Carl G. Hempel. A Tribute on the Occasion of His Sixty-Fifth Birthday.* 1969.
25. P. V. Tavanec (ed.), *Problems of the Logic of Scientific Knowledge.* 1969.
26. Marshall Swain (ed.), *Induction, Acceptance, and Rational Belief.* 1970.
27. Robert S. Cohen and Raymond J. Seeger (eds.), *Ernst Mach: Physicist and Philosopher.* Boston Studies in the Philosophy of Science, Volume VI. 1970.
28. Jaakko Hintikka and Patrick Suppes, *Information and Inference.* 1970.
29. Karel Lambert, *Philosophical Problems in Logic. Some Recent Developments.* 1970.
30. Rolf A. Eberle, *Nominalistic Systems.* 1970.
31. Paul Weingartner and Gerhard Zecha (eds.), *Induction, Physics, and Ethics.* 1970.
32. Evert W. Beth, *Aspects of Modern Logic.* 1970.
33. Risto Hilpinen (ed.), *Deontic Logic: Introductory and Systematic Readings.* 1971.
34. Jean-Louis Krivine, *Introduction to Axiomatic Set Theory.* 1971.
35. Joseph D. Sneed, *The Logical Sstructure of Mathematical Physics.* 1971.
36. Carl R. Kordig, *The Justification of Scientific Change.* 1971.
37. Milic Capek, *Bergson and Modern Physics.* Boston Studies in the Philosophy of Science, Volume VII. 1971.
38. Norwood Russell Hanson, *What I Do Not Believe, and Other Essays* (ed. by Stephen Toulmin and Harry Woolf). 1971.
39. Roger C. Buck and Robert S. Cohen (eds.), *PSA 1970. In Memory of Rudolf Carnap.* Boston Studies in the Philosophy of Science, Volume VIII. 1971.
40. Donald Davidson and Gilbert Harman (eds.), *Semantics of Natural Language.* 1972.
41. Yehoshua Bar-Hillel (ed.), *Pragmatics of Natural Languages.* 1971.
42. Sören Stenlund, *Combinators, λ-Terms and Proof Theory.* 1972.
43. Martin Strauss, *Modern Physics and Its Philosophy. Selected Papers in the Logic, History, and Philosophy of Science.* 1972.
44. Mario Bunge, *Method, Model and Matter.* 1973.
45. Mario Bunge, *Philosophy of Physics.* 1973.
46. A. A. Zinov'ev, *Foundations of the Logical Theory of Scientific Knowledge (Complex Logic).* (Revised and enlarged English edition with an appendix by G. A. Smirnov, E. A. Sidorenka, A. M. Fedina, and L. A. Bobrova.) Boston Studies in the Philosophy of Science, Volume IX. 1973.
47. Ladislav Tondl, *Scientific Procedures.* Boston Studies in the Philosophy of Science, Volume X. 1973.
48. Norwood Russell Hanson, *Constellations and Conjectures* (ed. by Willard C. Humphreys, Jr.). 1973.

49. K. J. J. Hintikka, J. M. E. Moravcsik, and P. Suppes (eds.), *Approaches to Natural Language.* 1973.
50. Mario Bunge (ed.), *Exact Philosophy – Problems, Tools, and Goals.* 1973.
51. Radu J. Bogdan and Ilkka Niiniluoto (eds.), *Logic, Language, and Probability.* 1973.
52. Glenn Pearce and Patrick Maynard (eds.), *Conceptual Change.* 1973.
53. Ilkka Niiniluoto and Raimo Tuomela, *Theoretical Concepts and Hypothetico-Inductive Inference.* 1973.
54. Roland Fraissé, *Course of Mathematical Logic* – Volume 1: *Relation and Logical Formula.* 1973.
55. Adolf Grünbaum, *Philosophical Problems of Space and Time.* (Second, enlarged edition.) Boston Studies in the Philosophy of Science, Volume XII. 1973.
56. Patrick Suppes (ed.), *Space, Time, and Geometry.* 1973.
57. Hans Kelsen, *Essays in Legal and Moral Philosophy* (selected and introduced by Ota Weinberger). 1973.
58. R. J. Seeger and Robert S. Cohen (eds.), *Philosophical Foundations of Science.* Boston Studies in the Philosophy of Science, Volume XI. 1974.
59. Robert S. Cohen and Marx W. Wartofsky (eds.), *Logical and Epistemological Studies in Contemporary Physics.* Boston Studies in the Philosophy of Science, Volume XIII. 1973.
60. Robert S. Cohen and Marx W. Wartofsky (eds.), *Methodological and Historical Essays in the Natural and Social Sciences. Proceedings of the Boston Colloquium for the Philosophy of Science 1969-1972.* Boston Studies in the Philosophy of Science, Volume XIV. 1974.
61. Robert S. Cohen, J. J. Stachel, and Marx W. Wartofsky (eds.), *For Dirk Struik. Scientific, Historical and Political Essays in Honor of Dirk J. Struik.* Boston Studies in the Philosophy of Science, Volume XV. 1974.
62. Kazimierz Ajdukiewicz, *Pragmatic Logic* (transl. from the Polish by Olgierd Wojtasiewicz). 1974.
63. Sören Stenlund (ed.), *Logical Theory and Semantic Analysis. Essays Dedicated to Stig Kanger on His Fiftieth Birthday.* 1974.
64. Kenneth F. Schaffner and Robert S. Cohen (eds.), *Proceedings of the 1972 Biennial Meeting, Philosophy of Science Association.* Boston Studies in the Philosophy of Science, Volume XX. 1974.
65. Henry E. Kyburg, Jr., *The Logical Foundations of Statistical Inference.* 1974.
66. Marjorie Grene, *The Understanding of Nature. Essays in the Philosophy of Biology.* Boston Studies in the Philosophy of Science, Volume XXIII. 1974.
67. Jan M. Broekman, *Structuralism: Moscow, Prague, Paris.* 1974.
68. Norman Geschwind, *Selected Papers on Language and the Brain.* Boston Studies in the Philosophy of Science, Volume XVI. 1974.
69. Roland Fraissé, *Course of Mathematical Logic* – Volume 2: *Model Theory.* 1974.
70. Andrzej Grzegorczyk, *An Outline of Mathematical Logic. Fundamental Results and Notions Explained with All Details.* 1974.
71. Franz von Kutschera, *Philosophy of Language.* 1975.
72. Juha Manninen and Raimo Tuomela (eds.), *Essays on Explanation and Understanding. Studies in the Foundations of Humanities and Social Sciences.* 1976.

73. Jaakko Hintikka (ed.), *Rudolf Carnap, Logical Empiricist. Materials and Perspectives.* 1975.
74. Milic Capek (ed.), *The Concepts of Space and Time. Their Structure and Their Development.* Boston Studies in the Philosophy of Science, Volume XXII. 1976.
75. Jaakko Hintikka and Unto Remes, *The Method of Analysis. Its Geometrical Origin and Its General Significance.* Boston Studies in the Philosophy of Science, Volume XXV. 1974.
76. John Emery Murdoch and Edith Dudley Sylla, *The Cultural Context of Medieval Learning.* Boston Studies in the Philosophy of Science, Volume XXVI. 1975.
77. Stefan Amsterdamski, *Between Experience and Metaphysics. Philosophical Problems of the Evolution of Science.* Boston Studies in the Philosophy of Science, Volume XXXV. 1975.
78. Patrick Suppes (ed.), *Logic and Probability in Quantum Mechanics.* 1976.
79. Hermann von Helmholtz: *Epistemological Writings. The Paul Hertz/Moritz Schlick Centenary Edition of 1921 with Notes and Commentary by the Editors.* (Newly translated by Malcolm F. Lowe. Edited, with an Introduction and Bibliography, by Robert S. Cohen and Yehuda Elkana.) Boston Studies in the Philosophy of Science, Volume XXXVII. 1977.
80. Joseph Agassi, *Science in Flux.* Boston Studies in the Philosophy of Science, Volume XXVIII. 1975.
81. Sandra G. Harding (ed.), *Can Theories Be Refuted? Essays on the Duhem-Quine Thesis.* 1976.
82. Stefan Nowak, *Methodology of Sociological Research. General Problems.* 1977.
83. Jean Piaget, Jean-Blaise Grize, Alina Szeminska, and Vinh Bang, *Epistemology and Psychology of Functions.* 1977.
84. Marjorie Grene and Everett Mendelsohn (eds.), *Topics in the Philosophy of Biology.* Boston Studies in the Philosophy of Science, Volume XXVII. 1976.
85. E. Fischbein, *The Intuitive Sources of Probabilistic Thinking in Children.* 1975.
86. Ernest W. Adams, *The Logic of Conditionals. An Application of Probability to Deductive Logic.* 1975.
87. Marian Przelecki and Ryszard Wójcicki (eds.), *Twenty-Five Years of Logical Methodology in Poland.* 1977.
88. J. Topolski, *The Methodology of History.* 1976.
89. A. Kasher (ed.), *Language in Focus: Foundations, Methods and Systems. Essays Dedicated to Yehoshua Bar-Hillel.* Boston Studies in the Philosophy of Science, Volume XLIII. 1976.
90. Jaakko Hintikka, *The Intentions of Intentionality and Other New Models for Modalities.* 1975.
91. Wolfgang Stegmüller, *Collected Papers on Epistemology, Philosophy of Science and History of Philosophy.* 2 Volumes. 1977.
92. Dov M. Gabbay, *Investigations in Modal and Tense Logics with Applications to Problems in Philosophy and Linguistics.* 1976.
93. Radu J. Bogdan, *Local Induction.* 1976.
94. Stefan Nowak, *Understanding and Prediction. Essays in the Methodology of Social and Behavioral Theories.* 1976.
95. Peter Mittelstaedt, *Philosophical Problems of Modern Physics.* Boston Studies in the Philosophy of Science, Volume XVIII. 1976.

96. Gerald Holton and William Blanpied (eds.), *Science and Its Public: The Changing Relationship.* Boston Studies in the Philosophy of Science, Volume XXXIII. 1976.
97. Myles Brand and Douglas Walton (eds.), *Action Theory.* 1976.
98. Paul Gochet, *Outline of a Nominalist Theory of Proposition. An Essay in the Theory of Meaning.* 1980.
99. R. S. Cohen, P. K. Feyerabend, and M. W. Wartofsky (eds.), *Essays in Memory of Imre Lakatos.* Boston Studies in the Philosophy of Science, Volume XXXIX. 1976.
100. R. S. Cohen and J. J. Stachel (eds.), *Selected Papers of Léon Rosenfeld.* Boston Studies in the Philosophy of Science, Volume XXI. 1978.
101. R. S. Cohen, C. A. Hooker, A. C. Michalos, and J. W. van Evra (eds.), *PSA 1974: Proceedings of the 1974 Biennial Meeting of the Philosophy of Science Association.* Boston Studies in the Philosophy of Science, Volume XXXII. 1976.
102. Yehuda Fried and Joseph Agassi, *Paranoia: A Study in Diagnosis.* Boston Studies in the Philosophy of Science, Volume L. 1976.
103. Marian Przelecki, Klemens Szaniawski, and Ryszard Wójcicki (eds.), *Formal Methods in the Methodology of Empirical Sciences.* 1976.
104. John M. Vickers, *Belief and Probability.* 1976.
105. Kurt H. Wolff, *Surrender and Catch: Experience and Inquiry Today.* Boston Studies in the Philosophy of Science, Volume LI. 1976.
106. Karel Kosík, *Dialectics of the Concrete.* Boston Studies in the Philosophy of Science, Volume LII. 1976.
107. Nelson Goodman, *The Structure of Appearance.* (Third edition.) Boston Studies in the Philosophy of Science, Volume LIII. 1977.
108. Jerzy Giedymin (ed.), *Kazimierz Ajdukiewicz: The Scientific World-Perspective and Other Essays, 1931-1963.* 1978.
109. Robert L. Causey, *Unity of Science.* 1977.
110. Richard E. Grandy, *Advanced Logic for Applications.* 1977.
111. Robert P. McArthur, *Tense Logic.* 1976.
112. Lars Lindahl, *Position and Change. A Study in Law and Logic.* 1977.
113. Raimo Tuomela, *Dispositions.* 1978.
114 Herbert A. Simon, *Models of Discovery and Other Topics in the Methods of Science.* Boston Studies in the Philosophy of Science, Volume LIV. 1977.
115. Roger D. Rosenkrantz, *Inference, Method and Decision.* 1977.
116. Raimo Tuomela, *Human Action and Its Explanation. A Study on the Philosophical Foundations of Psychology.* 1977.
117. Morris Lazerowitz, *The Language of Philosophy. Freud and Wittgenstein.* Boston Studies in the Philosophy of Science, Volume LV. 1977.
118. Stanislaw Leśniewski, *Collected Works* (ed. by S. J. Surma, J. T. J. Srzednicki, and D. I. Barnett, with an annotated bibliography by V. Frederick Rickey). 1980. (Forthcoming.)
119. Jerzy Pelc, *Semiotics in Poland, 1894-1969.* 1978.
120. Ingmar Pörn, *Action Theory and Social Science. Some Formal Models.* 1977.
121. Joseph Margolis, *Persons and Minds. The Prospects of Nonreductive Materialism.* Boston Studies in the Philosophy of Science, Volume LVII. 1977.
122. Jaakko Hintikka, Ilkka Niiniluoto, and Esa Saarinen (eds.), *Essays on Mathematical and Philosophical Logic.* 1978.
123. Theo A F. Kuipers, *Studies in Inductive Probability and Rational Expectation.* 1978.

124. Esa Saarinen, Risto Hilpinen, Ilkka Niiniluoto, and Merrill Provence Hintikka (eds.), *Essays in Honour of Jaakko Hintikka on the Occasion of His Fiftieth Birthday*. 1978.
125 Gerard Radnitzky and Gunnar Andersson (eds.), *Progress and Rationality in Science*. Boston Studies in the Philosophy of Science, Volume LVIII. 1978.
126. Peter Mittelstaedt, *Quantum Logic*. 1978.
127. Kenneth A. Bowen, *Model Theory for Modal Logic. Kripke Models for Modal Predicate Calculi*. 1978.
128. Howard Alexander Bursen, *Dismantling the Memory Machine. A Philosophical Investigation of Machine Theories of Memory*. 1978.
129. Marx W. Wartofsky, *Models: Representation and the Scientific Understanding*. Boston Studies in the Philosophy of Science, Volume XLVIII. 1979.
130. Don Ihde, *Technics and Praxis. A Philosophy of Technology*. Boston Studies in the Philosophy of Science, Volume XXIV. 1978.
131. Jerzy J. Wiatr (ed.), *Polish Essays in the Methodology of the Social Sciences*. Boston Studies in the Philosophy of Science, Volume XXIX. 1979.
132. Wesley C. Salmon (ed.), *Hans Reichenbach: Logical Empiricist*. 1979.
133. Peter Bieri, Rolf-P. Horstmann, and Lorenz Krüger (eds.), *Transcendental Arguments in Science. Essays in Epistemology*. 1979.
134. Mihailo Marković and Gajo Petrović (eds.), *Praxis. Yugoslav Essays in the Philosophy and Methodology of the Social Sciences*. Boston Studies in the Philosophy of Science, Volume XXXVI. 1979.
135. Ryszard Wójcicki, *Topics in the Formal Methodology of Empirical Sciences*. 1979.
136. Gerard Radnitzky and Gunnar Andersson (eds.), *The Structure and Development of Science*. Boston Studies in the Philosophy of Science, Volume LIX. 1979.
137. Judson Chambers Webb, *Mechanism, Mentalism, and Metamathematics. An Essay on Finitism*. 1980.
138. D. F. Gustafson and B. L. Tapscott (eds.), *Body, Mind, and Method. Essays in Honor of Virgil C. Aldrich*. 1979.
139. Leszek Nowak, *The Structure of Idealization. Towards a Systematic Interpretation of the Marxian Idea of Science*. 1979.
140. Chaim Perelman, *The New Rhetoric and the Humanities. Essays on Rhetoric and Its Applications*. 1979.
141. Wlodzimierz Rabinowicz, *Universalizability. A Study in Morals and Metaphysics*. 1979.
142. Chaim Perelman, *Justice, Law, and Argument. Essays on Moral and Legal Reasoning*. 1980.
143. S. Kanger and S. Öhman (eds.), *Philosophy and Grammar. Papers on the Occasion of the Quincentennial of Uppsala University, Sweden*. 1980.
144. Tadeusz Pawlowski, *Concept Formation in the Humanities and the Social Sciences*. 1980.
145. Jaakko Hintikka and David Gruender (eds.), *Theory Change, Ancient Axiomatics, and Galileo's Methodology. Proceedings of the 1978 Pisa Conference on the History and Philosophy of Science*. 1980. (Forthcoming.)
146. Jaakko Hintikka and David Gruender (eds.), *Probabilistic Thinking, Thermodynamics, and the Interaction of the History and Philosophy of Science. Proceedings of the 1978 Pisa Conference on the History and Philosophy of Science*. 1980. (Forthcoming.)

147. Uwe Mönnich, *Aspects of Philosophical Logic.* 1980. (Forthcoming.)

148. Dov M. Gabbay, *Semantical Investigations in Heyting's Intuitionistic Logic.* 1981. (Forthcoming.)

149. Evandro Agazzi, *Modern Logic - A Survey. Historical, Philosophical, and Mathematical Aspects of Modern Logic and its Applications.* 1980. (Forthcoming.)

150. A. F. Parker-Rhodes, *The Theory of Indistinguishables. A Search for Explanatory Principles below the Level of Physics.* 1981. (Forthcoming.)